Your Career in

Administrative Medical Services

Roberta C. Weiss, LVN, Ed.D.

Allied Health Curriculum Specialist
Vocational Education Teacher and Trainer

Your Career in

Administrative Medical Services

W.B. Saunders Company
A Division of Harcourt Brace & Company
Philadelphia London Toronto
Montreal Sydney Tokyo

W.B. SAUNDERS COMPANY

A Division of Harcourt Brace & Company

The Curtis Center
Independence Square West
Philadelphia, Pennsylvania 19106

Library of Congress Cataloging-in-Publication Data

Weiss, Roberta C.
    Your career in administrative medical services / Roberta C. Weiss.—
1st ed.
      p.   cm.
    ISBN 0-7216-6077-0
    1. Medical assistants—Vocational guidance. 2. Medical
secretaries—Vocational guidance. I. Title.
    [DNLM: 1. Hospital Administration. 2. Career Choice. 3. Allied
Health Personnel. WX 155 1996]
R728.8.W45   1996
610.69'53—dc20
DNLM/DLC   95-17451

YOUR CAREER IN ADMINISTRATIVE MEDICAL SERVICES   ISBN 0-7216-6077-0

Copyright © 1996 by W.B. Saunders Company

All rights reserved. No part of this publication may be reproduced or transmitted in any form or by any means, electronic or mechanical, including photocopy, recording, or any information storage and retrieval system, without permission in writing from the publisher.

Printed in the United States of America

Last digit is the print number: 9   8   7   6   5   4   3   2   1

This book is very lovingly dedicated to my mom; my sister, Mickie; my brother, Jeff; my extended family, Lynn, Larry, Melody, and Damon; and all those people in my life whose love and guidance brought me to the place I am today.

Preface

All the major technological breakthroughs, the ever-changing and increasing costs in medical care, and the constant need to fill new positions in all health-related disciplines mean the delivery of health care, as we once knew it, no longer exists. Instead, we are rushing to meet the demands of a society, that, because its members live longer, is in greater need of being cared for by professionals who have been trained in a number of occupations within the same educational discipline. According to the U.S. Department of Labor, if current trends continue, health care will continue to be one of the fastest growing industries in our economy. By the year 2005, employment in health-related occupations is projected to grow from 8.9 million in 1994 to 12.8 million. And, most employers agree that those individuals trained in more than one occupation will have the greatest opportunity to fill the millions of new jobs.

Your Career in Administrative Medical Services is a textbook based upon the concept of training a professional in more than one occupation within the same health care discipline. It is written in simple yet comprehensive language, and is accompanied by photographs and basic illustrations in order to provide the reader with a well-rounded introduction to the principles and techniques involved in the various allied health occupations within the field of administrative medical and allied health services.

While this author and educator firmly believes that there are many adequate textbooks currently in use that provide information for the person interested in either advanced training in administrative medical services or a specific area, no textbook is available to cover more than one occupation in the field.

This text consists of 27 chapters that cover concepts and hands-on skills and techniques needed in the study of basic health careers opportunities, scientific principles and medical terminology, administrative medical assisting, medical transcription, medical information coding, medical records clerking, health unit coordinating, and administrative and job-seeking skills for the administrative and medical clerical worker. To assist readers in their understanding of the subject matter, the beginning of each chapter has lists of performance objectives and terminology used, while at the end of each chapter is a short summary of the material covered. In addition, to assist both the student and the instructor, the text is also accompanied by a *Teacher's Resource Package*, which includes such concepts as teaching methodologies and strategies, resources and references, clinical and classroom activities, and an in-depth glossary of medical terms used in the study of administrative medical services.

Roberta C. Weiss, LVN, Ed.D.
North Hollywood, California

Acknowledgments

I would like to extend my deepest gratitude and appreciation to all those who have assisted me in the preparation and development of this textbook, including: the members of the staff at Huntington Memorial Hospital, Pasadena, CA, who caringly gave of their time, energy, and expertise in the preparation of this text; the individual patients undergoing cardiovascular treatments and procedures in Huntington Memorial Hospital's cardiology department, respiratory therapy department, clinical laboratory department; the employees in the business office, the medical records department, the admitting department, and central supply department, who gave of their time and provided that "personal touch" by allowing their photographs to be taken; my very dear friend and colleague, Ms. Marla Keeth, LVN, Coordinator of the Health Careers Academy at Blair High School, Pasadena Unified School District/Los Angeles County Regional Occupational Programs, for her assistance and continued sustenance in providing time for her health career students to be photographed and to work with actual patients in the preparation of this textbook; the highly professional health careers students enrolled in Blair High School's Health Careers Academy for their time, expertise, and participation while being photographed; Debbie Ogilvie, Career School Sales Representative for W.B. Saunders Company, for her time and energy in bringing me and my "new" concept to the right people at Saunders; and, finally, Lisa Biello, Vice-President and Editor-in-Chief for Health-Related Professions for W.B. Saunders Company, who was willing to take a chance on something that had never been done and who has provided me with many hours of guidance and inspiration throughout the development of this project.

Contents

Preface vii
Acknowledgments ix

Section I
Introduction to Health Careers 1

Chapter 1
Career Opportunities in Administrative Medical Services 3

Performance Objectives, Terms and Abbreviations 3
History of Administrative Medical Services 4
Training and Education 4
Administrative Medical Careers and Employment Opportunities 5
Hospital Organization and Medical Specialties 6
Summary 6
Review Questions 7

Chapter 2
Health, Wellness, and Illness 9

Performance Objectives, Terms and Abbreviations 9
Clinical Model of Health Care Delivery 10
Ecological Model of Health Care Delivery 10
Role Performance Model of Health Care Delivery 10
Adaptive Model of Health Care Delivery 10
Eudaimonistic Model of Health Care Delivery 10
Developing a Personal Definition of Health 11
Wellness, Disease, and Illness 11
Relationship of Health, Illness, and Disease 12
Understanding the Concept of Holistic Health 12
Health Beliefs and Behaviors 12
Stages of Illness 13
Sick Role Behavior 14
Patterns and Trends in Health and Illness 15
Summary 16
Review Questions 16

Chapter 3
Medical Law and Ethics 17

Performance Objectives, Terms and Abbreviations 17
Understanding Medical Ethics and Medical Law 18
The Patient and the Health Care Worker 18
The Good Samaritan Law and Medical Malpractice 19
Patient Medical Records and the Law 21
Summary 22
Review Questions 22

Chapter 4
Health and Safety in Administrative Medical Services 23

Performance Objectives, Terms and Abbreviations 23
Asepsis and Infection Control 24
Providing for a Healthy and Safe Environment 27
Summary 31
Review Questions 32

Chapter 5
Medical Terminology and the Medical Record 33

Performance Objectives, Terms and Abbreviations 33
Using Medical Abbreviations 33
Body Structure and Medical Terminology 34
Understanding the Patient's Medical Record 36
Summary 37
Review Questions 38

Chapter 6
Basic Structure and Function of the Human Body 39

Performance Objectives, Terms and Abbreviations 39
Systems of the Human Body 39
Summary 51
Review Questions 52

Section II
Administrative Medical Assisting: Basic Concepts and Applications 55

Chapter 7
Administrative Management and Communications 57

Performance Objectives, Terms and Abbreviations 57
Receiving and Dealing with Patients, Family Members, and Visitors 57
Communication Systems Used in the Health Care Environment 58
Processing Written Communications within the Health Care Environment 60
Office Machines Used in the Health Care Environment 62
Summary 62
Review Questions 63

Chapter 8
Admissions, Transfers, and Discharges 64

Performance Objectives, Terms and Abbreviations 64
The Admission Process 64
The Transfer Process 65
The Discharge Process 72
Summary 72
Review Questions 73

Chapter 9
Filing Procedures and Medical Records Management 74

Performance Objectives, Terms and Abbreviations 74
Indexing Medical Information 75
Filing and Records Management 75
Storing Files and Medical Records 77
Medical Record Assembly and Analysis 78
Chart Forms 79
Importance of Providing Complete Data 81
Summary 81
Review Questions 82

Chapter 10
Data Abstracting 83

Performance Objectives, Terms and Abbreviations 83
Abstracting Data 83
Vital Statistics and Health Care 84
Summary 85
Review Questions 86

Chapter 11
Insurance Coding and Indexing 87

Performance Objectives, Terms and Abbreviations 87
Careers in Insurance Coding 88
Health Insurance Programs 88
Understanding and Using Coding Systems 88
Processing Insurance Claims 90
Summary 90
Review Questions 92

Chapter 12
Release of Information and the Law 93

Performance Objectives, Terms and Abbreviations 93
Patient Privacy and the Law 93
Informed Consent and Liability 94
Patient Consent Forms and Documentation 95
Summary 95
Review Questions 95

Chapter 13
Providing Quality Assurance in the Health Care Environment 96

Performance Objectives, Terms and Abbreviations 96
Understanding Quality Assurance 97
Quality Assurance and the Medical Record 99
Summary 99
Review Questions 99

Chapter 14
Basic Computer Applications 100

Performance Objectives, Terms and Abbreviations 100
Defining Computers and Their Components 101
Computer Programs and Applications 102
Computer Problems and Resolutions 103
Summary 103
Review Questions 103

Section III
Medical Transcription: Basic Concepts and Applications 105

Chapter 15
Introduction to Medical Transcription 107

Performance Objectives, Terms and Abbreviations 107
The Role of the Medical Transcriptionist 107
Training and Career Opportunities 108
Summary 109
Review Questions 109

Chapter 16
Working as a Medical Transcriptionist 110

Performance Objectives, Terms and Abbreviations 110
Basic Medical Transcription Equipment 110
Putting Your Equipment Together with Your Knowledge 112
Punctuation and Capitalization 113
Transcribing and Understanding the Use of Numbers 114
Summary 115
Review Questions 115

Chapter 17
Sentence Structure and Letter Writing 117

Performance Objectives, Terms and Abbreviations 117
Understanding Basic Sentence Structure 117
Proofreading and Sentence Structure 118
Letter Writing Skills and the Medical Transcriptionist 118
Typing the Envelope 120
Summary 125
Review Questions 125

Chapter 18
Transcribing Medical Reports and Records 126

Performance Objectives, Terms and Abbreviations 126
Using Reference Materials 126
Transcribing Patient History and Physical Reports 127
Transcribing Pathology and Radiology Reports 127
Transcribing Consultation Reports 132
Transcribing Discharge Reports 132
Transcribing Other Medical and Hospital Reports 132
Summary 132
Review Questions 142

Section IV
Medical Records Clerk: Basic Concepts and Applications 143

Chapter 19
Introduction to Medical Records 145

Performance Objectives, Terms and Abbreviations 145
The Medical Records Team 146
The Medical Records Department 147
The Tumor Registry 148
Utilization Review and Quality Assurance 149
Working in the Medical Records Department 149
Medical Records and the Law 150
Summary 152
Review Questions 152

Chapter 20
Processing Medical Records 154

Performance Objectives, Terms and Abbreviations 154
Processing the Medical Record 154
Processing Admissions and Discharges 155
Assembling the Medical Record 156
Analysis of the Medical Record 157
Summary 180
Review Questions 180

Chapter 21
Advanced Functions of the Medical Records Clerk 181

Performance Objectives, Terms and Abbreviations 181
Meeting Accreditation Standards 182
Advanced Functions of the Medical Records Clerk 182
Completing Legal Documents 183
Working with Other Departments 184
Summary 184
Review Questions 184

Section V
Medical Coding and Claims Worker: Basic Concepts and Applications 185

Chapter 22
Introduction to Medical Coding 187

Performance Objectives, Terms and Abbreviations 187
Purpose and Function of Coding 188
International Classification of Diseases 189
CPT Coding 190
Using Modifiers in CPT Coding 193
Terminology and Reference Materials 194
Summary 198
Review Questions 198

Chapter 23
Coding Diseases 200

Performance Objectives, Terms and Abbreviations 200
Diseases of the Integumentary and Musculoskeletal Systems 201
Diseases of the Digestive, Urinary, and Reproductive Systems 202
Diseases of the Cardiovascular System and Blood Forming Organs 203
Diseases of the Nervous System and Sensory Organs 204
Disorders of the Endocrine System, Infectious Diseases, and Neoplasms 205
Coding Unusual Situations and Nondefined Conditions 205
Summary 206
Review Questions 206

Chapter 24
Medical Coding and Claims Processing 208

Performance Objectives, Terms and Abbreviations 208
Processing Insurance Claims and the Law 208

Obtaining Patient Information 209
Billing for Services Rendered 210
Processing Insurance Claims 211
Processing Medicare and Medicaid Claims 211
Processing Workers' Compensation
    Insurance Claims 216
Coordinating Benefits 218
Summary 219
Review Questions 219

Section VI
Health Unit Coordinator: Basic Concepts and Applications 221

Chapter 25
Introduction to Health Unit Coordinating 223

Performance Objectives, Terms and Abbreviations 223
The Role of the Health Unit Coordinator 224
Carrying Out Basic Procedures 225
Health Unit Coordinator Tasks
    and Procedure Skills 229
Summary 232
Review Questions 232

Chapter 26
Transcribing Medical Orders 234

Performance Objectives, Terms and Abbreviations 234
Maintaining the Medical Chart 235
Transcribing Data from the Medical Chart 235
Understanding Basic Pharmacology and Transcribing
    Medication Orders 236
Transcribing Orders Related to Nursing Care
    and Treatments 237
Transcribing Orders Related to Diagnostic
    Treatments 239
Cancelling, Renewing, and Stopping Orders 240
Summary 241
Review Questions 241

Section VII
Job Seeking Skills and Employment Opportunities in Administrative Medical Services 243

Chapter 27
Job Seeking Skills and Employment Opportunities 245

Performance Objectives, Terms and Abbreviations 245
Initiating the Job Hunt 246
The Cover Letter 247
Preparing the Resume 247
Taking Part in the Interview Process 251
After the Interview, What's Next? 252
Summary 252
Review Questions 254

Appendix A
Medical Abbreviations and Symbols 255

Appendix B
Root Words, Prefixes, and Suffixes 261

Appendix C
Billing Forms 265

Index 291

Section I

Introduction to Health Careers

1
Career Opportunities in Administrative and Health Information Medical Services

2
Health, Wellness, and Illness

3
Medical Law and Ethics

4
Health and Safety in Administrative Medical Services

5
Medical Terminology and the Medical Record

6
Basic Structure and Function of the Human Body

Career Opportunites in Administrative and Health Information Medical Services

Performance Objectives

Upon completion of this chapter, you will be able to:

1. Discuss career opportunities available in administrative medical services.
2. Describe the training required at various levels of administrative medical services.
3. Distinguish among divisions within health care facilities.
4. Discuss some of the desirable personal characteristics and technical skills required of an administrative medical services worker.
5. Identify several potential duties of the administrative medical services worker.
6. List potential employers of administrative medical services workers.

Terms and Abbreviations

AAMA abbreviation for the American Association of Medical Assistants, Inc., a national association for medical assistants, medical assistant students, and medical assisting instructors.

American Guild for Patient Account Managers a national association for administrative medical services workers specializing in managing patient accounts.

AAMT abbreviation for the American Association for Medical Transcription, a national association for medical transcriptionists.

ART abbreviation for Accredited Record Technician.

CCS abbreviation for Certified Coding Specialist.

CMA abbreviation for Certified Medical Assistant, which is a certification earned through examination and after appropriate training or work experience as a medical assistant in certain institutions.

CMT abbreviation for Certified Medical Transcriptionist.

Diagnostic services services within a medical facility which deal with identifying pathological conditions or diseases.

Environmental services services within a medical facility which provide properly furnished facilities for safe use by patients and staff.

General services services within a medical facility which include admitting, feeding, and medicating patients and preserving records concerning both patients and medical personnel.

HMO abbreviation for Health Maintenance Organization, which is an organization comprised of various medical professionals and services to which a patient can subscribe.

Medical record a document that includes a patient's evaluation, testing, and treatment received during his or her health care.

MRT abbreviation for Medical Record Technician.

NAHUC abbreviation for the National Association of Health Unit Clerks-Coordinators, which is a national organization for health unit clerks and coordinators.

Patient care services services within a medical facility that manages the activities of daily living for patients.

RRA abbreviation for Registered Record Administrator.

Therapeutic services services within a medical facility which treats pathological conditions and diseases.

We give periods of time descriptive names in order to indicate something important about that particular time. For example, when we speak of the "stone age," we are referring to a time when humans had stone tools and simple survival skills. Now commentators are saying that we live in the "information age." Computers and electronic technology make global transfer of data instantaneous. Thus, workers in the health care industry must be versed in information systems for recording, storing, and retrieving data related to the health and wellness of those entrusted to their care. In addition to grasping an understanding of basic medical concepts and allied health skills, health care providers involved in the field of administrative medical services must also be able to use various types of machines, equipment, and computers and information systems, since some of their responsibilities also require interpretation of data.

History of Administrative Medical Services

Early in the 20th century, the administrative aspects of a medical practice were quite simple. A doctor performed his or her own paperwork, or, in some cases, had only one assistant who was responsible for assisting the physician in the practice. More recently, administrative medical assistants have become a necessity due to the enormous volume of paperwork produced by changing technology. The administrative medical services worker has consistently gained more and more authority, along with increased responsibility. There are now exciting possibilities for both variety in the workplace and specialization.

Careers in administrative and health information services can be found in both the business world and the health care industry. In business, these positions exist in large law firms, as well as in insurance companies. Careers in the health care industry are found in government agencies, physician's offices, the business and insurance offices of clinics, long-term care facilities, and hospitals. Locations within the hospital, where data are managed, include the admitting office, business office, medical records room, and each nursing station with its unit secretary or health unit coordinator.

Secretaries, accountants, and administrative personnel are also concerned with patient information and data processing. They must understand and have access to records about a patient's health and treatment. However, this chapter will deal with those members of the administrative medical services and health information team who specialize in running the administrative center of the physician's office or medical clinic, as well as those administrative professionals responsible for gathering, classifying, storing, interpreting, and protecting data in the health care industry.

Training and Education

No matter what aspect of administrative medical services you choose to work in, the fact that you have selected an occupation in the health care industry says a lot about you and your interest in helping others. Unfortunately, however, not all people are cut out to work in an area which may require dealing with mounds of paperwork and smiling, even if you're having a bad day.

If you are to be successful in your position, it is important that you not only possess certain educational and technical skills but, just as important, have the personal qualifications necessary to work well with others. In addition to being a compassionate person, you must also be dependable and punctual. You must be flexible and like being organized. You should be able to work well under pressure, and take pride in both yourself and the job you are doing. And, above all, you must like working with people, and, when doing so, you must be able to be tactful, accept criticism, and follow rules and detailed instructions (Figure 1-1).

Most administrative medical services workers have a minimum of a high school diploma or its equivalent. There are exceptions to this, but in most cases, completion of high school coursework is required by most employers. In addition, to be successful in administrative services, you should also have received some advanced technical training in the field you plan to work in. This generally includes some computer training, understanding of office machines and medical terminology, a basic understanding of anatomy and physiology, coursework concentrating on office practices, and a solid foundation related to medical office business, billing, accounting, and insurance processing. Most workers receive this training as part of an overall education program in a vocational or adult school, community college, or as part of a regional occupational program.

Administrative Medical Services

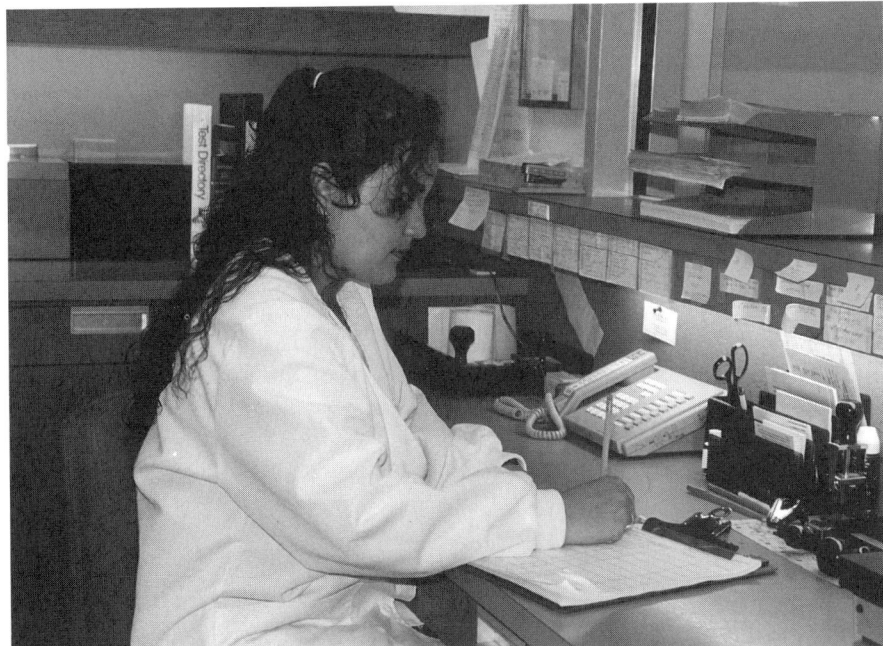

Figure 1-1
The administrative medical services worker.

Administrative Medical Careers and Employment Opportunities

Health care facilities and private medical practices use many different names and titles for their administrative services workers. Doctors' offices and clinics often hire *Administrative Medical Assistants*, *Medical Secretaries*, and *Medical Receptionists*. Larger health care institutions often have job titles such as *Medical Office Clerk*, *Hospital Admitting and Discharge Clerk*, *Medical Records Clerk*, *Insurance Billing and Coding Clerk*, *Medical Transcriptionist*, *Health Unit Coordinator* and *Unit Secretary*, among others. No matter what your job title may be, there are certain tasks that are usually required of all these jobs. These include filing, typing, computer data entry, composing and transcribing correspondence, mail processing, billing and collection procedures, public and patient relations, telephone and appointment scheduling techniques, and basic banking and bookkeeping skills.

During the normal workday, the administrative medical services worker comes into contact with a wide variety of people. In addition to dealing with the many varieties of patients and their individual problems and personalities, you will also interact with members of the nursing staff, doctors, pharmacists, laboratory and radiology technicians, supervisors, housekeepers, hospital volunteers, and many other people who are members of the hospital staff, as well as visitors seeking understanding and answers to questions about their loved ones. You must always be prepared to communicate with surgeons, who may use technical terms, as well as patients, some of whom may speak little English. Such diversity in your work creates a work environment filled with challenges, some of which may seem very frustrating while others are very rewarding.

Employment Opportunities

Over the past several decades, employment opportunities for the administrative medical services workers have expanded far beyond the one-person office assistant who is responsible only for completing basic paperwork and answering the telephone. While once a physician's office offered the only jobs available to someone interested in the field, today the options are much more extensive.

Medical care has never been more expensive or competitive. Insurance companies are not willing to pay for unnecessary procedures. Many different institutions are competing with one another to provide patient services. In today's economy, the health care institution, as well as the private medical practice, must run in a cost-efficient manner if it is to survive.

Today, there are many different types of medical facilities and agencies which can be very rewarding places of employment for the administrative medical services worker. These include general acute care hospitals, where patients are hospitalized for a short period of time (anywhere from a day to a few weeks), specialized hospitals, which have facilities that provide care for specific problems (such as psychiatric illnesses or chronic diseases), long-term care facilities (also known as convalescent hospitals), which concentrate on providing care to the elderly, and outpatient clinics, which generally house several doctors with varied specialties in a combined practice. Other types of facilities with employment opportunities for the administrative medical services worker include physicians' and dentists' private practices, rehabilitation centers, health maintenance organizations, and home health care agencies.

In addition to working in a health care facility, the administrative medical services worker may also seek employment in university medical schools, research centers, pharmaceutical and medical supplies companies, laboratories, and insurance companies, and, in some cases, may be self-employed.

Hospital Organization and Medical Specialties

As a member of the administrative medical services team, if you choose to work in the hospital environment, it is very important for you to understand how today's modern medical facilities function. These facilities must be well organized, with a "chain of command" to tell each employee who his or her immediate supervisor is. Such organization provides an efficient way for the facility to fulfill its mission and purpose.

An example of a hospital organization chart is shown in Table 1-1. As you can see, each service has specialized departments. These departments are determined by the type of service they provide.

Medical Specialties

If you are working in the hospital or clinic environment, it is more than likely you will come into contact with many different medical specialties. A medical specialist is a physician who devotes him or herself to a single branch of medical knowledge. You may find yourself working in an area that deals often with one branch of medicine, such as gastroenterology or hematology. Table 1-2 provides definitions of most of the present-day medical specialties and the name of the physician practicing those specialties.

Summary

In this chapter, we discussed the very rapidly changing field of administrative medical services. We determined that there were many varied career opportunities available to graduates of this field, both in the business world and in the health care industry. We also talked about some of the personal characteristics and technical skills which are necessary in order to be successful as a professional in this field and described the basic level of training for these occupations. Finally, we defined the various types of medical specialties and the name of each of the medical specialists involved in the various branches of medicine, noting that all of these were available to the administrative medical services worker as a means to employment.

Table 1-1
Organizational Chart of a Health Care Facility

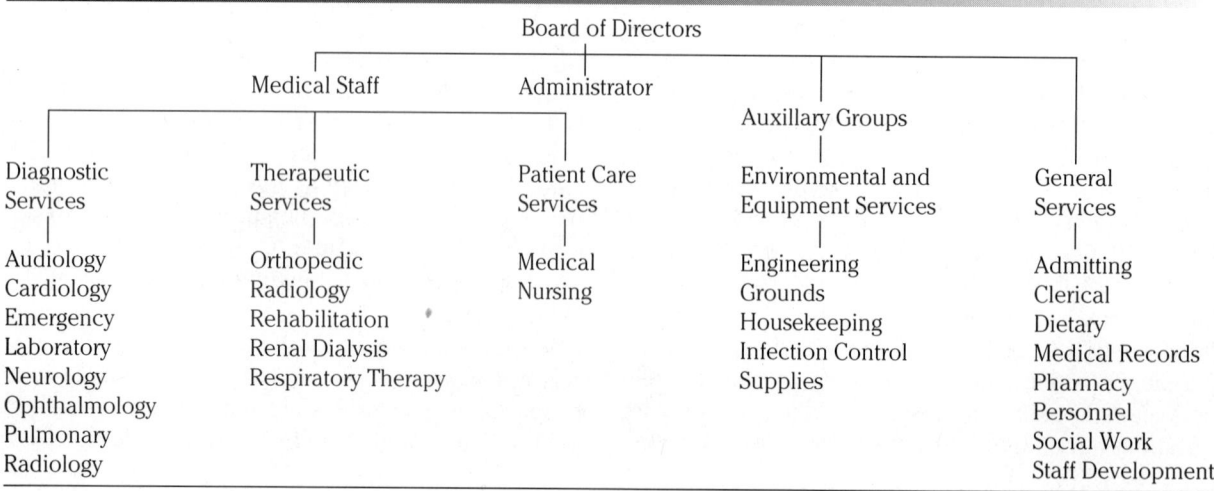

Administrative Medical Services

Table 1-2
Medical Specialties

Specialty	Name of Physician	Description of Specialty
Allergies	Allergist	Deals with diagnosis and treatment of body reactions resulting from sensitivity to foods, pollens, dust, medicine, or other substances
Anesthesiology	Anesthesiologist	Administers various forms of anesthesia in surgery or diagnosis to cause loss of feeling or sensation
Cardiology	Cardiologist	Deals with the diagnosis and treatment of diseases of the heart
Dermatology	Dermatologist	Deals with the diagnosis and treatment of diseases of the skin
Endocrinology	Endocrinologist	Deals with the diagnosis and treatment of diseases of the endocrine system and the hormones produced by the ductless glands
Family Practice	Family Practitioner	Diagnoses and treats diseases by medical and surgical methods for all members of the family
Gastroenterology	Gastroenterologist	Deals with the diagnosis and treatment of diseases of the digestive system
Gynecology	Gynecologist	Deals with the diagnosis and treatment of disorders of the female reproductive system
Internal Medicne	Internist	Diagnoses and treats diseases of adults
Neurology	Neurologist	Diagnoses and treats diseases of the nervous system and brain
Obstetrics	Obstetrician	Cares for women during pregnancy, childbirth, and the interval immediately following
Oncology	Oncologist	Diagnoses and treats cancer
Ophthalmology	Ophthalmologist	Diagnoses and treats disorders of the eye and prescribes glasses
Orthopedics	Orthopedist	Diagnoses and treats disorders of the muscular and skeletal systems
Otolaryngology	Otolaryngologist	Diagnoses and treats disorders of the eyes, ears, nose, and throat
Pathology	Pathologist	Studies and interprets changes in organs, tissues, cells and changes in the body's chemistry to aid in diagnosing disease and determining the type of treatment which may be necessary
Pediatrics	Pediatrician	Deals with the prevention, diagnosis, and treatment of children's diseases
Proctology	Proctologist	Diagnoses and treats diseases of the rectum
Psychiatry	Psychiatrist	Diagnoses and treats mental disorders
Radiology	Radiologist	Uses radiant energy, including x-rays, radium, cobalt, etc., in the diagnosis of diseases
Urology	Urologist	Diagnoses and treats diseases of the kidneys, bladder, ureters, and urethra and of the male reproductive system

Review Questions

1. Define the following abbreviations:
   a. AAMA _____
   b. AAMT _____
   c. ART _____
   d. CMT _____
2. Briefly define what a *medical record* is.
3. What does the abbreviation *NAHUC* mean?
4. What departments fall under *general services* within a health care facility?
5. What departments fall under *diagnostic services* within a health care facility?
6. What departments fall under *patient care services* within a health care facility?

7. What departments fall under *therapeutic services* within a health care facility?

8. What departments fall under *environmental services* within a health care facility?

9. What do the following abbreviations mean?

a. RRA _____

b. MRT _____

c. CCS _____

10. Briefly explain the function of an *HMO*.

Health, Wellness, and Illness

Performance Objectives

Upon completion of this chapter, you will be able to:

1. Define health, wellness, illness, and disease and briefly explain the relationship of each to the other.
2. Briefly discuss nontraditional views of health and illness.
3. Identify and briefly discuss health beliefs and behaviors.
4. Identify, and briefly discuss, the five basic stages of illness.
5. Identify, and briefly discuss, the 11 sequential stages of illness.
6. Describe what is meant by sick role behavior.
7. Explain the effects illness has on family members.
8. Discuss the effects of hospitalization.
9. Identify patterns and trends which influence health and illness.

Terms and Abbreviations

Compliance the extent to which a person's behavior coincides with that recommended by health care providers.
Disease an alteration in the body's functions which results in a reduction in capacities or a shortening of the normal life span.
Health the state of complete physical, mental, and social well-being, and not merely to the presence of disease or infirmity.
Health status the state or health of an individual at a given time.
Illness a highly personal state in which the individual feels unhealthy or ill.
Model an abstract outline or a theoretical depiction of a complex phenomenon; also known as a *paradigm*.
Morbidity pertains to illness; the *morbidity rate* is the ratio of sick to well people in a given population.
Wellness the active process through which an individual becomes aware of and makes choices that lead to a more successful existence.

Health is an ever-changing, evolving concept, which is basic both to the study of medicine and to the delivery of administrative and hands-on care giving. For centuries, the concept of disease was the yardstick by which all health was measured, and it wasn't until as late as the 19th century that health care providers began to deal with the causes of diseases. More recently, there has been an increased emphasis on health.

There is no general consensus regarding the definition of health. There is, however, knowledge as to how one can attain a certain level of health, but health itself cannot be measured. According to the World Health Organization (WHO), *health* can be defined as a state of complete physical, mental, and social well-being, and not merely as the absence of disease or infirmity. Because this definition was proposed in the late forties, when thousands of soldiers were returning home from World War II with series injuries and traumas, many health care providers of the time thought this definition to be impractical. Today, however, some view it as a possible goal for all people, while others consider complete well-being unobtainable.

Because health is a complex concept, much research has done in an attempt to measure it. Out of this research, specific models or paradigms have been developed to explain health and, in some instances, its relationship to illness. The five models of health care delivery that are most widely accepted include the *clinical* model, the *ecologic* model, the *role performance* model, the *adaptive* model, and the *eudaimonistic* model.

Clinical Model of Health Care Delivery

It appears that the narrowest interpretation of health care delivery is the clinical model. According to this model of health care delivery, all people are viewed as physiological systems with individual and related functions. In this model, health is seen as a concept which is used to identify the absences of signs and symptoms of disease or injury.

In the clinical model of health care delivery, disease is seen as the opposite of health and a relatively passive state of freedom from wellness. The extreme of health, according to this model, is the absence of any signs or symptoms of disease, and the presence of such signs or symptoms is considered to be an indication of extreme illness. Today, the focus of many medical practices is to relieve the signs and symptoms of disease and thus eliminate the malfunction and pain of patients. When the signs and symptoms of the disease are no longer present in a person, the physician often considers that individual's health has been restored. These practices, therefore, base the care and treatment of their patients on the clinical model of health care delivery.

Ecological Model of Health Care Delivery

The ecological model of health care delivery is based upon the relationship of all humans to their internal and external environments. The model covers three concepts (Figure 2-1): a *host*, which is a person or group of people who may or may not be at risk of acquiring an illness or a disease; an a*gent*, which can be any factor within either the internal or the external environment that, by its very presence or absence, can lead to illness or disease; and the *environment*, internal or external, which may or may not predispose the person to the development of a disease or illness.

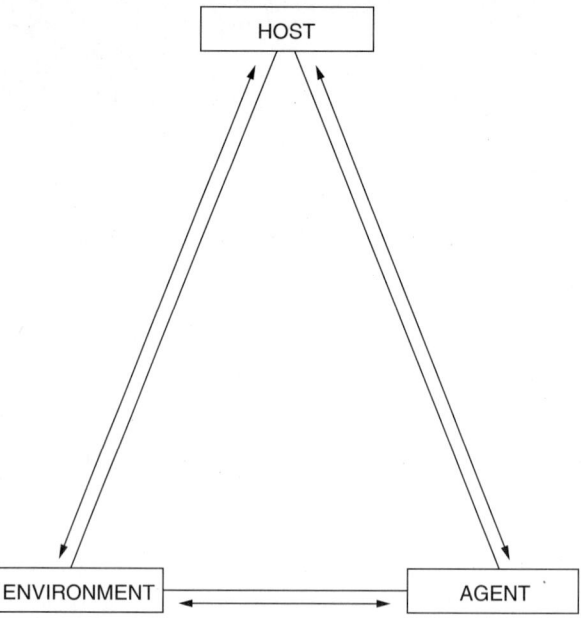

Figure 2-1
Ecological model of health care delivery.

Role Performance Model of Health Care Delivery

The role performance model of health care delivery adds the social and psychological implications and standards to the concept of health. In this concept of health care delivery, health is defined in terms of the person's ability to fulfill his or her role in society, that is, work. According to this model, people who can fulfill their roles are considered healthy even if they appear clinically ill. The emphasis of the delivery of healthcare, according to this model, therefore, should be placed upon a person's capacity rather than his or her committment to roles and tasks.

Adaptive Model of Health Care Delivery

The focus on the adaptive model of health care delivery is based upon adaptation. Diseases, according to this model, are seen as a failure in the person's ability to adapt, and the treatment is seen as an ability to restore the person to a state in which he or she can cope with illness or disease.

Eudaimonistic Model of Health Care Delivery

The eudaimonistic model of health care delivery incorporates the most comprehensive view of health. Health, according to this model, is seen as an actualization or realization of a person's total potential, and the highest aspiration of an individual is his or her fulfillment and complete development. Illness, according to the eudaimonistic model of health care delivery, is a condition or situation which prevents one from being able to self-actualize.

Developing a Personal Definition of Health

As you begin your journey into a career as a provider of health care, you will soon realize that perceptions of health are highly individualized. Meanings and descriptions of health vary considerably, and an individual's personal definition of health may not necessarily agree with that of other health care professionals.

There are several factors which influence one's personal definition of health. These include one's developmental status, social and cultural influences, previous experiences which may influence one's perception of health, and expectations of one's self.

Wellness, Disease, and Illness

The concept of wellness has received increasing attention in recent years. Some people believe that wellness and health are one and the same, while others believe that the two differ. In one respect, wellness is very similar to self-actualization, as it is defined by the eudaimonistic model of health care delivery. Wellness, like good health, can only exist as a relatively passive state of freedom from illness in which the individual is at peace with his or her environment.

Wellness can also be defined as an active process through which an individual becomes completely aware of choices that can lead to a more successful existence. These choices are influenced by the person's self-concept, his or her culture, and the environment. Wellness, therefore, is seen through a continuum, the extremes of which are total wellness and premature death (Figure 2-2).

Disease and Illness

Disease is a medical term that is used to describe any alteration in a person's body functions that results in a reduction in his or her capacities or a shortening of the normal life span. Intervention by physicians has always been the goal of eliminating or decreasing the disease process. Primitive people thought that disease was caused by outside forces or spirits. Later on, however, this belief was replaced by the single-cause theory. Increasingly, a number of factors are considered to interact to cause disease and determine a person's response to a given treatment.

Illness, while it may or may not be related to disease, differs from it in that it is defined as a highly personal state in which a person feels unhealthy or ill. Someone who feels pain or nausea tends to modify his or her behavior and thus may consider him- or herself ill. An individual could have a disease, for example, a growth in the intestine, and not feel ill. According to researchers in the theory of illness and disease, for a person to feel ill, the physician must be able to determine if that person meets three distinct criteria. First, there must be symptoms present, such as an elevated temperature or pain. Second, the person must be able to describe his or her perceptions of the illness, for example, good, bad, or sick. And third, the ability to carry out daily activities, such as a job or schoolwork, must have been affected.

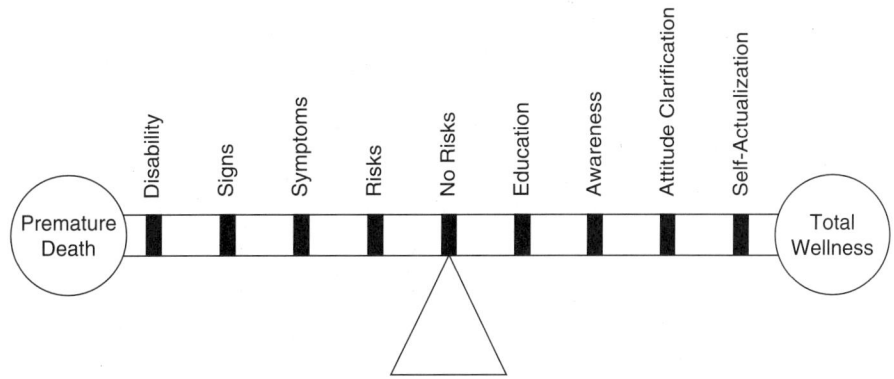

Figure 2-2
Continuum between total wellness and premature death.

Relationship of Health, Illness, and Disease

Health and Illness

A person's health and illness can be considered points along one continuum, related but separate entities, or completely separate entities. A continuum of well-being or health constitutes a grid or graduated scale, in which peak wellness is at one end and death is at the other. Such a continuum accommodates a wide range of normal health, which reflects both that no one ever attains perfect health and not everyone becomes sick.

Illness and Disease

Traditionally, health care providers have dealt with diseases at a subsystem level. Subsystems are those aspects of the human body which are subsumed by the larger system of the whole body. A subsystem, for example, may be a cell, an organ, or an organ system. Only recently have medical practitioners started looking at the person as an entity, or as a whole being. Health care providers, such as cardiovascular technicians, by contrast, have traditionally viewed the person as an entity, meaning that they have viewed the patient from a holistic point of view. Cardiovascular providers today base the scope of their practice on the multiple-causation theory of health problems or illness. For example, unemployment, pollution, lifestyle, and stressful events may all contribute to illness. In effect, these can all be considered subsystem problems, that is, problems which stem from systems in which the patient is a subsystem (Figure 2-3). Therefore, the concept of illness must include all aspects of the total person as well as the biological and genetic factos that contribute to disease.

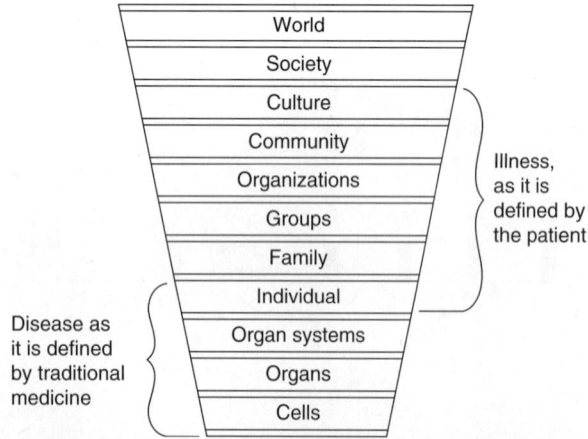

Figure 2-3
Systems hierarchy differentiating illness from disease.

Illness, then, is influenced by a person's family, his or her social network, the environment, and culture.

Understanding the Concept of Holistic Health

The concept of *holistic health* is based upon the belief that the whole is greater than the sum of its parts. When using a holistic health approach to health care, you must consider the events affecting the whole person. Illness is viewed as an opportunity for growth; health is a dynamic state of being that, for each person, moves back and forth along a continuum. According to holistic health, the extremes of the continuum are highest health potential and death; good health, normal health, mild illness, illness or poor health, and critical illness are points along the continuum (Figure 2-4).

In the holistic health model of health care delivery, wellness is seen as ever-evolving growth toward fulfilling a person's potential. It considers the person's needs, abilities, and disabilities. A critical assumption in holistic health is that the perception of health involves an individual decision, encouraging self-responsibility and self-control. Health can exist in the presence of illness. For example, a man who is feeling pain in his heart may perceive himself as being near or at the point of highest health potential along the highest health potential-death continuum if he feels he is functioning at his highest potential relative to his pain.

Health Beliefs and Behaviors

The *health status*, or state, of a person is the health of that person at a given time. In its general meaning, the term may refer to anxiety, depression, or acute illness and thus describes the individual as a whole. Health status can also describe such specifics as a person's vital signs, for example, pulse rate and body temperature. The *health beliefs* of an individual involve those concepts pertaining to his or her health which that individual person believes to be true. Such beliefs may or may not be founded on fact.

Health behaviors are defined as those actions which a person takes in order to understand his or her state of health, maintain an optimal state of health, prevent illness and injury, and reach his or her maximum physical and mental potential. Behaviors such as eating wisely, exercising, paying attention to signs of illness, following treatment advice, and avoiding known health hazards such as smoking are all examples. The ability to relax, achieve emotional maturity, lead a productive life, and give way to self-expression also affects one's state of health.

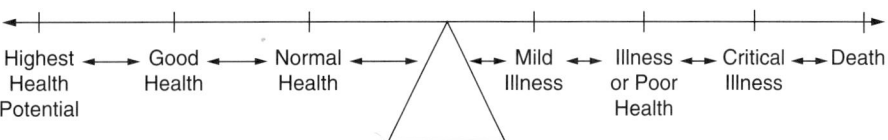

Figure 2-4
Health-illness continuum representing the holistic health model.

Influences on One's State of Health

There are many variables that affect or influence one's state of health. Some of these are internal factors, such as a person's genetic make-up, and others are external, such as a person's culture and physical environment.

Of the many internal and external factors influencing how one achieves a state of health, there are 13 which have a direct effect on a person. These include a person's genetic make-up, race and culture, sex, age, developmental level, mind-body relationship, (how a person allows his or her emotional responses to affect body function), lifestyle, physical environment, occupation, standard of living, family make-up and support network, self-concept, and geographic area of residence.

Factors Influencing One's Health Behavior

In addition to those factors that influence one's state of health, there are other factors influencing one's health behavior. Culture and family influences are two examples. Persons can usually control their health behaviors and can choose healthy or unhealthy activities. In contrast, persons have little or no choice over their genetic make-up, age, sex, physical environments, culture, or area of residence.

Individual behavior is based upon individual perceptions and modifying factors that influence one's own's life.

Individual perceptions generally include:

- The importance of health.
- Perceived control over life.
- Any perceived threat of a specific illness.
- Perceived susceptibility toward a specific illness or disease.
- Perceived seriousness of a specific illness or disease.
- Perceived benefits of preventive action toward a specific illness or disease.
- Perceived value of early detection.

Individual factors that modify a person's perceptions generally include:

- Demographic variables such as age, sex, and race.
- Interpersonal variables, such as level of concern from a significant other, family patterns of health care, and interactions with health care professionals.
- Situational variables, such as cultural acceptance of specific health-related behaviors.

Stages of Illness

Throughout the evolution of medicine and health care, various scientists and researchers have described the various stages of illness. These stages, according to most researchers, are classified according to *basic*, or physical stages, and *sequential* stages.

Basic Stages of Illness

There are five basic stages of illness. The first stage of illness is the time during which most persons come to believe something is wrong with them. We refer to this as the *symptom experience* stage. The basic stage at which a patient signals an acceptance of his or her illness is called *acceptance of the sick role*. At this time, the patient has generally decided this his or her symptoms or concerns are sufficiently severe to suggest presence of illness.

Medical care contact is the third basic stage of illness. During this stage, the patient seeks the advice of a health care provider, either on his or her own initiative or at the urging of a significant other. Persons going through this stage are usually seeking three types of information: validation of a real illness, explanation of the symptoms in understandable terms, and either reassurance that they will recover or a prediction of the probable outcome.

During the fourth basic stage of illness, known as the *dependent patient role* stage, the patient has already gone through the process of having his or her illness validated by the health care professional. The person now sees him- or herself as a patient, dependent on the professional for help. Often during this stage the person becomes reluctant to accept a professional's recommendation.

The final basic stage of illness is the *recovery* or *rehabilitation* stage. During this time, the patient learns to give up the sick role and return to former roles and functions. For persons suffering from acute illnesses, the time as an ill person is generally short, and recovery is usually rapid. Thus, they find it relatively easy to return to their former lifestyles.

Sequential Stages of Illness

There are 11 sequential or consecutive stages of illness. Many of these stages vary in duration, and some may occur simultaneously. They usually include the following:

- *Symptoms experience:* During this time the patient begins to experience actual symptoms, such as pain; he or she may also become more aware that there may be a problem and begin to respond with fear or anxiety, while giving the symptoms a label and a meaning.
- *Self-treatment:* During this time, the patient often tries to treat him- or herself, especially if it is believed that the symptoms are not serious.
- *Communication to significant others:* During this time, the patient begins to communicate and confide their symptoms to support persons, a significant other, or health care providers.
- *Assessment of symptoms:* During this time, an assessment of the patient's symptoms may be made by the patient, support persons, or a health care professional.
- *Assumption of the sick role:* At this point, the person accepts his or her illness, is declared sick, and accepts the role of a sick person.
- *Expression of concern:* During this stage, support persons and friends begin to offer the patient sympathy and express concern; some persons may even recommend practitioners to their sick friend.
- *Assessment of probable efficacy or appropriateness of treatment:* During this time, the person may begin to assess the variety of treatments available; a number of variables may affect this assessment, including previous experience and availability of information.
- *Selection of a treatment plan:* During this time, a treatment plan may be selected with or without the advice of a health care professional; such factors as cost, time, knowledge, and effects are often involved during the selection process.
- *Implementation of treatment:* During this stage, the patient allows the health care practitioner to implement the prescribed treatment.
- *Evaluation of the effects of treatment:* During this stage, the patient begins to see and accept the possible outcomes of treatment, including a full recovery with a change in treatment, recovery with some disability, and no recovery; at this time, the patient may also choose to exercise his or her right to completely reject or terminate all treatment.
- *Recovery and rehabilitation:* During this final stage, the patient begins to gradually return to usual social roles.

Sick Role Behavior

Sick role behavior has to do with activities that are undertaken by those who consider themselves ill for the purpose of getting well. Such behavior usually includes four aspects: the patient's inability to be held responsible for his or her own condition; allowing the patient to be excused from certain social roles and tasks; encouraging the patient to try to get well as quickly as possible; and assisting both the patient and family members in seeking competent health care and treatment.

Effects of Illness on Family Members

A person's illness often affects the family or a significant other. The type of effect and its extent depend chiefly on three factors: which member of the family is ill, how serious and long the illness is, and what cultural and social customs the particular family holds.

Changes that most frequently occur in a family include role changes, task reassignments, increased stress due to anxiety about the outcome of the illness and conflict about unaccustomed responsibilities, financial problems, loneliness as a result of separation and pending loss, and change in social customs.

Each member of the family may be affected differently, depending, for example, on whether a grandparent, the father of a nuclear family, or a teenager is ill. Each person plays a different role in the family, and each supports the family in different ways. Parents of young children, for example, have greater family responsibilities than parents of grown children.

The degree of change that family members experience is often related to their dependence on the sick person. For example, when a child is ill, there are few changes other than added responsibilities directly related to the child's illness. When the mother is ill, however, many changes are often necessary because other family members must assume her functions.

Effects of Hospitalization

Normal behavior patterns generally change with illness; with hospitalization, however, the change can be even greater. Hospitalization usually disrupts a person's privacy, autonomy, lifestyle, roles, and economics.

When a person enters a hospital or other health care institution, such as a skilled nursing care facility, the greatest loss comes to the patient's privacy. *Privacy* has often been described as a comfortable feeling reflecting a deserved degree of social retreat. It is a personal, internal state that cannot and should not be imposed on without the agreement of the person seeking the privacy. Its boundaries are considered to be highly individual. People need varying degrees of privacy and establish their boundaries for it. When these boundaries are crossed, they feel invaded. Hospital personnel sometimes show little concern for a patient's privacy. Patients are asked to provide information they often consider private, they may share a room with a stranger, and their health is frequently discussed with many health care providers.

A person's *autonomy* deals with those activities a person performs which allow him or her to be completely independent and self-directed without outside control. People vary in their sense of autonomy; some are accustomed to functioning independently in most of their life activities, while others are accustomed to direction from others. Hospitalized patients frequently give up much of their autonomy. Decisions about meals, hygienic practices, and sleeping habits are frequently made for them. The loss of individuality is often difficult to accept, often leading to a patient feeling dehumanized.

Hospitalization often makes a drastic change in a person's lifestyle and social role. Many hospitals determine when people wake up and when they should go to sleep. Food in a hospital is usually mass produced, and individual preferences in taste are not always accommodated.

Hospitalization not only affects a person's lifestyle; it frequently affects his or her life roles. A woman or man may no longer be capable of earning wages; a single parent may be unable to fulfill his or her normal parental responsibilities. Such changes not only affect the patient's physical and emotional well-being, but just as important hospitalization can often place a genuine financial burden on patients and their families. Even though many patients have health insurance, it may not reimburse all costs; in addition, many patients lose wages while they are hospitalized.

As a health care provider, you can become aware of costs sustained by the patient and, by doing so, provide care and treatment that is as economical as is safely possible. For example, you can use only the minimum number of supplies necessary for providing a patient's breathing treatment. You can also support activities that promote health and that return patients to their normal activities as soon as possible.

Patterns and Trends in Health and Illness

The health of most Americans throughout the United States is steadily improving. This is primarily due to early preventative efforts based on new knowledge obtained through research; improvements in sanitation, housing, nutrition, and immunization essential to disease prevention; and individual members of society taking measures to promote health and prevent disease. According to the U.S. Surgeon General, there are other factors, or health risks, for which individuals are responsible, which make them more susceptible to illness and disease. These factors include *cigarette smoking*, *alcohol and drug abuse*, *injuries* sustained from automobile and other types of accidents, and *occupational injuries* suffered as a result of not following safety standards in the work place. All of these risky behaviors present challenges to the health care worker, especially members of the cardiovascular team, whose goal is to promote and provide the patient with care and treatment relative to his or her cardiovascular system.

Measuring Trends in Health Care Delivery

Evidence of current health status and changes in the last few decades are measured in various ways. Measurements are made of longevity or life expectancy, the ways people feel about their health, mortality rates and causes, morbidity rates and causes, the types of health behaviors people practice, and the amount and kind of health services used.

Life Expectancy

In 1960, life expectancy at birth was about 70 years. By 1983, it reached nearly 75 years. Differences, however, exist among races and between males and females. In 1982, for example, the life expectancy for a white male was 71.5 years. For a white female, it was 78.8 years. For an African-American male, it was 64.9 years, while it was 73.5 years for an African-American female.

How People Feel about Their Health

The way a person feels about his or her health is one way in which a person's health status can be ascertained. Over the past several years, persons of all ages have generally held the same perception about their health status. In 1981, for example, when most individuals were asked how they felt about the state of their health compared to other persons of the same age, approximately 87.3 percent said they felt good or excellent. What is most interesting is that about 70 percent of persons who stated they felt good or excellent were elderly persons.

Summary

In this chapter, we discussed the concepts of health, wellness, illness, and disease, and explained their relationship to one another. We also talked briefly about nontraditional views of health and illness, as well as about health beliefs and behaviors, the basic and sequential stages of illness, what is meant by sick role behavior, and the effects illness has on both the patient and family members. Finally, we discussed the effects of hospitalization and the role it plays in caring for an ill person and identified various patterns and trends that influence both health and illness.

Review Questions

1. Briefly explain the meaning of *disease*.
2. Briefly define the meaning of *health*.
3. Briefly define the meaning of *wellness*.
4. Briefly define the meaning of *illness*.
5. The _____ rate is the ratio of sick to well people in a given population.
6. The term _____ pertains to the extent to which a person's behavior coincides with that recommended by health care providers.
7. What model of health care delivery incorporates the most comprehensive view of health?
8. What model of health care delivery sees disease as a failure in the person's ability to adapt?
9. What model of health care delivery is based upon the relationship of all humans to their internal and external environments?
10. Health _____ are defined as those actions which a person takes in order to understand his or her health and prevent illness and injury.

Medical Law and Ethics

Performance Objectives

Upon completion of this chapter, you will be able to:

1. Delineate the difference between medical law and medical ethics.
2. Discuss the Code of Ethics which administrative medical services assistants and workers are morally bound to follow.
3. Identify at least four examples of ethical behavior for the administrative medical services worker.
4. Discuss the Patient's Bill of Rights.
5. Discuss the purpose of licensing medical personnel.
6. Explain the Rule of Personal Liability.
7. Briefly discuss the Good Samaritan Act.
8. Demonstrate an understanding of specific patient consent forms.
9. Discuss the legal implications of a patient's medical record.

Terms and Abbreviations

Battery any unlawful touching of another person without his or her consent, regardless of resultant injury.

Civil law differentiated from criminal law in that it enforces private rights and liabilities.

Defamation any attack on a person's reputation; called "libel" when the written word is used and "slander" when the spoken word is used.

Duty of care a lawful obligation which protects the patient by requiring the health care worker to provide safe care to the patient.

Ethics moral standards and principles.

Good Samaritan Act a law, differing in various states, that protects a health care worker from a malpractice suit when coming to the aid of a person in an emergency situation.

Informed consent the act of informing the patient of the risks and alternatives to a medical procedure; the patient is asked to sign a consent form acknowledging the assumed risks.

Malpractice literally, "bad practice" by a professional; any care below an expected standard which results in injury.

Medical Practice Acts statutes dealing with licensing of medical personnel.

Negligence occurs when any member of the health care delivery system fails to do something for a patient that another person of the same training and position would do under the same or ordinary circumstances.

Patient's Bill of Rights a document produced and adopted by the American Hospital Association which states what a patient has the right to expect during his or her medical treatment.

Privileged information information given by a patient to a medical care provider which cannot be disclosed without the consent of the patient.

Reasonable care defined as acts or services performed by a health care professional who is then protected by law if it is proven that he or she acted reasonably compared to fellow workers.

Rule of Personal Liability dictates that all health care workers be held responsible for their own conduct.

The laws and ethical codes of conduct for the medical profession must be understood by the administrative medical services worker. These laws and ethics codes protect all members of the medical profession, as well as the patient.

Medical ethics has been important in the study of medicine as far back as 400 B.C., when Hippocrates wrote the Hippocratic Oath, a document that first developed the standards of medical conduct and ethics. Today, many individual health care professional associations have adopted their own codes of ethics in order to govern their health care workers. Health unit coordinators, medical records technicians and clerks, medical transcriptionists, and administrative medical assistants all have ethical codes that provide guidelines for professional behavior.

The practice of medicine today exists within a framework of laws. Such laws vary from state to state, so it is important to know your state's laws governing the administrative medical services worker, as well as federal and local statutes.

Understanding Medical Ethics and Medical Law

Medical ethics are concerned with whether the administrative medical services worker's actions are right or wrong, whereas *medical law* is more concerned with focusing on the actions' legality. *Ethical behavior* deals with behavior that is specific and represents the ideal conduct for a certain group. Each group of health care professionals that requires its members to obtain a license to function in their position has drafted and adopted a specific code of ethics. These ethics are based upon moral principles and practices. If you are accused of unethical behavior by your professional association, you can be issued a warning or expelled from the association.

Medical law is concerned with the legal conduct of the members of the medical profession. There are federal, state, and local laws that must be followed, and violation of such laws may subject the offender to civil or criminal prosecution. Professional licenses can be taken away, fines can be levied, and prison sentences can be imposed. Often, the line between what is unethical and what is illegal can be unclear. If you, for example, as a member of the health care delivery system, decided you were going to be rude to a patient, that would be unethical behavior. But if you decided to inform the patient's neighbor of his or her disease, that would be illegal behavior. In addition, you can also be sued by a patient for *defamation*, which is defined as an attack on a person's reputation. When the defamation is written, it is called *libel*, and when spoken, *slander*.

Another example of the difference between unethical and illegal behavior concerns reading a friend's chart in a part of the hospital where you do not work. If you read it out of curiosity, the behavior is unethical. If you talk about it, it is illegal. If an action is illegal, it is always unethical. However, it can be unethical without being illegal. Remember, ethics represent the highest standards of behavior.

Ethical Behavior

As an administrative medical services worker, you are responsible for displaying ethical behavior at all times. This means maintaining the highest level of ethical conduct. To accomplish such a task, you should always:

- Respect the rights of all patients to have opinions, lifestyles, and beliefs that are different from your own.
- Remember that everything seen, heard, or read about a patient is considered confidential and should not leave the job site.
- Be conscientious in doing your work, doing the best you can at all times.
- Be ready to be of service to patients and co-workers at any time during the workday.
- Let the patient know that it is a privilege for you to assist him or her.
- Follow closely the specific rules of ethical conduct prescribed by your employer.

The Patient and the Health Care Worker

The Patient's Bill of Rights

Awareness of the patient's rights is the responsibility of all members of the health care team. Because this is such a vital and important aspect of providing care and treatment to the patient, the American Hospital Association (AHA), in its wisdom, felt it was necessary to establish a document which identified what the patient could expect from those individuals responsible for patient care. As a result, the *Patient's Bill of Rights* was adopted by the organization. The intent of this document is to make both members of the health care system and patients aware of what the patient has a right to expect. According to the AHA, the patient has a right to:

- Considerate and respectful care.
- Obtain from the attending physician complete current information concerning diagnosis, treatment, and prognosis, in terms that they can reasonably be expected to understand.
- Informed consent, which should include knowledge of the proposed procedure, along with its risks and probable duration of incapacitation; in addition, the patient has a right to information regarding medically significant alternatives.
- Refuse treatment to the extent permitted by law and be informed of the medical consequences of such action.
- Have case discussion, consultation, examination, and treatment conducted discretely, with those not directly involved in the patient's care, obtaining permission from the patient for their presence.
- Expect that all communication and records pertaining to his or her care be treated in a confidential manner.
- Expect the hospital to make a reasonable response to a request for services and provide evaluation, service, and referrals as they may be indicated by the urgency of the case.
- Obtain (1) information regarding any relationship the hospital may have with other health care and educational institutions affecting his or her care, and (2) information about the relationshp among named individuals who are treating the patient.
- Be advised if the hospital proposes to engage in or perform human experimentation affecting his or her care or treatment, and be given the right to refuse participation in such research projects.
- Expect reasonable continuity of care.
- Examine and receive an explanation of the bill regardless of the source of payment.
- Know what hospital rules and regulations apply to his or her conduct as a patient.

Licensing of Health Care Workers

The medical profession and many of the individual health care professions are legally regulated throughout the United States by the issuing of licenses and certificates. All 50 states require the licensing of hospitals. The statues dealing with the individual licensing and certification of the individual health care professionals are commonly referred to as *medical practice acts*.

Licenses can be revoked or suspended when a medical professional has been found guilty of violating various statutes involved in the licensing process. Grounds for losing a medical license include serious crimes such as murder, rape, and arson. Crimes of "moral turpitude," such as tax crimes, minor sexual offenses, and false statements while applying for a license, can also be grounds for loss of the license. Other crimes which may cause the worker to have his or her license suspended or revoked include incapacity due to insanity or excessive use of alcohol or drug addiction.

Protection Under the Law

In the health care environment, both the patient and the health care worker must be assured protection under the law. For the patient, this means being assured of safe care. For the health care professional, it means being protected from irresponsible law suits.

The patient is protected by a process known as *duty of care*. This entitles the patient to safe care by making it mandatory that he or she be treated according to the common or average standards of practice expected in the community under similar circumstances. The duty of care also provides that the patient be treated with *reasonable care*, that is, protection of the health care professional by law if it can be proven that he or she acted reasonably compared to fellow workers of the same or similar training in a situation of the same nature. If it is proven that the health care worker failed to meet such a standard, resulting in harm to the patient, negligence may be proven.

Negligence is the failure to give reasonable care or the giving of unreasonable care. The patient is harmed because the health professional did something wrong or failed to do something that he or she should have done under the circumstances.

The Good Samaritan Law and Medical Malpractice

The *Good Samaritan Law* is a law which, in its entirety, addresses the problem of medical malpractice suits for a physician or any trained health care professional who comes upon an accident scene and attempts to render aid to the victim. The law, which has been enacted in all 50 states, encourages members of health care professions to offer treatment without fearing the possibility of a malpractice suit. Laws throughout the country do differ, so it is wise to check the law in your own state in order to determine what professional liability may exist during an emergency situation.

The term *malpractice*, unfortunately, is a term that is familiar to all of us, because of the excessive number of lawsuits being filed and settled throughout the country. For the medical worker, it seems that the higher the educational level and requirements of the worker, the greater the likelihood that they may be held responsible for their actions.

When used in the medical professions, malpractice refers to any misconduct or lack of skill that results in the patient's injury. A patient who thinks that his or her physician has been negligent in diagnosing and treating an illness or accident may file a medical malpractice claim. Most claims are made against physicians, although any employee working in the health care environment can be named in a malpractice lawsuit. Most insurance companies who issue medical malpractice policies on physicians take into account that the policy will also cover the physician's employees, but medical office or hospital employees may wish to purchase their own insurance policy, which is usually quite inexpensive.

Physicians are liable for the actions of their employees while the employee is on duty, for example, if an administrative medical assistant accepts a lab specimen from two patients and then mislabels the specimens, consequently causing one patient to be told that his specimen is normal and the other to be administered antibiotics. If subsequently, the patient who has been told that everything is fine develops an infection which could have been prevented if the antibiotics had been administered, the doctor could be sued for the negligence of the employee mislabeling the specimens.

In health care, under the *Rule of Personal Liability*, all individuals are held responsible for their own personal conduct. In a medical malpractice suit, such as in the example just described, the physician and the administrative medical assistant could be held jointly liable for medical malpractice.

An administrative medical assistant can also be held jointly responsible if he or she works for a doctor who is involved in an illegal act, is aware of the crime, but fails to report it. Physicians must also report crimes that they learn about when practicing medicine, such as a shooting, child or elder abuse, or rape. An administrative medical services workers can also be held jointly responsible with the physician if he or she fails to report such crimes. In some cases, protecting patient confidentiality and the patient's right to complete privacy regarding their records does not apply. Births, deaths, communicable diseases, and crimes are all examples of times when the physician is bound by law to report what has occurred. Failing to report these types of cases may result in the medical assistant also being held liable.

In some instances, the law is very specific regarding confidentiality and reporting of information. In cases dealing with patients suffering from AIDS, for example, there are laws and regulations considering the confidential nature of the patient's illness. However, these regulations seem to be constantly changing, therefore it is important to keep current regarding information about the laws and regulations governing confidentiality in the treatment of this disease in the state in which an individual works.

Obtaining Patient Consent

When a doctor makes a diagnosis and recommends a specific mode of treatment, the patient has the responsibility to accept or refuse such diagnosis and treatment. The physician has the responsibility of informing the patient in words that he or she can understand as to the risks and alternatives of any suggested procedure. The patient has the responsibility of deciding whether or not to accept all the explained risks. Once it is ascertained that the physician has properly explained the procedure and the patient fully understands it and the risks, a consent form must be signed by the patient, indicating that he or she fully accepts the risk of the procedure. This process is called *informed consent*. As one of your responsibilities, you may be asked to prepare the consent form (Figure 3-1) for any type of procedure, whether it is to be performed in the office or in the hospital. Consents are also required before an experimental procedure or prior to any other unusual procedure. Consent forms are also used prior to the administration of any experimental drugs or medication.

In cases in which the patient may have difficulty understanding or speaking English, the consent form must be translated or prepared in the patient's native language. A patient who has not been properly informed through the "informed consent" process can sue the physician for medical malpractice.

Specific guidelines have been established with regard to the details and the signing of the consent form. These include the following:

- Always make sure that the patient fully understands the consent form and realizes what he or she has signed; patients who are mentally handicapped should be given an explanation that can be understood completely, with as few confusing terms as possible.
- The patient must never be forced to sign the consent form, and must not be allowed to sign it under the influence of alcohol or drugs.
- All signatures must be witnessed, dated, and signed in ink, with the full legal names used.
- Any adult over 18 years of age may sign his or her own consent unless the patient is incompetent and has a guardian or there is an

Administrative Medical Services

Patient or someone acting for the patient agrees to the following terms of hospital admission.

1) **MEDICAL TREATMENT:** Patient will be treated by his/her attending doctor or specialists. Patient authorizes Hospital to perform services ordered by the doctors. Special consent forms may be needed. Many doctors and assistants (such as those providing x-rays, lab tests, and anesthesiology) may not be Hospital employees and are responsible for their own treatment activities.
2) **GENERAL DUTY NURSING:** Hospital provides only general nursing care. If the patient needs special or private nursing, it must be arranged by the patient or by the doctor treating the patient.
3) **MONEY AND VALUABLES:** The Hospital has a safe in which to keep money or valuables. It will not be responsible for any loss or damage to items not deposited in the safe. The Hospital will not be responsible for loss or damage to items such as glasses, dentures, hearing aids and contact lenses.
4) **TEACHING PROGRAMS:** The Hospital participates in programs for training of health care personnel. Some services may be provided to the patient by persons in training under the supervision and instruction of doctors or hospital employees. These persons may also observe care given to the patient by doctors and hospital employees. Photos or video tapes may be made of surgical procedures.
5) **RELEASE OF INFORMATION:** The Hospital may disclose all or any part of the patient's medical and/or financial records (INCLUDING INFORMATION REGARDING ALCOHOL OR DRUG ABUSE), to the following:
   a. **Third Party Payors:** Any person or corporation, or their designee, which is or may be liable under a contract to the hospital, the patient, a family member, or employer of the patient, for payment of all or part of the hospital's charges, including but not limited to, insurance companies, utilization review organizations, workman's compensation payors, hospital or medical service companies, welfare funds, governmental agencies or the patient's employer;
   b. **Medical Audit:** The Hospital conducts a program of medical audit and the patient's medical information may be reviewed and released by employees, members of the medical staff or other authorized persons to appropriate agencies as part of this program.
   c. **Medical Research:** Information may be released for use in medical studies and medical research.
   d. **Other Health Care Providers:** Information may be released to other health care providers in order to provide continued patient care.

I understand that the authorization granted in items 5. a, b, c and d may be revoked by me at any time, except to the extent to which action has been taken in reliance upon it. The authorization will stay in effect as long as the need for information in items 5. a, b, c and d exist.

I have read and understand this Admissions Agreement, have received a copy and I am the patient, the parent of a minor child or the court appointed guardian for the patient and am authorized to act on the patient's behalf to sign this Agreement.

_____ _____
WITNESS   PATIENT / PARENT OF MINOR CHILD / COURT APPOINTED GUARDIAN
                           (PLEASE CIRCLE THE CORRECT TITLE)

_____
DATE

MEDICAL POWER OF ATTORNEY A.R.S. §14-5501: I appoint _____ as my agent to act in all matters relating to my health care, including full power to give or refuse consent to all medical, surgical and hospital care. This power of attorney shall be effective upon my disability or incapacity or when there is uncertainty whether I am dead or alive and shall have the same effect as if I were alive, competent, and able to act for myself.

_____ _____
WITNESS   PATIENT

FINANCIAL AGREEMENT

I agree that in return for the services provided to the patient, I will pay the account of the patient, and/or prior to discharge make financial arrangements satisfactory to the hospital for payment. If the account is sent to an attorney for collection, I agree to pay reasonable attorney's fees and collection expenses. The amount of the attorney's fee shall be established by the Court and not by a Jury in any court action. A delinquent account may be charged interest at the legal rate.

If any signer is entitled to hospital benefits of any type whatsoever under any policy of insurance insuring patient, or any other party liable to patient, the benefits are hereby assigned to hospital for application on patient's bill. However, IT IS UNDERSTOOD THAT THE UNDERSIGNED AND PATIENT ARE PRIMARILY RESPONSIBLE FOR PAYMENT OF PATIENT'S BILL.

IN GRANTING ADMISISON OR RENDERING TREATMENT, THE HOSPITAL IS RELYING ON MY AGREEMENT TO PAY THE ACCOUNT. EMERGENCY CARE WILL BE PROVIDED WITHOUT REGARD TO THE ABILITY TO PAY.

_____ _____
PATIENT   OTHER PARTY AGREEING TO PAY

_____ _____
WITNESS   RELATIONSHIP TO PATIENT

_____
DATE

Figure 3-1
Consent form. (Reprinted with permission from LaFleur, M. W. and Starr, W. K. *Health Unit Coordinating.* Philadelphia: W. B. Saunders, 1986, p 483.)

emergency, in which case two physicians must sign the consent form.
- Married minors may sign their own consent forms for treatment.
- Unmarried minors must have a consent signed by one parent or legal guardian. However, consent of both parents is usually suggested; a stepparent may not sign a consent form.
- An emancipated minor, who is under the age of 18 and who has been declared by a court of law to be legally responsible for him- or herself, may sign his or her own consent form.
- Because any break in the skin may be considered an operation, a consent form must be signed in order to avoid liability for battery.
- Telephone consents are valid in an emergency

situation, provided that the telephoned consent is witnessed by two people and is immediately followed by a written confirmation.
- A consent is valid for a reasonable time after signing, so long as there is no change in the anticipated procedure.

Patient Medical Records and the Law

The patient's medical record is considered a legal document and, as such, is the property of the physician, if the patient is an out-patient, and the property of the hospital, if the patient has been hospitalized. It is extremely important that these records be as accurate, complete, up to date, and neat as possible, both to protect members of the medical staff from any future litigation, and to serve as evidence of the truth if there is a lawsuit or court case regarding the patient's care or treatment. While all patient records are considered confidential, any or all parts of one may be subpoenaed and used during a court action. Therefore, most hospitals and private medical practices make it a standard practice to obtain a signed "release of information" form from the patient when they are first seen or admitted into the hospital.

If you are required to write in the patient's medical records, always remember to use only permanent ink and never erase an entry. If an error has been made, simply cross the error out, using only one line, initial it, and rewrite the correct entry above or next to the original entry. No documentation written in pencil or with erasures is acceptable, and any record with either of these can be automatically rejected as legal evidence.

Summary

In this chapter, we discussed some of the very important concepts dealing with medical law and ethics, including the differences between each and what constitutes medical negligence and medical malpractice. We also talked about ethical behavior on the part of the administrative medical services worker, as well as the role of the Patient's Bill of Rights and what the patient has the right to expect from those health care professionals caring for him or her. In addition, we talked about the need to license medical personnel, as well as the purposes of the Rule of Personal Liability and the Good Samaritan Act. Finally, we talked about patient consent forms and discussed the legal implications of the patient's medical record.

Review Questions

1. Briefly explain what is meant by *duty of care*.

2. Briefly explain the concept of *medical ethics*.

3. A patient's _____ is a document which states what a patient has the right to expect during his or her medical treatment.

4. _____ pertains to any attack on a person's reputation.
   a. ethics
   b. negligence
   c. defamation

5. _____ pertains to any care which is provided to a patient which is below an expected standard and which can result in injury.

6. When the written word is used to defame someone, it is called _____; when the spoken word is used to defame someone, it is called _____.

7. Give at least two examples of ethical behavior:
   a. _____
   b. _____

8. Briefly define the *Good Samaritan Act*.

9. _____ pertains to any member of the health care delivery team failing to do something for a patient that another person of the same training and position would do under the same or ordinary circumstances.
   a. negligence
   b. defamation
   c. malpractice

Health And Safety in Administrative Medical Services

Performance Objectives

Upon completion of this chapter, you will be able to:

1. Identify factors associated with making some people more susceptible to infection than others.
2. Explain the difference between medical and surgical asepsis.
3. Describe and be able to demonstrate how to perform correct handwashing techniques.
4. Explain the relationship between good body alignment and practicing good body mechanics.
5. Identify specific types of activities that will ensure a safe environment for both patients and members of the administrative medical services team.
6. Identify potential emergency situations and how to respond to them.
7. Discuss the skills necessary for providing privacy for patients seen in the hospital and in the administrative medical services environment.
8. Discuss the role of the administrative medical services worker in maintaining treatment rooms within the health care environment.

Terms and Abbreviations

Alignment having parts in their proper relationship to one another.
Asepsis the absence of disease-producing organisms.
Bacteria one-celled microorganisms which are capable of causing fermentation, decay, and, in some cases, disease.
Bacteriology the study of bacteria.
Balance keeping an object in a steady position, or stable.
Base of support the area on which an object rests; when the body is upright, the feet form the base of support for the body.
Body mechanics the efficient use of the body during activity.
Center of gravity the point at which the mass of the body is centered.
Infection process which occurs when pathogens attack a person and produce signs and symptoms of an illness.

Line of gravity the imaginary vertical line that passes through the center of gravity.
Medical asepsis medical techniques used to reduce the number of and prevent the spread of pathogens.
Musculoskeletal system comprised of the bones, joints, and muscles of the body.
Nosocomial infection an infection acquired by the patient while being cared for in the hospital.
Pathogen any disease-producing organism.
Posture the position of the body parts in relation to one another.
Susceptible refers to one having a very low resistance to disease.

For as many years as people have lived on earth, they have suffered from infectious diseases. For a very long time, however, it was thought that an infection might be part of the actual healing process. No one knew how or why infections were transmitted. Physicians would move from one patient to another without ever washing their hands or changing their dirty lab coats. And who ever heard of wearing sterile gloves or isolation gowns? It wasn't until as recently as the middle of the nineteenth century that the germ theory of disease was even suggested.

Asepsis and Infection Control

Asepsis is defined as the absence of microorganisms that are capable of causing disease. As it is now practiced, asepsis is based upon the many discoveries of scientists from around the world. One of the greatest of these was Louis Pasteur. Born in France, Pasteur was able to make tremendous strides in the prevention and treatment of disease. It is for his entire body of work that he is considered the founder of the study of bacteriology. In fact, today there are millions of people around the globe who owe their lives to this man. It was his work that ultimately proved that germs could be spread through the air, but that the spread of such germs could be controlled. In addition, he also proved that these germs could be killed. Because of his knowledge and his work in the study of bacteriology, Pasteur was also one of the very first practitioners to practice surgery with the use of antiseptics to prevent infection.

Susceptibility and Infection

Some people tend to be more susceptible to infection than others. One factor that seems to give rise to this fact has to do with the age of the person. The very young and the very old seem to develop infections easily. This is because a small child does not have the mature immune system that is needed to fight off infections, nor does the elderly person, simply because of his or her age.

People with disabilities or who have body systems which have become weakened because of their disabilities also seem to be more susceptible than others to infection. One example of this is the patient who suffers from a spinal cord injury and becomes more susceptible to urinary bladder infections because of the inability to empty the bladder effectively. Another example is the patient who has been diagnosed with multiple sclerosis who is also more susceptible to pneumonia because of impaired lung capacity and muscle paralysis.

People who are well nourished are less susceptible to infection than the malnourished. Nutrition plays an important role in the prevention of infection and in recovery from illness. Without proper nourishment, our bodies would not be able to function properly. A good diet, therefore, can provide the substances the body needs to repair itself.

Ingestion of medications can also increase one's incidence of infections. Drugs used in the treatment of leukemia, for example, decrease the number of white blood cells being manufactured by the body. These cells are part of the body's immune system and are necessary to fight infection.

Patients who have been diagnosed with acquired immune deficiency syndrome, or AIDS, also have a weakened immune system. These patients are very susceptible to all kinds of infections.

The Infection Cycle

Microorganisms move from place to place in a cycle (Figure 4-1). If the cycle is broken, the microorganisms cannot grow, spread, or cause disease. A person with an infection acts as a reservoir for the microorganisms, allowing them to grow and multiply. The infection may be spread by such means as the hands, bed clothes or linens, and equipment. Microorganisms can even be transmitted by a simple cough or a sneeze. The person exposed and receiving the organism is called the *host*. The infectious agent may enter the host through an opening in the skin, mouth, or nose. If the host is unable to fight off the infecting organism, that person will begin to show signs or symptoms of a disease or an infection.

Germs and microorganisms are everywhere: in the air, in the soil, in the food and water we consume, and even on other people. The skin is called our first line of defense and if this is broken, germs and organisms may enter the body.

In order for an infection to occur, the environment for the microorganisms to grow and multiply must be just right. This means that food and moisture are needed. Oxygen is also needed, except in those cases in which the microorganisms are *anaerobic*, that is, do not require oxygen for growth. There are a few bacteria which fall into this category. One common example of an anaerobic bacteria is the one that causes tetanus. It generally grows well in a puncture wound because it is anaerobic. Bacteria needing oxygen for growth are called *aerobic* bacteria. Normal body temperature is best for most bacteria to

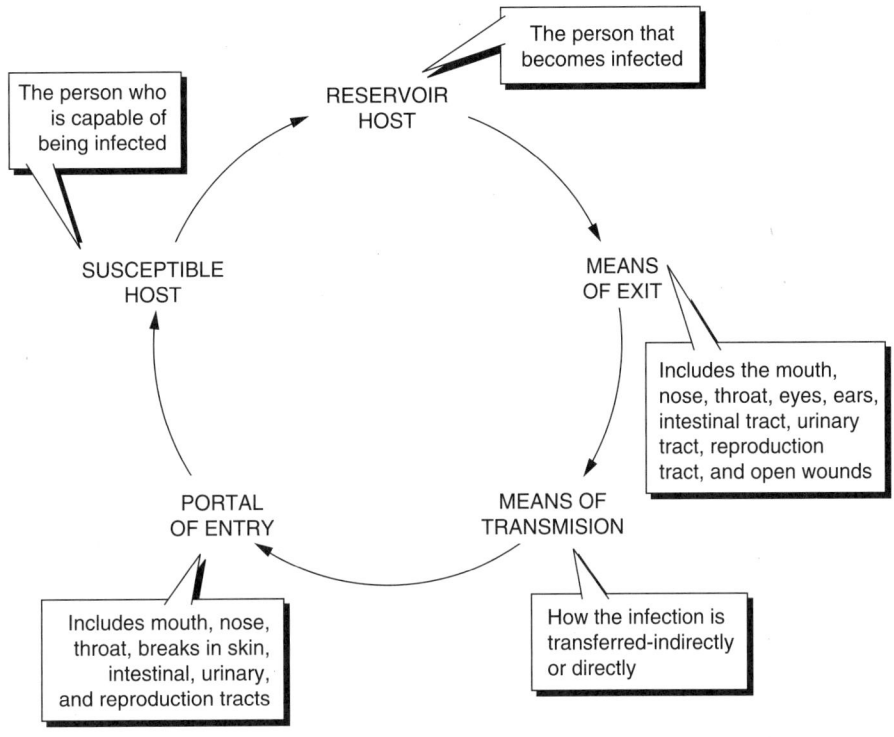

Figure 4-1
The infection cycle.

grow and multiply. High temperature kills most microorganisms, while low temperatures retard and tend to slow the growth rate. Finally, many bacteria need darkness to grow and multiply. When exposed to light, they die. Food, moisture, oxygen, warm temperature, and darkness all encourage the growth of bacteria. To break the infection cycle, at least one or more of these factors must be modified.

Practicing Medical and Surgical Asepsis

The most common means by which organisms are transmitted are the hands, since they carry germs to the mouth, eyes, and nose and to other people. The best way to inhibit this transmission is through the practice of medical and surgical asepsis.

Medical asepsis is the practice of reducing the numbers and preventing the spread of microorganisms. It is a practice which can be accomplished in a number of ways, such as covering your nose and mouth when you sneeze or cough, washing your hands before you handle food and after you have used the restroom, washing your hands before and after you have treated a patient, wearing a clean uniform or lab coat each day to work, practicing good body hygiene, and keeping dirty items, such as linen and equipment, separate from clean items in your office or your department.

Surgical asepsis is the practice of eliminating all microorganisms, both pathogenic or disease-producing and nonpathogenic. Examples of surgical asepsis include using sterile equipment for giving injections, performing venipuncture, starting IVs, and surgery, and using sterile dressings and bandages to cover open wounds.

Whether you work in the clinical environment or the administrative services area, one of the most important tasks you will be faced with in health care is to provide the very best possible surroundings for you and the patients intrusted to your care. A clean, dry, light, and airy department will help prevent the growth of pathogens.

One of the easiest and best ways of preventing the spread of bacteria is through using correct *handwashing* techniques. Figure 4-2 shows the correct procedure for handwashing. It is a skill used not only for your protection, but also for those around you—patients, co-workers, and visitors. You should wash your hands before and after doing treatments and at any other time your hands become dirty.

When washing your hands, always remember to use plenty of flowing water, a good germicidal soap, and friction. It is the friction, or strong rubbing movement of both hands against one another, that loosens the bacteria from the skin. The water is directed such that it flows from the wrist down toward the fingers. This carries the suds and dirt down the drain. Never lean against the sink or allow your uniform or lab

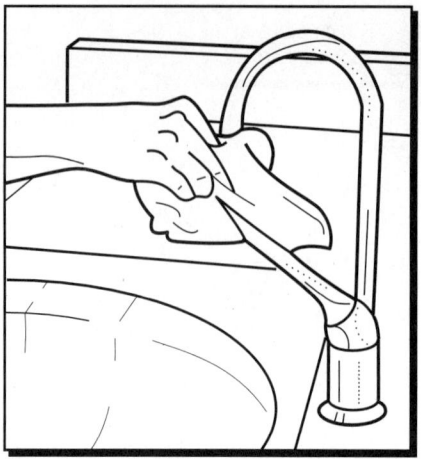
Step 1: Never touch sink with your uniform.

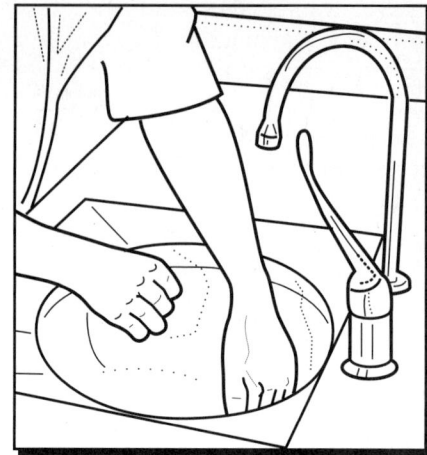
Step 2: Wet hands.

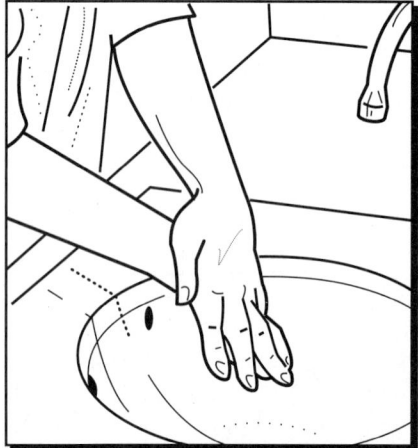
Step 3: Use strong rubbing movements.

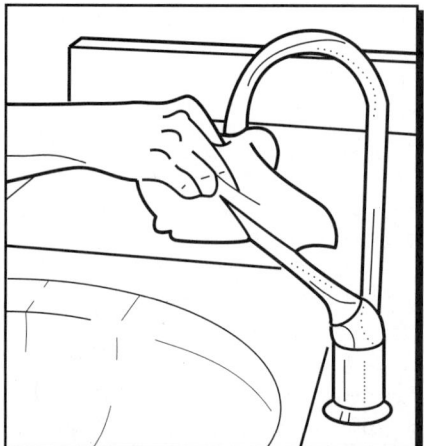
Step 4: Rinse well, with hands lower than wrist.

Figure 4-2
Steps in handwashing to prevent spread of bacteria.

coat to touch it. And, after drying your hands with a paper towel, you must always remember to use the towel to turn off the faucet before you discard it in the trash. After you have finished washing your hands, you may also apply some lotion to them to prevent any chapping and to keep the skin soft.

Medical Asepsis in the Administrative Medical Services Department

In most large medical facilities, as well as in the majority of individual medical practices and out-patient clinical settings, administrative medical services are usually self-contained business or clerical departments. This makes it a great deal easier to keep these areas clean and thus prevent the growth and spread of bacteria and microorganisms. Tables, desks, chairs, and business equipment should have adequate space between each item so that they may be kept spotlessly clean.

Any article which may have been used by an individual patient must be cleaned after each use. This prevents the possible spread of pathogens between patients and provides protection to your co-workers from the possibility of contracting an infection from a patient. In the past, many hospitals placed articles in the direct sunlight in order to destroy pathogens. Since this is not very practical today, any equipment which may have been used by a patient while in your department must be soaked in an antiseptic solution or wiped off with a disinfectant.

Antiseptics are used to inhibit and stop the growth of pathogens, yet they are mild enough to be used on the skin. A *disinfectant* is used to stop the growth of the pathogenic organisms, but it is harmful if used directly on the skin. *Sterilization* is a process used to kill all microorganisms. Items which have

Administrative Medical Services

been sterilized are usually marked or wrapped in special containers in order to maintain their cleanliness until they are to be used.

As with all members of the health care team, it is also the responsibility of the administrative medical services worker to be on guard continuously against the spread of infections. If you start to perform a procedure which requires direct hands-on contact with a patient and suddenly remember that you failed to wash your hands since working with your last patient, you must immediately stop what you're doing and wash your hands. Any health care professional, whether a clinician or an administrative medical services worker, who observes that medical aseptic techniques are not being followed and does nothing to correct it is not safe to practice as a member of the health care profession. Nothing should ever be assumed when practicing asepsis.

Following Universal Precautions

In today's medical and health care facilities, all members of the health care professions are required to practice universal precautions to prevent the spread of communicable diseases and infections. All patients and staff are considered possible carriers of potentially harmful pathogens. Blood and other body fluids carry many pathogens. Body fluids that may be infected include blood, sputum, urine, stool, and drainage from open wounds. In order to decrease the risk of accidental exposure to body fluids, cardiovascular departments have strict infection control procedures that must be followed by all members of the staff. This is especially true when working with patients who have open wounds. These precautions are for all patients, and they stress "prevention" from exposure and possible infection. As a member of the administrative medical services team, you have a responsibility to make yourself aware of the infection control procedures followed by your department and facility.

Providing for a Healthy and Safe Environment

Health care workers are active people. In the performance of your work, you will engage in a variety of movements. Even though your role may not require you to come into contact with patients in the clinical environment, you may be required to reach, pull, lift, stoop, sit, stand, carry objects, push wheelchairs and stretchers, and stand at the treatment area or a patient's bedside. Your body must be in good physical condition. This means being constantly aware of how to most effectively and efficiently use your muscles. Doing so will prevent injury to yourself and your patients.

Practicing Good Body Mechanics

Body mechanics is the efficient use of the body during an activity. Good posture is insurance for health and happiness. *Posture* pertains to the position of your body and the relationship of its individual parts to one another. Once you know your faulty posture habits and have a corrected posture image, you can begin to control alignment. Your movements will be safer and more efficient. You will feel tall, poised, and graceful. You should always remember to practice good posture in your walking, sitting, and working. It will decrease the fatigue you may be experiencing during the day. Figure 4-3 shows the correct alignment of body parts.

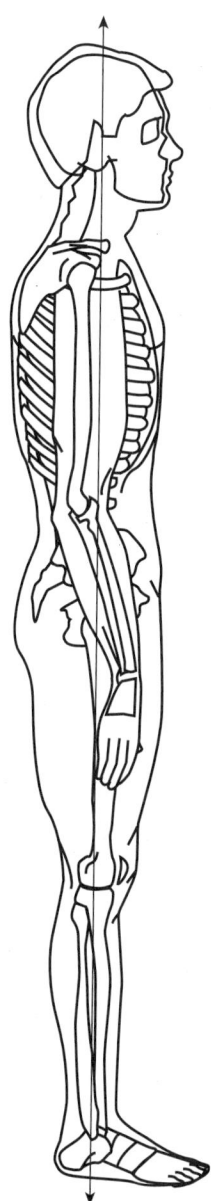

Figure 4-3
Correct postural body alignment.

The most common complaint of all health care workers is low back pain. Often this is precipitated by lifting a load while twisting around at the waist. It may result in severe muscle strain or in some cases even disc herniation or rupture. It is costly in loss of salary and medical expenses. The important point is that it is preventable.

When the parts of the skeletal system are aligned, balance can be maintained (Figure 4-4). *Balance* is the maintaining of an object in a steady position so that it does not tip or fall. To maintain balance, the body must have a wide base of support. In a standing position, the feet are the base of support. The *center of gravity* is located at the center of the pelvis just below the umbilicus. To maintain balance, the center of gravity must fall within the base of support (be-

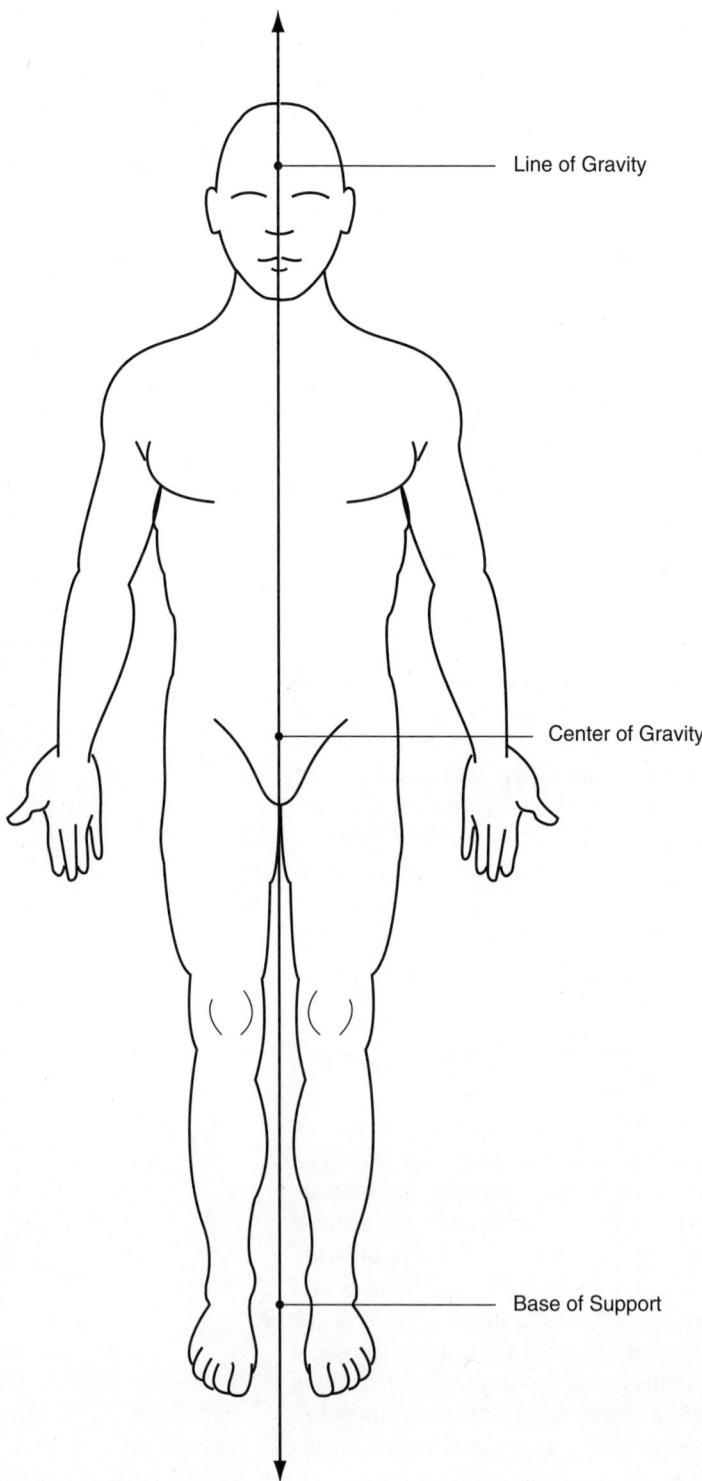

Figure 4-4
Maintaining balance through body alignment.

tween the feet). When the center of gravity falls outside of the base of support, one loses stability or balance.

In order to use your muscles most efficiently and prevent the possibility of muscle strain or other injury, there are certain guidelines you can follow when lifting objects or helping patients. These include:

- Always remember to size up the load and never lift more than you feel comfortable lifting. When in doubt, get help!
- Always remember to stand with your feet shoulder width apart or with one foot slightly in front of the other.
- If you have to squat, always remember to bend at your knees and never at the waist, which will add more stress to your back.
- Always remember to bring the load close to your body, using a firm grip with both hands.
- When lifting an object, always remember to tighten your abdominal muscles and then lift the load using your leg muscles.

It's important to remember that even lifting very light loads can result in injury if you do not use proper body mechanics. Using good posture and body mechanics in your everyday work will decrease the amount of energy expanded and provide more ease and grace in your movements. Knowledge of proper body mechanics and alignment will help you prevent injury to yourself and others (Figure 4-5).

Providing a Safe Environment

The make-up of all departments within the health care environment is important to the well-being of the patient. The goal throughout your facility should be to provide a therapeutic or healing environment. Therefore, it is the responsibility of all members of the administrative medical services team to provide continuously for safety, privacy, and order.

Even though your responsibilities do not require that you provide direct patient care, like some of your colleagues, you must still be alert to providing safe care in every treatment and procedure done for each and every patient. A safe environment will help prevent injuries and potential lawsuits. Falls, which are considered to be the leading cause of most lawsuits in the hospital environment, can be prevented by taking a few well thought-out precautions, including:

- Keeping the floor clean and dry and putting things away as soon as you are through using them.
- Cleaning up all spills immediately after they occur.
- Checking with the nursing station before you perform a procedure on a patient you are unsure of regarding his or her level of consciousness or orientation.
- Giving extra attention to patients with vision or hearing impairments.

Any time you are required to use an electrically operated piece of equipment, you must make sure it is in proper working condition. Make sure there are no frayed wires or faulty plugs. Report equipment problems to your supervisor, and move faulty equipment out of patient care areas so that it will not be used accidently.

Another way to provide a safe environment is to provide for your patient's privacy. Privacy is the legal right of every patient, and anytime you are required to perform a procedure, you must also provide for the patient's privacy. Always remember to knock before entering a patient's room or a treatment area. If you are required to be in an area in which a procedure is being performed that requires the patient to disrobe, always make sure that you make every effort to assist the clinician in exposing only that part of the body needed for the procedure. And if you must obtain information from the patient directly, in his or her hospital room, make sure you remember to have the privacy curtain pulled around the patient. Remember, as health care workers, we are accustomed to situations that would be embarrassing to the average person. Always be careful to provide for the privacy of each of your patients.

Finally, another very important aspect of providing for a safe environment has to do with knowing your hospital floor plan regarding exits, locations of fire extinguishers, and how to use them. Make a point of finding out how to report a fire. Many clinics and hospitals have special code words which alert the entire building of fires and emergencies. You should know these codes and be aware of your responsibilities if an emergency does occur in your facility.

Common Emergencies

Part of providing a safe environment means being aware of potential crises or emergency situations and then responding appropriately. Patients don't care whether or not you work in the clinical area or in the administrative services department. They just want to feel that you will know what to do if an emergency does occur. Naturally, any emergency should be reported to your supervisor immediately.

A common emergency is *fainting*, which is the temporary loss of consciousness. This occurs quite often and is not considered a serious malady. It is generally associated with illness or emotional stress. The patient shows pallor, sweating, and a slow, weak

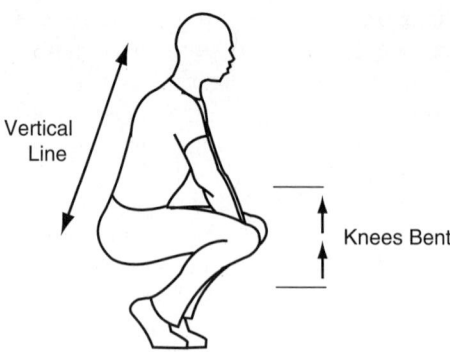

A. Lifting Heavy Objects

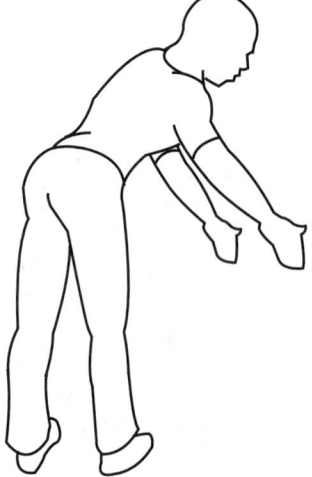

B. Twisting Incorrectly

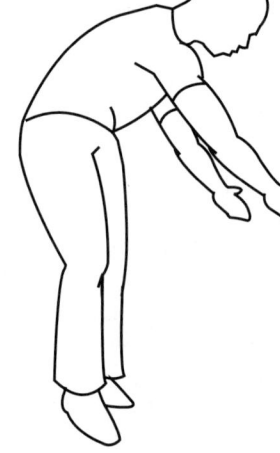

Twisting Correctly

Figure 4-5
Using correct body alignment to prevent injuries.

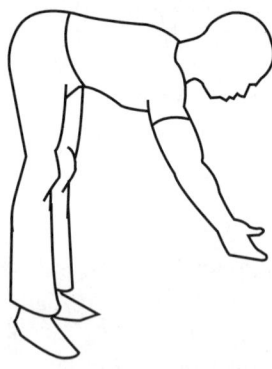

C. Bending Incorrectly

Bending Correctly

pulse that later becomes rapid. To treat the patient, first check the pulse, respirations, and blood pressure and make sure the airway is open. Then, elevate the patient's feet and provide supportive care.

Another common emergency which you may encounter is a *convulsive seizure*. These are disorders of the central nervous system which are characterized by recurrent attacks. They generally involve changes in the patient's level of consciousness, motor activity, or sensory phenomena. The onset is usually sudden and lasts only a brief time. The individual convulses, falling to the floor, and in some cases may exhibit violent, involuntary contractions of the muscles.

The aim of treatment during a seizure is to prevent the person from injurying him- or herself. Move furniture out of the way, loosen the clothing around the person's neck, and place a pillow or soft roll under the person's head. After the seizure is over, the patient will be sleepy and lethargic. Rest should be encouraged.

Falls, which may lead to contusions, sprains, or dislocations, are another common emergency, and, as we have already noted, are one of the leading causes of lawsuits in the hospital environment. A contusion or bruise is an injury to the muscle and soft tissues as a result of a blunt force. Discoloration occurs when small capillaries break and leak into the tissues. The area is usually quite painful and generally causes swelling.

To treat a sprain, gently apply cold compresses or ice packs to the area. This will cause vasoconstriction, which slows the internal bleeding and thus decreases the pain. The injured part should be elevated and wrapped in an elastic bandage for compression and support.

An emergency which could have a deady result, is *cardiac arrest*. In this very critical crisis situation, there is a sudden stoppage of the heart. The patient loses consciousness, stops breathing, and has no pulse and heart sounds. Time is the most important factor in resuscitation. You must quickly assess the situation and call for help. A delay of more than four-to-six minutes could result in irreversible damage to the brain. As a member of the administrative medical services team, you must be familiar with standard cardiopulmonary resuscitation (CPR) techniques and, if necessary, begin rescue breathing or CPR immediately.

Maintaining the Treatment Area

By keeping your department neat, you will be contributing significantly to keeping things running smoothly. You will also find that your own work, as well as everyone else's, is much easier once there is a system for keeping everything in order. Many individual medical offices and out-patient clinical facilities generally require the administrative or "front office" person to accept responsibility for keeping the facility neat, clean, and orderly. Therefore, as soon as possible, you should learn exactly where different articles and equipment and supplies are kept and always return them to the same place. This routine keeps the shelves, cabinets, and treatment rooms in a state of readiness. Shelves should be labeled so that everyone will know at a glance where a particular item is located. Table tops should be kept clear and clean so that there is always work space. Wastebaskets should be emptied regularly, and the rest of the department should be kept clean and in perfect working order.

Equipment should be checked often and cleaned with the recommended disinfectant after being used. Moveable equipment should be placed where it will not tilt or roll. Water should be drained from the autoclave or sterilizer after each use. In addition, it is a good idea to check frequently with your supervisor for any ideas or suggestions regarding changes in the normal clean-up routine of the department.

An inventory is important in any department. A good reference file, listing the names, addresses, and telephone numbers of the companies from which supplies and equipment can be ordered, should be kept. Prices and order numbers of supplies and equipment are usually listed with information supplied from the company. Supplies on hand should be checked often so that new supplies can be ordered before they are needed. Do not order more supplies than will be used in a reasonable length of time. Storage space should be available for anything that is ordered. If you are required to keep the inventory for your department, always make sure that you have your list approved by the department supervisor before you make out a requisition or take an order.

Any equipment needing repairs should be reported to your supervisor as soon as possible. Never try to repair the equipment or take advice from another employee who is not qualified to repair the equipment because more damage and greater costs may result. Most equipment requires certification as being "safe" before it can be used for patient care.

Medical supplies are very expensive. Therefore, you should only use what is necessary and return reusable supplies to their proper places. In some institutions, supplies are ordered for each individual receiving treatment. This is especially true for medications, dressings, and bandages. Wise use of medical supplies in your department will decrease the cost of the patient's medical care, which will ultimately decrease the costs for the medical facility, thus putting additional monies back into your pocket in the form of increases in salary.

Summary

In this chapter, we discussed important aspects of health and safety in the health care environment. We talked about medical and surgical asepsis and infection control procedures, noting the importance of each in the protection of both the patient and the health care worker. We also discussed awareness of posture and maintenance of good body alignment and use of proper body mechanics. We explained the role safety plays in the health care environment and discussed some of the more common emergencies which can be avoided through the proper maintenance of a safe clinical environment. Finally, we discussed proper maintainance of the administrative medical services department and its individual treatment areas, noting that it is the responsibility of all members of the department to maintain the department in a state of readiness and order.

Review Questions

1. Briefly define the concept of *medical asepsis*.
2. What does the term *alignment* mean?
3. _____ pertains to keeping an object in a steady position, or stable.
4. A(n) _____ is a process which occurs when pathogens attack a person and then produces signs and symptoms of an illness.
5. What is the name of the process used to kill all microorganisms?
   a. disinfection
   b. sterilization
   c. sanitization
6. The point at which an object rests, when the body is upright, is called:
   a. center of gravity
   b. line of gravity
   c. base of support
7. What type of infection is acquired by the patient while being cared for in the hospital?
   a. viral
   b. bacterial
   c. nosocomial
8. A one-celled microorganism which is capable of causing fermentation, decay, and in some cases, disease, is called _____.
9. Briefly explain the *infection cycle*.

Medical Terminology and the Medical Record

Performance Objectives

Upon completion of this chapter, you will be able to:
1. Identify specific prefixes, suffixes, and root words related to medical terms.
2. List common medical abbreviations used in the health care environment.
3. Identify body structure terms related to position, direction, anatomical planes, posture, and types of movement.
4. Explain the importance and uses of the medical record.
5. Distinguish among subjective, objective, assessment, and plan information on a patient's medical record.

Terms and Abbreviations

Combining form created by joining word elements and the combining vowel together.
Combining vowel a situation in which an "o" is added for ease in pronounciation.
Prefix always placed at the beginning of a word; a word element which, when combined with a root, changes or adds to the root word's meaning.
Root the body or main part of the word, which carries the primary meaning of the word.
Suffix always placed at the end of a root; a word element used to change or add to the meaning of the word.
Word elements the five word elements that are the basis of medical terminology: the prefix, suffix, root, combining form, and combining vowel.

The discussion of medical terminology will help you to understand the many terms commonly used in medicine. The prefixes and suffixes of many medical terms give definite information about the meaning of the term. If you know these prefixes and suffixes, it will be a lot easier to understand many medical words. A *prefix* is a word fragment placed in front of the basic or root word (Appendix B). A *suffix* is a word fragment added at the end of the basic or root word (Appendix B). The *root word* is the main body of the word, that is, the part that usually gives the meaning to the word (Appendix B).

There are no specific rules governing the pronounciation of medical terms. A medical dictionary will give you some suggestions as to the pronounciation of words. However, in many hospitals and medical facilities, these pronounciations will vary among professionals and individual departments.

Using Medical Abbreviations

Many abbreviations for words and phrases are used in the treatment of patients to save time and space. As an important member of the health care team, you must learn these abbreviations to follow directions and communicate with other health care workers (Appendix A). Since medicine has adopted many abbreviations that are commonly used by all health care professionals in writing their notes in the patient's medical record, it is also very important that you study and learn these common abbreviations so that you will be able to easily recognize their meanings.

Body Structure and Medical Terminology

The human body may be compared to a smooth-running machine. It has many parts which must work together in order to promote good health, growth, and life. The body is a combination of organs and systems which are supported and protected by a framework of bones known as the skeleton. The muscles working upon the skeleton provide for the movements as we work and play throughout each day. All of this is protected by an external covering known as the skin, which is considered the largest of all our body organs. You will need to know the various parts of the body and how to describe them in medical terms.

The human body is divided into five specific *cavities* or compartments: thoracic, abdominal, pelvic, cranial, and spinal (Figure 5-1). Within the thoracic cavity are the lungs, heart, aorta, and thymus gland. The brain lies within the cranial cavity, while the spine is located within the spinal cavity. The abdominal cavity contains the stomach, liver, gallbladder, small intestine, colon or large intestine, spleen, and pancreas. The reproductive organs and the urinary bladder are all located within the pelvic cavity. Various structural units make up the body. Cells, tissues, and organs are organized into individual systems. These systems include the skeletal, muscular, nervous, circulatory, digestive, respiratory, urinary, reproductive, and endocrine.

Anatomical Position

Many terms are used to describe the body and identify the position, direction, and location of its parts. In addition, these terms are also used to describe various medical characteristics of the body, such as the location of incisions on or injuries to the body. In order for the health care professional to be able to read the patient's medical record and thus have a mental picture of the patient's condition, all descriptive terms are based on an accepted standard position. This standard is known as the *anatomical position* (Figure 5-2). In this position, the person is standing erect, facing forward, with the head and trunk aligned, the arms straight by the sides with

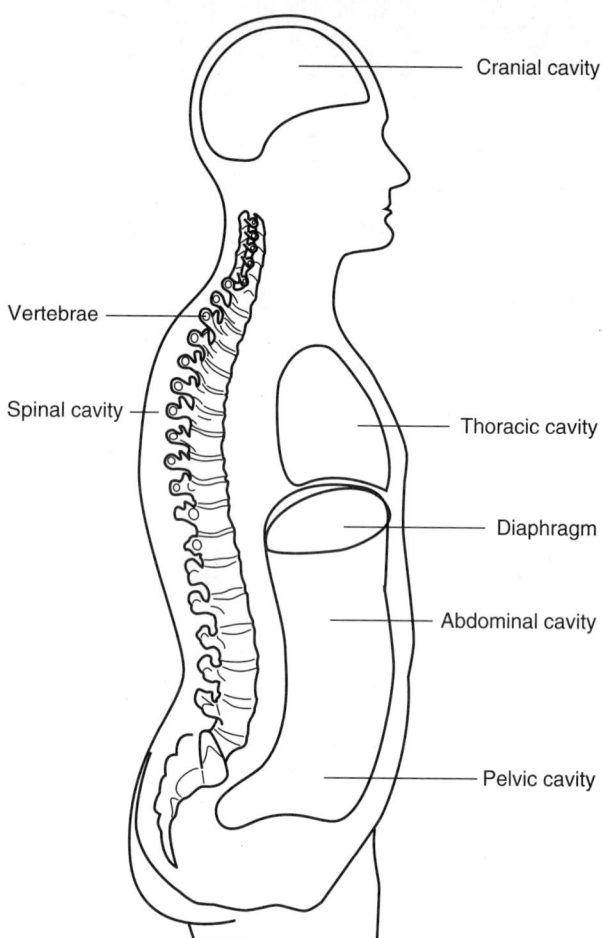

Figure 5-1
Body cavities.

palms facing forward, and the legs straight with the feet together.

Defining Position and Direction

Once you know the definition of the anatomical position, you can begin to use specific terms to describe position, direction, and location. While Figure 5-3 shows the relationship of some of these terms to one another, following is a list of all of the terms which are used to describe the body's position and direction:

- *Anterior*   Toward the front or in front of; ventral
- *Posterior*   Toward the back or in back of; dorsal
- *Medial*   Nearest the midline
- *Lateral*   Away from the midline or toward the side

Administrative Medical Services

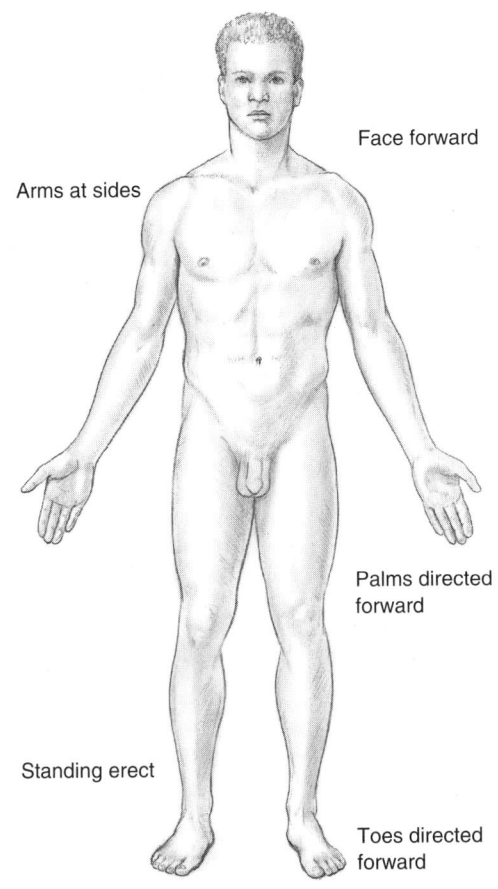

Figure 5-2
Anatomical position. (Reprinted with permission from Applegate, E. J. *The Anatomy and Physiology Learning System*. Philadelphia: W. B. Saunders, 1995, p 14.)

- *Internal* Inward or inside
- *External* Outward or outside
- *Proximal* Nearest the point of reference
- *Distal* Farthest away from the point of reference
- *Superior* Above
- *Inferior* Below
- *Cranial* Toward the head
- *Caudal* Toward the tail

Defining Anatomical Planes

There are specific terms used to describe and identify structures, areas of the body, and certain types of movement of the extremities (Figure 5-4). These include:

- *Sagital* An imaginary plane that runs parallel to the long axis of the body dividing it into right and left sections.

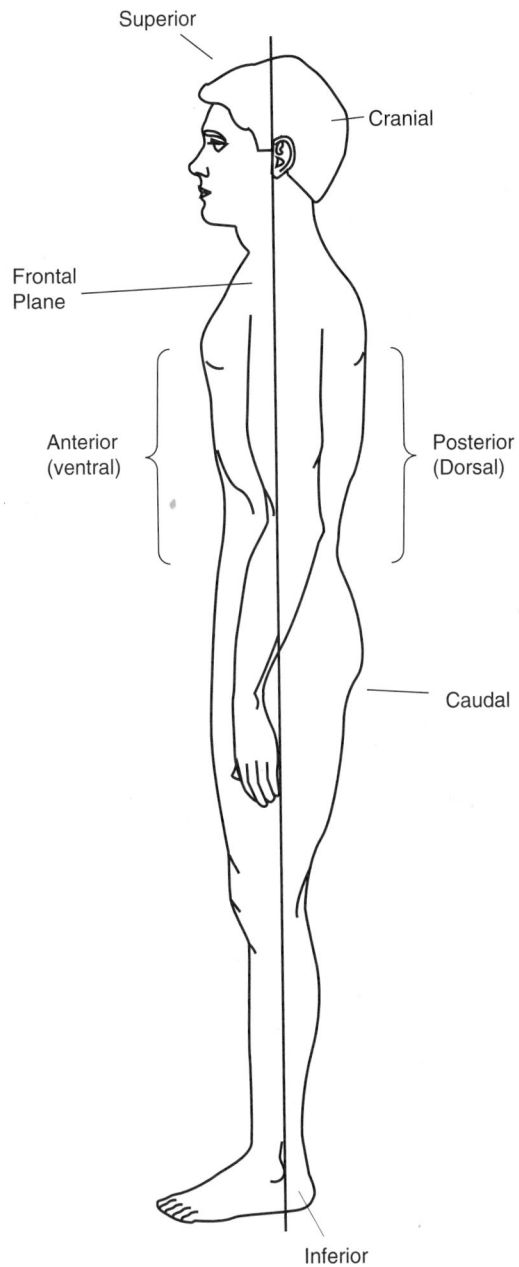

Figure 5-3
Body positions and directions.

- *Frontal* An imaginary plane that runs through the side of the body dividing it into anterior (front) and posterior (back) sections.
- *Transverse* An imaginary plane dividing the body into superior (upper) and inferior (lower) sections.

Defining Anatomical Postures

There are specific terms used to describe anatomical postures. These include:

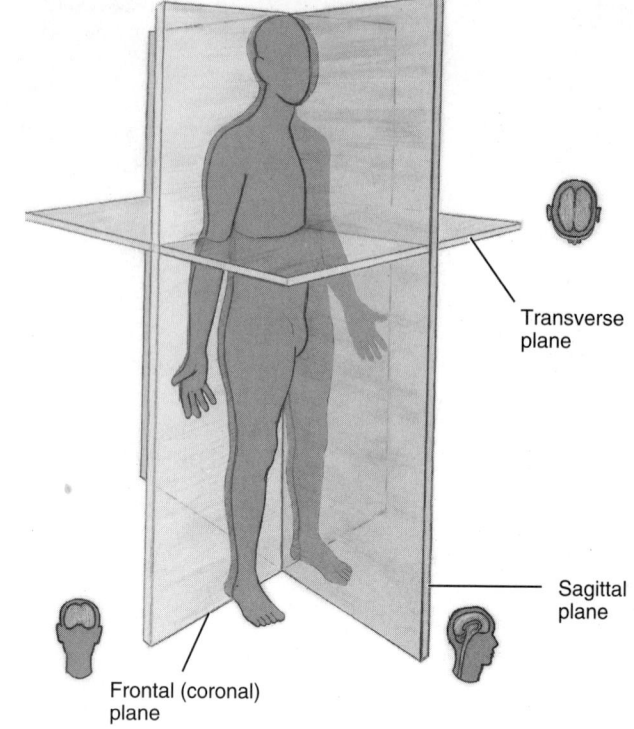

Figure 5-4
Anatomical planes.
(Reprinted with permission from Applegate, E. J. *The Anatomy and Physiology Learning System.* Philadelphia: W. B. Saunders, 1995, p 14.)

- *Erect*   Standing position
- *Supine*   Lying down
- *Prone*   Lying flat on the stomach with the face down
- *Side-lying* Lying with the body positioned on either the left or right side

Defining Types of Movement

There are specific terms used to describe various types of body movement. These include:

- *Flexion*   Bending at a joint
- *Extension*   Straightening at a joint; unbending
- *Abduction*   Moving away from the center of the body
- *Adduction*   Moving toward the center of the body
- *Rotation*   Rolling a part on its own axis, such as turning the head
- *Pronation*   Moving the palm from the anatomical position into a position with the palm facing posteriorly or backward
- *Supination*   Moving the palm into the anatomical position facing anteriorly or forward
- *Eversion*   Turning the foot outward
- *Inversion*   Turning the foot inward

It's very important to remember that not all joints can perform all motions. The immovable joints are, of course, incapable of performing any of them. Of the freely moving joints, only the ball-and-socket joints, such as the hip and the shoulder, can perform all motions. The hinge joints, such as those found at the knee and the elbow, are only able to flex and extend.

Understanding the Patient's Medical Record

Memory, at its best, is often fleeting and inaccurate. Medical records fill the need by containing accurate and detailed facts. These facts serve as a basis for the study, evaluation, and review of the patient's medical care.

Patient's medical records have been found among the earliest writings of Egyptians and in Hindu literature. Until modern times, however, no effort was ever made to keep patients' histories in any type of systematic order. These records are of the greatest importance to the physician while the patient is under his or her care. They are also extremely valuable if the patient returns for treatment after several years. For the patient, the record has a historical value, as it is the document forming his or her case history.

Patients' histories also have great statistical value. For instance, they may be of use to the physician in evaluating a specific type of treatment, or in finding out the incidence of a particular disease. They may also be used as the basis for a lecture, an article, or a textbook. Further, patients' histories have legal importance. They may be summoned as legal documents and used in a court of law to uphold the rights of the doctor if he or she is involved in litigation. They may also be used to confirm the claim of the patient if the doctor is called as a witness.

The medical record is used by health care practitioners to exchange information or as a means of communication. The physician, nurses, and other personnel contribute information to the record. A doctor may change a patient's care if the nurse has observed a change in the patient's condition. Likewise, after reading the notes written by the ECG technician, a nurse may plan specific nursing care that will allow the patient the greatest degree of comfort toward the prevention of any future cardiac symptoms.

There are many different types of printed forms available for keeping medical records. Some hospitals print their own forms. Preprinted forms are useful because they save time. They also assist the recorder in remembering questions he or she wants to ask. They are a way to accurately record the details of the patient's care and treatment.

The physician is always the person who initiates the treatment by checking the appropriate referral blanks or writing directions or "orders" for treatment in a narrative form. A request for a lab test or an electrocardiogram, for example, is sent to the cardiovascular department or the laboratory, as soon as possible, so that the procedure or lab test can be scheduled and carried out. Usually, the hospitalized patient and the chart are brought to the department from the nursing unit. Information relating to how the testing or procedural goals are being accomplished, and the patient's response to the procedure, is then recorded in the medical record after the test or procedure has been completed.

Charting Notes in the Medical Record

Most large health care facilities and individual medical practices document the patient's care and treatment in the medical record by using a four-point system known as *SOAP notes*. The SOAP abbreviation refers to the four methods *according to which* the patient's care is identified, assessed, and ultimately, carried out. The *S* stands for *subjective* symptoms which the patient may be presenting, and includes any information the patient supplies, family remarks made regarding the patient, and any other information stated by health care providers. The *O* refers to any *objective* information, tests, or treatments which may have been provided for the patient, as well as any observations or measurements made by a member of the health care team. The *assessment*, or how well the patient is responding to a given treatment, is abbreviated by the *A*. It includes any professional opinions or goals for a specific type of therapy or treatment ordered by the health care provider. The *P*, which stands for *plan*, refers to what needs to be done or what will be done for the patient, based on the ojective and subjective findings and the assessment.

Not all hospitals or individual departments use the SOAP note format. Your department may choose to set up its own method of record keeping. Some departments use 5-by-8-inch file cards and record only the visits of the patients. Others may use an 8-by-11-inch file folder which encloses individual progress notes. In any case, the SOAP format is a good way of organizing written data, no matter what type of forms your department uses.

Writing in the Medical Record

Since the patient's medical record is considered a legal document, it should always be written in ink. All information should be factual and have meaning. Notes must be accurate, without any spelling or grammatical errors. Any errors must be properly corrected by drawing one line through the error, with the person who made it writing his or her initials to verify that an error has been made. You must never scratch out mistakes or use correction fluids. Remember, too, that all information within the medical record is confidential. This means that the record should not be used for any reason other than as a means of exchanging information between members of the professional medical staff.

Summary

In this chapter, you were introduced to medical terminology and the medical record. In doing so, we determined that a basic knowledge of medical terms will make you more knowledgeable about your patient's condition and thus allow you to read medical reports and records with greater understanding. As you continue to gain experience in the health care industry, you will recognize the importance of using proper terms and abbreviations to save time and space when documenting patients' responses to medical treatments.

Review Questions

1. A _____ is always placed at the end of a word.
   a. prefix
   b. root
   c. suffix

2. A _____ is always placed at the beginning of a word.
   a. prefix
   b. root
   c. suffix

3. A _____ pertains to the body or main part of a word.
   a. prefix
   b. root
   c. suffix

4. A _____ _____ always requires an "o" for ease in pronunciation.
   a. combining vowel
   b. combining form

5. What does the term *lateral* mean?

6. What does the term *posterior* mean?

7. Define the following suffixes:
   a. ology
   b. ectomy
   c. itis

8. Define the following prefixes:
   a. erythro
   b. leuko
   c. anti

Basic Structure and Function of the Human Body

Performance Objectives

Upon completion of this chapter, you will be able to:
1. Identify the various systems of the human body and briefly explain the individual structures, components, and functions of each.
2. Identify common disorders of each of the systems of the body.

Terms and Abbreviations

Aneurysm abnormal dilation or ballooning of a blood vessel.
Arthritis disease of the musculoskeletal system in which there is an inflammation of bones at the joints with accompanying pain and swelling of the joints.
Constipation abnormal delay in, or infrequent, bowel movements.
Diarrhea abnormally frequent, watery bowel movements.
Dysentery inflammation of the lining of the intestines with diarrhea.

Fracture a common bone injury in which there is a break in a bone; may be a simple fracture, a compound fracture, in which there is a break in the skin, or a greenstick fracture, in which the fracture is incomplete and which is most often seen in children.
Incontinence inability to retain urine in the bladder.
Ligaments tissues that hold bones together.
Prostatism any condition that results in the obstruction of the prostate gland along with retention of urine in the bladder.

The human body is a miraculous piece of work consisting of many individual structures, organs, and systems, all of which work both independently and as part of a "team," whose overall goal is the preservation of life. While you may not be involved in the actual "hands-on" care and treatment of patients, as an administrative member of the professional health care team, it is important for you to have a basic understanding of the individual systems that make up the human body so that you can better comprehend why patients seek medical care in the first place. A basic knowledge of these very important concepts will almost guarantee you the ability to perform your job in a much more efficient and effective manner.

Systems of the Human Body

The human body is comprised of nine body systems. These include the musculoskeletal system, the digestive system, the circulatory system, the respiratory system, the nervous system, the integumentary system, the urinary system, and the reproductive system.

The Musculoskeletal System

The musculoskeletal system, which is made up of all the bones and muscles of the body provides us with the ability to stand erect and protects us from falling and injuring ourselves (Figures 6-1 and 6-2). The three types of muscles found in this system include *skeletal*, or voluntary muscles, which are connected to bone and make voluntary movements possible, such as walking or picking up an object off the floor; *smooth*, or involuntary muscles, which cannot be consciously controlled but are controlled by the autonomic nervous system, and which include such muscles as those found in the digestive and respiratory systems; and *cardiac* muscle, which makes up the heart wall.

All the bones of the body together make up the skeleton. The skeleton has three general functions: movement, because skeletal muscles are attached to bones; support of the body; and protection of our vital organs, such as the heart or liver.

The point at which two bones come together is called a *joint*. There are many different kinds of joints, each with a different function. Some joints allow us to bend our elbows, fingers, or knees. These are called *hinge* joints. Another type of joint is called an immovable joint, an example of which is the junction of the bones of the adult skull. This joint helps protect the brain.

Tendons and ligaments are actually parts of the musculoskeletal system. *Tendons* attach muscles to bones, while *ligaments* hold the bones together at the joints.

Common Disorders of the Musculoskeletal System

The most common injury or disorder of the musculoskeletal system is a *fracture*, which is any break in a bone. While there are several different types of fractures, the most commonly seen include *simple fractures*, which occur without resulting in a break in the skin; *compound fractures*, which involve breakage of

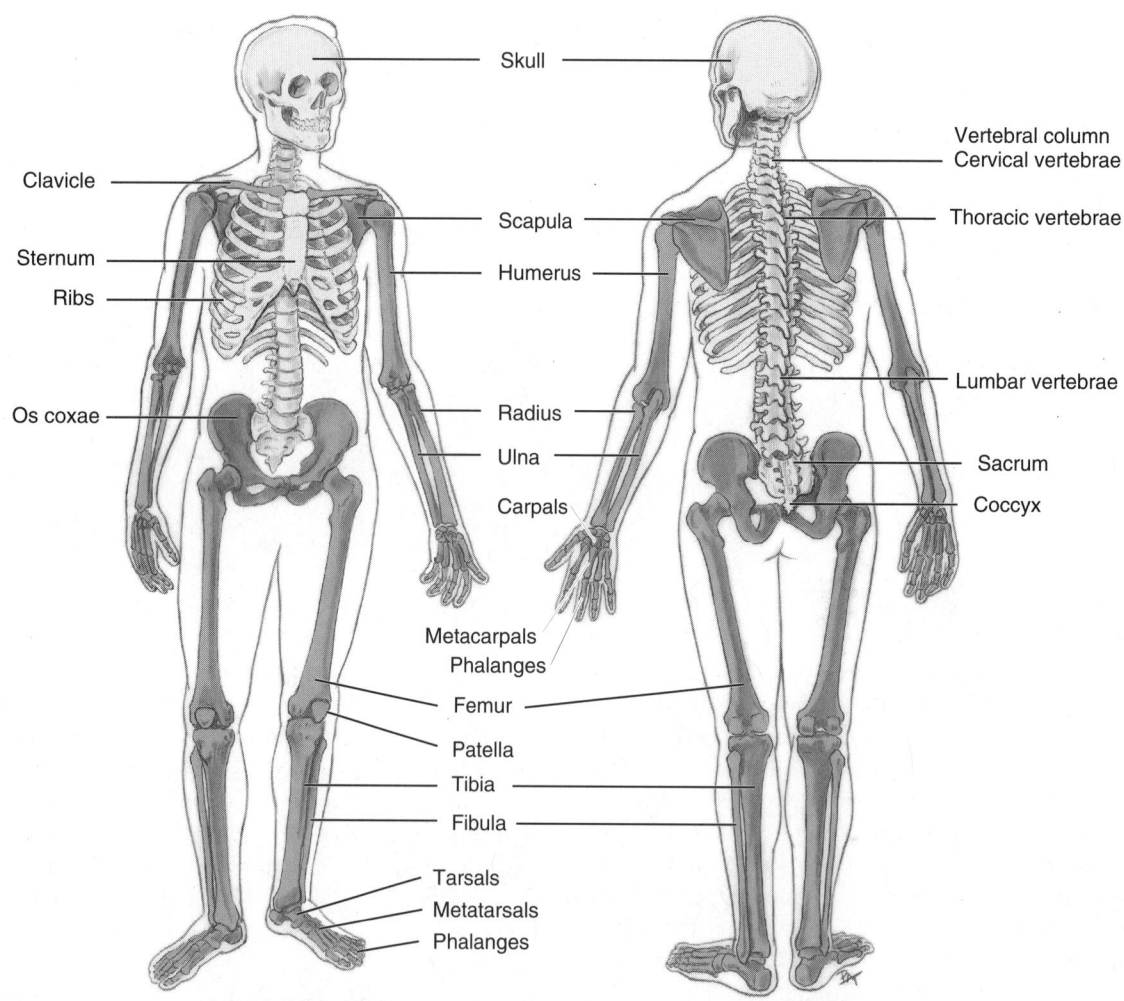

Figure 6-1
The skeletal system. (Reprinted with permission from Applegate, E. J. *The Anatomy and Physiology Learning System.* Philadelphia: W. B. Saunders, 1995, p 101.)

soft tissue, resulting in the broken bone pertruding through the skin; and *greenstick fractures*, which are incomplete breaks in the bone fibers and which most often occur in children.

In addition to fractures, the most common disorders and injuries to the musculoskeletal system include dislocations, sprains, and strains. A *dislocation*, which is almost always accompanied by torn or stretched ligaments, is a bone injury in which a bone is moved from its normal position in a joint. A *sprain* is an injury that occurs when a joint moves too far, thus resulting in over stretching or tearing of the ligaments. Back sprains or sprained ankles are common types of injuries. Muscle *strains* result from simply overworking the muscles. They are usually accompanied by a great deal of soreness.

A disease considered quite common to the musculoskeletal system is *arthritis*. It is a disorder involving inflammation of the bones at the joints, which is often accompanied by pain and swelling of the joints. Arthritis is most often seen in older patients.

The Digestive System

The digestive system includes the mouth, throat, esophagus, stomach, small and large intestines, and closely associated glands and organs (Figure 6-3). Its function is to digest food or to change it from an insoluble to a soluble form. This is accomplished through a series of actions occurring involving the chemicals and the digestive juices.

The first part of the digestive tract is the *mouth*. Here the food is ground up and torn apart by the *teeth*. Digestion begins in the mouth as saliva mixes with it and then acts on the food. The food then passes through the throat or *pharynx* as it is swallowed and eventually goes through the *esophagus* to the *stomach*. Food remains in the stomach from four-to-six hours. During this time, it is churned by the action of smooth muscles in the stomach and is then mixed with digestive juices secreted by glands in the stomach.

Food passes from the stomach into the *small intestine*, which is about 20 feet in length and is arranged in coils and loops held in place by a thin sheet of tissue called the *mesentery*. The small intestine is divided into three parts. The first part is called the *duodenum*, which is 8-to-10 inches long. The second part is called the *jejunum*, which is six-to-nine feet long. The final part of the small intestine is called the *ileum*. Most digestion occurs in the duodenum. Bile from the liver and pancreatic juice from the pancreas empty into the duodenum. These juices, along with secretions from glands in the small intestine, carry on most of the digestion of the food. The digested food is then absorbed from the lining of the small intestine and enters the bloodstream.

After food has been carried through the small intestine, the undigested food residue eventually moves into the *large intestine*. The large intestine consists of the *cecum;* the *appendix;* the *ascending, transverse,* and *descending colons;* the *rectum;* and the *anus.* The last part is the external opening of the digestive system. Food is not digested in the large intestine, but water and minerals are absorbed there. The food is moved through the entire digestive tract by wavelike contractions of the intestines called *peristalsis*.

Disorders and diseases of the digestive system are very common. Many of these include inflammatory conditions. Those most often seen include *appendicitis*, which is an inflammation of the appendix, *cirrhosis*, which is a chronic disease of the liver often caused by excessive intake of alcohol, *constipation* and *diarrhea, gastritis,* an inflammatory condition of the lining of the stomach, and *heartburn,* which is a burning sensation experienced in the stomach. Other conditions seen frequently include *hepatitis*, which is an inflammation of the liver, *nausea,* and *ulcers*, which occur in an area of the stomach or small intestine where the tissues are gradually disintegrating.

The Circulatory System

The circulatory system is made up of the heart, blood vessels, and blood (Figure 6-4). Its main function is to send blood throughout all parts of the body, carrying with it digested food and oxygen, and then return it to the heart carrying waste products of metabolism.

The heart is a hollow, cone-shaped organ which is about the size of your fist when it is closed. It has four separate chambers and is located in the center of the chest cavity with the tip pointing slightly to the left. It is well protected by the ribs and sternum. The function of the heart is to pump blood to every part of the body.

Blood vessels of the circulatory system include the *arteries,* which carry blood away from the heart; *veins,* which return the blood to the heart; and *capillaries,* which are small vessels that connect arteries to veins. The blood vessels you are most likely to find discussed at your facility are the *aorta,* which is the largest artery of the body and is located along the spinal column, and the *coronary arteries* and *veins* that supply the heart.

The blood which is carried throughout the body is a very complex fluid. It is made up of red blood cells, called *erythrocytes;* white blood cells, called *leukocytes;* and *platelets.* The red blood cells carry

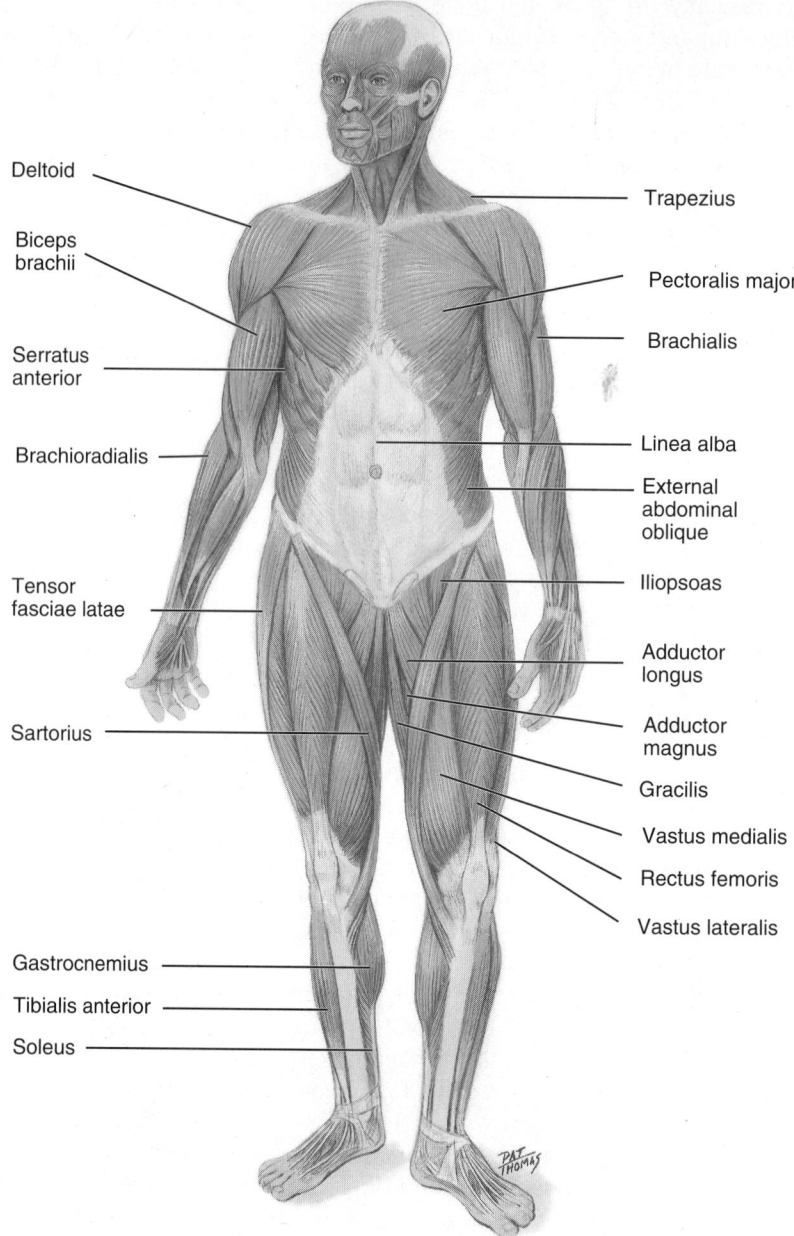

Figure 6-2
The muscular system. (Reprinted with permission from Applegate, E. J. *The Anatomy and Physiology Learning System.* Philadelphia: W. B. Saunders, 1995, p 145.)

oxygen to every cell in the body. The white blood cells destroy pathogenic bacteria and thereby help the body to combat diseases caused by bacterial infection. Platelets are involved in the clotting mechanism of blood.

The *lymphatic system* is also part of the circulatory system. Lymph is a clear, colorless fluid which is formed in tissue spaces throughout the body. It collects in tiny lymph capillaries and is eventually carried to even larger lymph vessels until it finally empties into the blood. The lymph vessels pass through the lymph nodes, which then filter out any foreign particles. Sometimes lymph nodes become infected because they filter out bacteria, which may then invade the node itself.

When discussing the circulatory system, we must also mention pulse and blood pressure. *Pulse* is the rhythmic throbbing which can be felt in an artery as a result of the heart beating. The pulse rate, rhythm, and strength are all indicators of health or the presence of disease. *Blood pressure* is the pressure that blood exerts on the inside walls of blood vessels. The *systolic* blood pressure is the pressure measured during the time in which the heart muscle is contracting. *Diastolic* blood pressure is the pressure measured during the time the heart muscle is relaxing. Blood

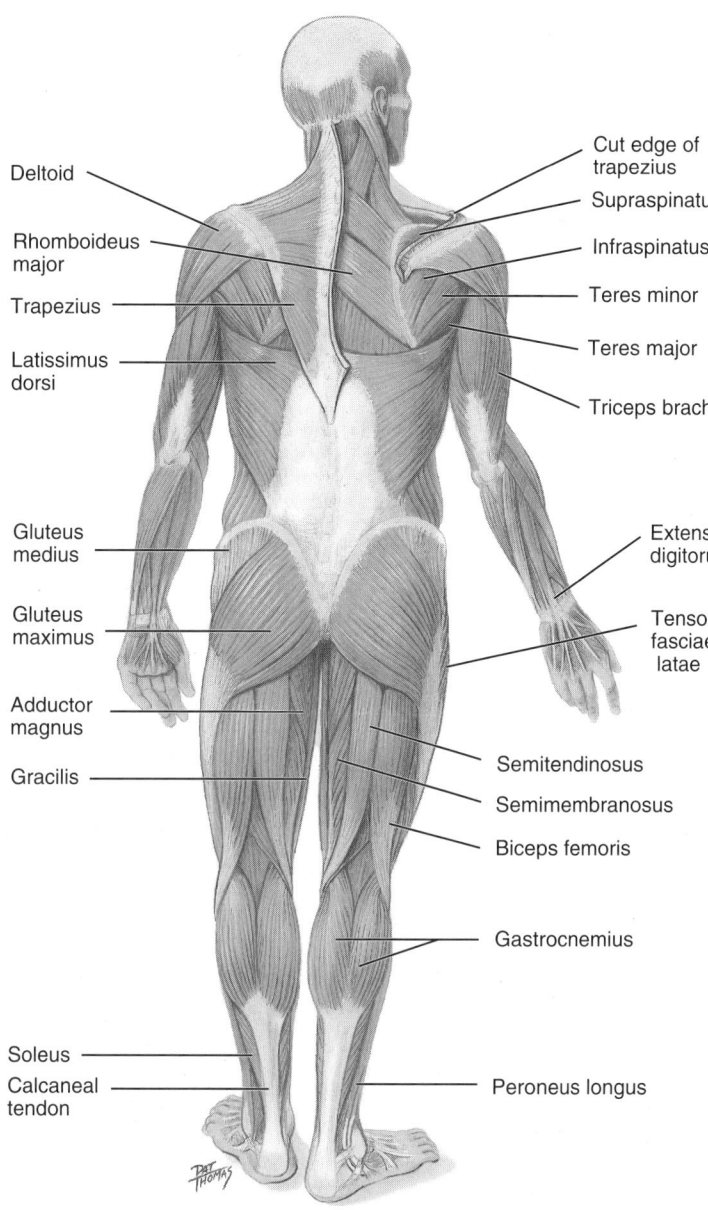

Figure 6-2 **Continued. (Reprinted with permission from Applegate, E. J. *The Anatomy and Physiology Learning System.* Philadelphia: W. B. Saunders, 1995, p 146.)**

pressure is reported as two figures, systolic pressure over diastolic pressure, such as 120/80.

Many patients coming into the medical office or admitted into the health care facility have diseases and disorders of the circulatory system. The most common of these include the following:

- *Anemia* A condition in which the blood is deficient in the number of red blood cells or in hemoglobin.
- *Arteriosclerosis* A condition in which there is a thickening, hardening, and loss of elasticity of the blood vessels.
- *Congestive Heart Failure* A condition which occurs when there is a failure of the heart to maintain an adequate output of blood in order to meet the demands of the body.
- *Coronary Occlusion* An obstruction of a coronary artery.
- *Myocardial Infarction* Also referred to as a heart attack, during which damage to the heart muscle has occurred as a result of diminished blood supply to the heart muscle from a coronary occlusion.
- *Hemorrhage* Extensive loss of blood.
- *Hodgkin's Disease* A disease of unknown cause which affects the lymph nodes.
- *Hypertension* An abnormally high blood pressure.
- *Heart Murmur* An abnormal heart sound.

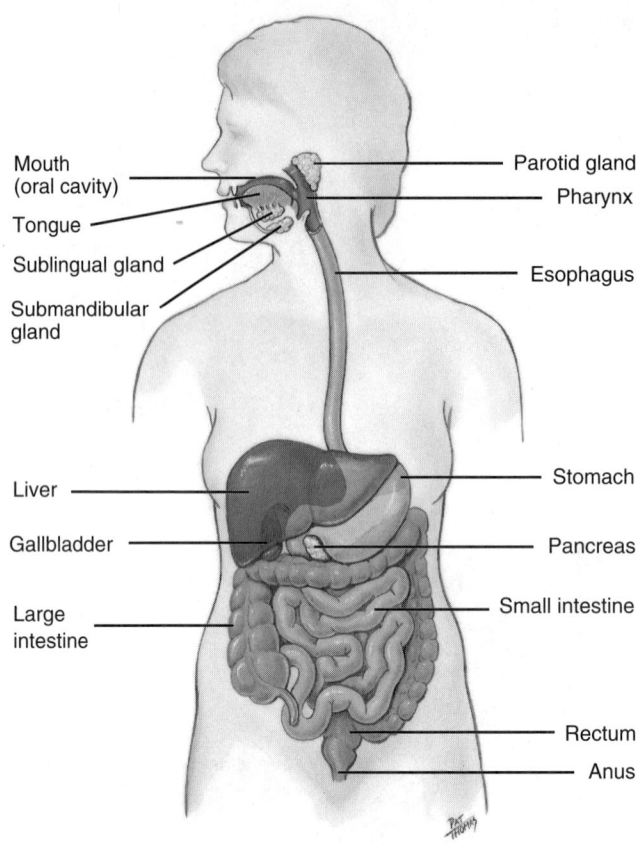

Figure 6-3
**The digestive system.
(Reprinted with permission from Applegate, E. J. *The Anatomy and Physiology Learning System.* Philadelphia: W. B. Saunders, 1995, p 328.)**

- *Phlebitis* An inflammation of a vein.
- *Vericose Veins* Dilation of veins which often occurs in the legs.

The Respiratory System

The function of the respiratory system is to bring air containing oxygen into the body and, at the same time, eliminate carbon dioxide and water. Each body cell must have a constant supply of oxygen and must also get rid of carbon dioxide.

The structures of the respiratory system include the *nasal cavities;* the *sinuses;* the *pharynx,* which is the throat; the *larynx,* or voice box, which contains the vocal chords and makes speech possible; the *trachea,* which is the short tube extending from the larynx to the bronchi; the *bronchi,* which lead to the *lungs,* which are a pair of lobed organs found in the thorax or chest cavity; and the *diaphragm,* which is a sheet of muscle that separates the abdominal and chest cavities (Figure 6-5). The contraction of the diaphragm forces air into the lungs. The pleura is the membrane that covers the lungs and lines the chest cavity.

Air is brought into the lungs as the diaphragm contracts. It then passes through the nasal cavity, pharynx, larynx, trachea, and bronchi and finally into the lungs. The actual exchange of oxygen with carbon dioxide takes place through the thin cell walls of the alveoli by the process of diffusion. Oxygen from the alveoli actually enters the blood and is carried to every cell. Carbon dioxide then returns to the alveoli and is eventually exhaled.

The most common disorders of the respiratory system include:

- *Asthma.* An allergic reaction in which the bronchioles swell making breathing difficult.
- *Bronchitis.* An inflammatory condition of the bronchi.
- *Common cold.* The most widespread of all communicable diseases, characterized by swollen and inflamed mucous membrane and discharge from the nose and throat.
- *Emphysema.* A condition causing swelling of the alveoli due to chronic bronchial obstruction, and which is common in heavy smokers.
- *Influenza.* A highly contagious disease characterized by inflammation of the upper respiratory tract, and generalized aches and pains.
- *Pleurisy.* Inflammation of the pleura.
- *Pneumonia.* Inflammation of the alveoli of the lung which may be caused by bacteria or viruses.

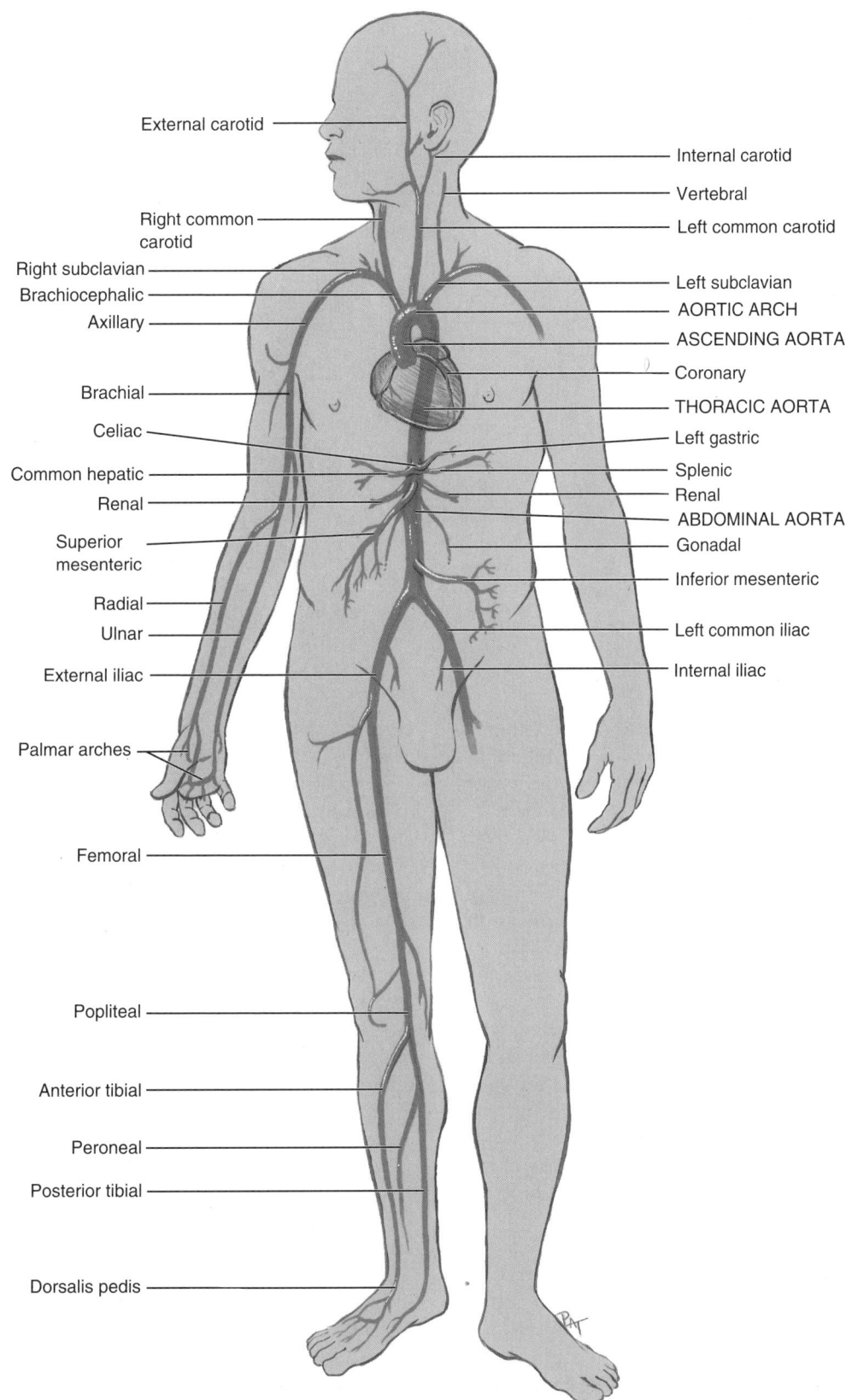

Figure 6-4
The circulatory system. (Reprinted with permission from Applegate, E. J. *The Anatomy and Physiology Learning System.* **Philadelphia: W. B. Saunders, 1995, p 272.)**

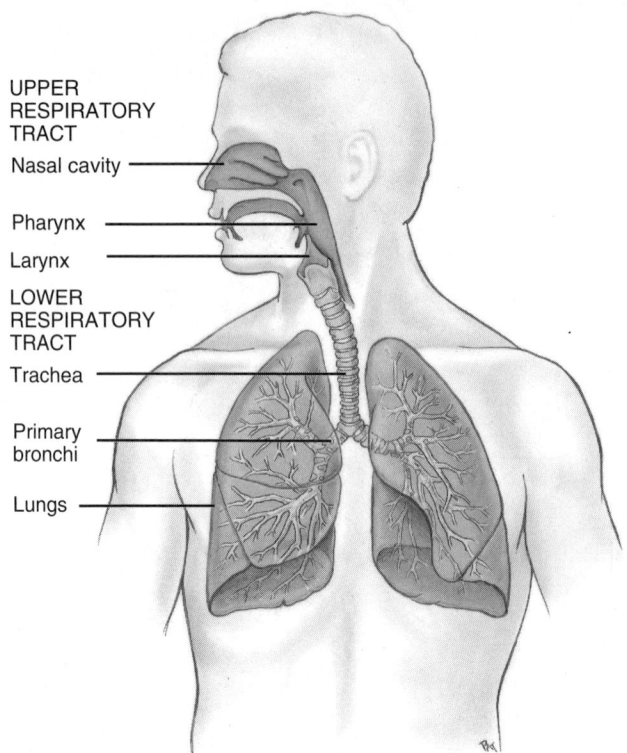

Figure 6-5
The respiratory system. (Reprinted with permission from Applegate, E. J. *The Anatomy and Physiology Learning System.* Philadelphia: W. B. Saunders, 1995, p 306.)

- *Tuberculosis.* A disease which can be either acute or chronic that is caused by the tubercle bacillus and usually affects the respiratory system.

The Nervous System

The nervous system is the body's main communication system. It consists of three parts: the central nervous system, which is made up of the brain and the spinal cord; the peripheral nervous system, consisting of the nerves extending to the outlying parts of the body; and the autonomic nervous system, which controls all of our involuntary functions, such as those of the digestive system or the respiratory system (Figure 6-6). The nervous system also includes the sensory organs. These are the structures which make up our eyes and ears, and control the sense of taste, smell, and touch; temperature; pain; pressure; and balance.

Common disorders of the nervous system include *cerebrovascular accidents,* often referred to as strokes, with which there is a destruction of brain tissue as a result of hemorrhage from blood vessels of the brain; *epilepsy,* which is a chronic disorder characterized by abnormal brain functions, sometimes referred to as convulsive seizures; *meningitis,* an inflammatory condition involving the membrane covering the brain and the spinal cord; *paralysis,* which is a temperary or permanent loss of functions, often accompanied by loss of sensation or voluntary motion; *poliomyelitis,* a less commonly seen acute viral disease which may destroy nerve cells and cause paralysis; *shingles,* a condition caused by a formation of blisters along the course of a nerve, most often affecting the intercostal nerves; and *vertigo,* often referred to as dizziness.

The Integumentary System

The skin, which is considered to be the largest organ of the body, is made up of a thin layer that covers the entire body (Figure 6-7). Sometimes called the integument or integumentary system, its functions include excreting excess waste materials, protecting deeper tissues, regulating body heat, and providing us with information about the environment, such as temperature, pain, or pressure, through its receptors. The integumentary system also has accessory organs. These organs include the hair, nails, sweat glands, and oil glands.

Common disorders of the integumentary system include wounds, or *lesions,* and skin diseases, such as *athlete's foot, acne,* and *fever blisters.*

The Urinary System

The primary function of the urinary system is excretion of the body's waste products. Figure 6-8 shows the parts of the urinary system. The *kidneys* are a pair of bean-shaped organs located on the back of

Administrative Medical Services

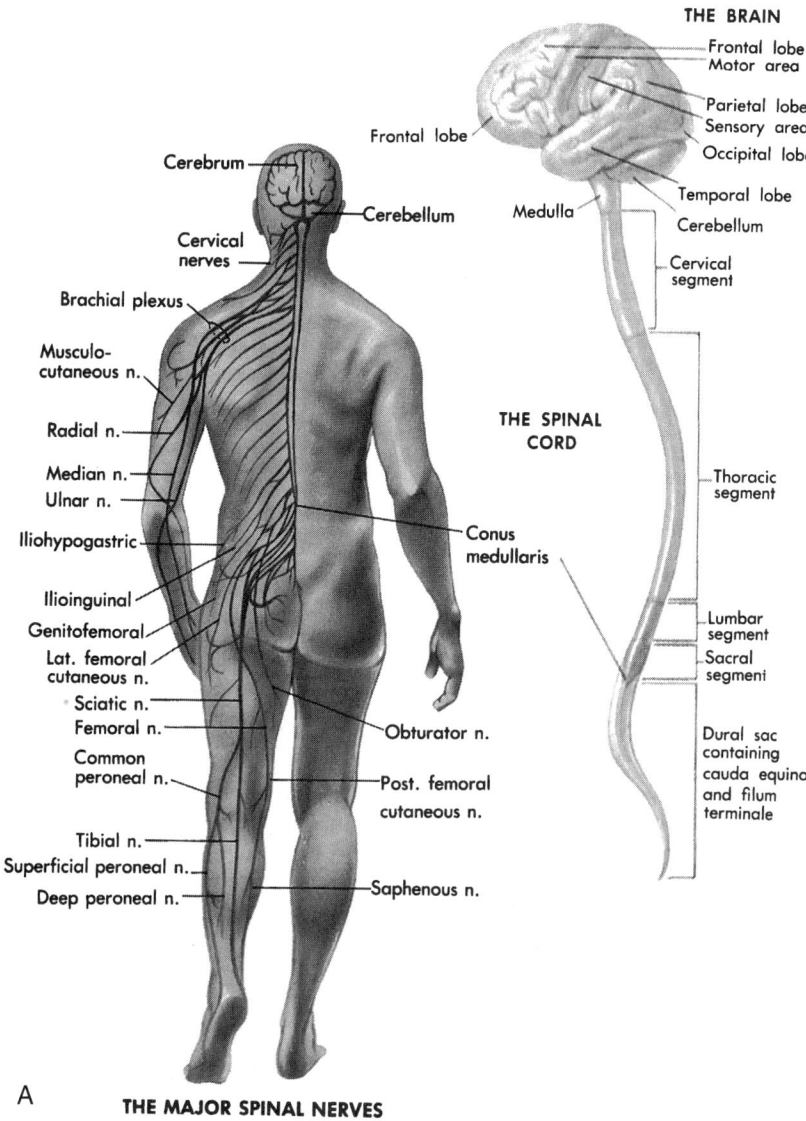

Figure 6-6
The nervous system: (A) brain and spinal nerves. (Reprinted with permission from Bonewit, K. *Clinical Procedures for Medical Assistants,* 3rd ed. Philadelphia: W. B. Saunders, 1995, plate 13.)

the abdominal wall. Their function is to filter waste materials from the blood. The *ureters* are tubes that extend from the kidneys to the *urinary bladder.* The urinary bladder is the sac in which urine is collected. The *urethra* is the tube extending from the urinary bladder to the exterior, or outside of the body.

The blood flows to the kidneys where waste products such as urea, uric acid, and various salts are filtered out. The waste product, urine, flows from the kidneys through the ureters, to the urinary bladder. The filtered and cleansed blood is then returned to the circulatory system. The urine is excreted from the bladder through the urethra.

Common disorders of the urinary system include *cystitis,* which is an inflammatory or infectious condition of the urinary bladder; *nephritis,* or *Bright's disease,* with which there is severe inflammation of the kidneys; *uremia,* a condition in which wastes normally excreted by the kidneys are retained in the blood; and *urinary calculi,* or formation of kidney stones.

The Endocrine System

The endocrine system is made up of several glands that produce hormones (Figure 6-9). *Hormones* are chemical substances that regulate the function of other organs.

The *pituitary gland* is located at the under-surface of the brain. It secretes several different hormones, each of which has a different function. Some of the functions are control of the activity of other glands, control of body growth, and contraction of involuntary muscles. Because the pituitary is involved in so many body functions, it is often referred to as the "master gland."

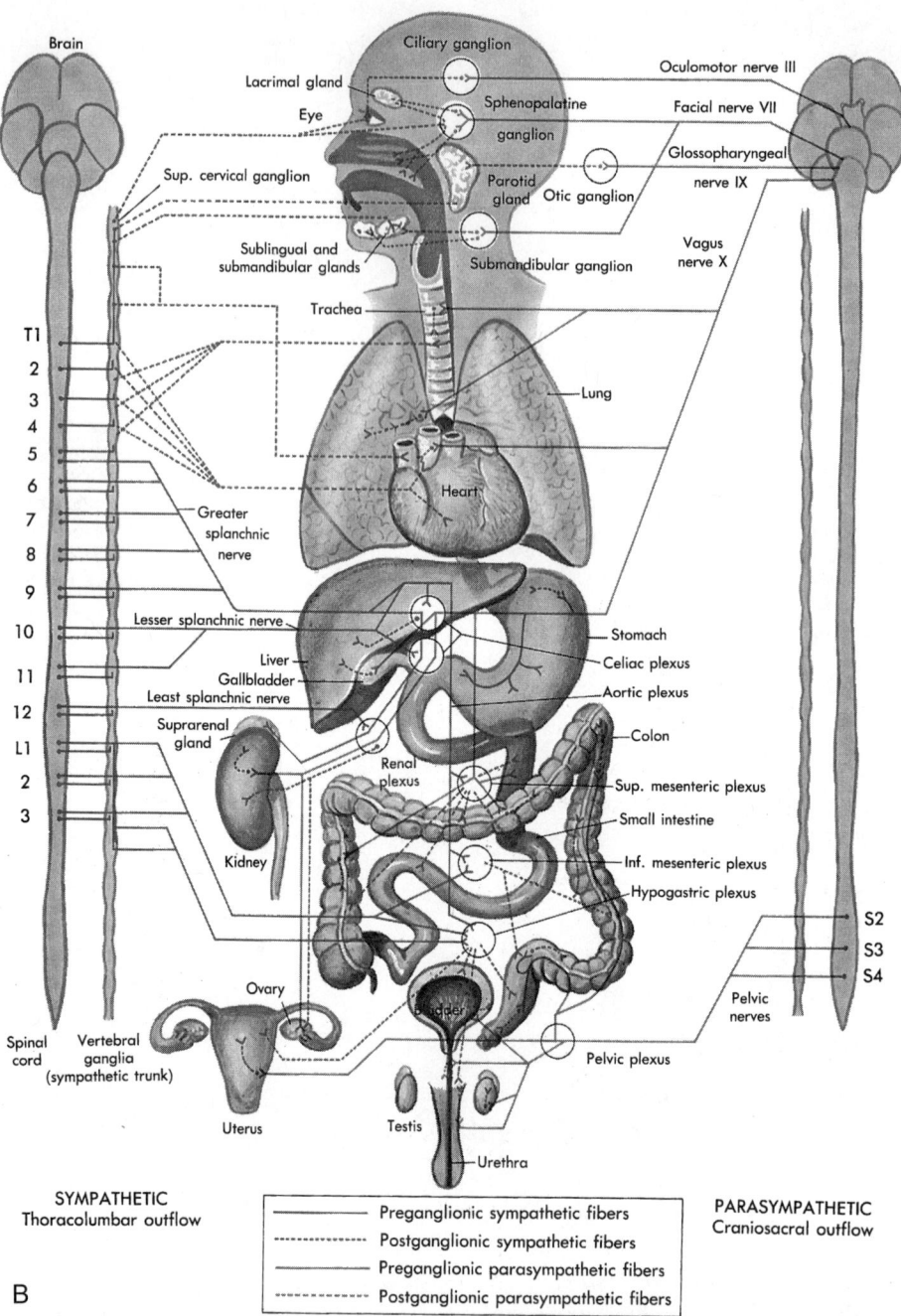

Figure 6-6
Continued. (B) autonomic nerves. (Reprinted with permission from Bonewit, K. *Clinical Procedures for Medical Assistants,* 3rd ed. Philadelphia: W. B. Saunders, 1995, plate 14.)

Acromegaly is a disease of adults that results from hyperactivity of the anterior pituitary. The bones of the face, hands, and feet widen. *Dwarfism,* a condition in which patients are abnormally small, and *giantism,* in which patients are abnormally large, may result from a deficiency or an excess of a hormone that controls body growth.

The *thyroid gland* is located just below the larynx in the neck. Its main function is to help control growth and metabolism. The secretion of the thyroid gland is called thyroxine. An enlargement of the thyroid gland is called a *goiter.* Goiters may result from a lack of iodine in the diet, which is necessary for the proper functioning of the thyroid. If there is an overexcretion of thyroxine, an *exophthalmic goiter* can result. In the patient suffering from this disease,

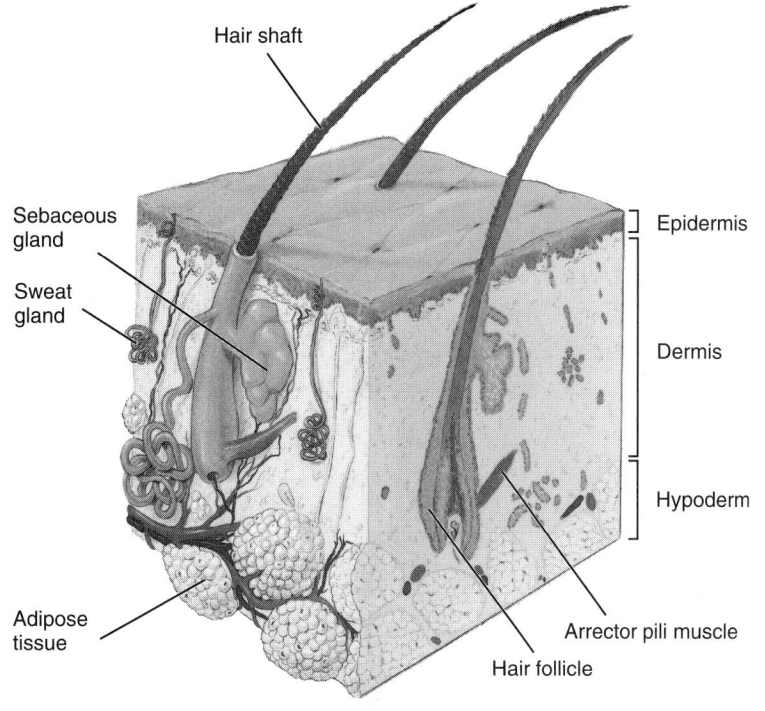

Figure 6-7
The integumentary system: layers of the skin. (Reprinted with permission from Applegate, E. J. *The Anatomy and Physiology Learning System.* Philadelphia: W. B. Saunders, 1995, p 85. Figure first appeared in Jarvis, C. *Physical Examination and Health Assessment.* Philadelphia: W. B. Saunders, 1992.)

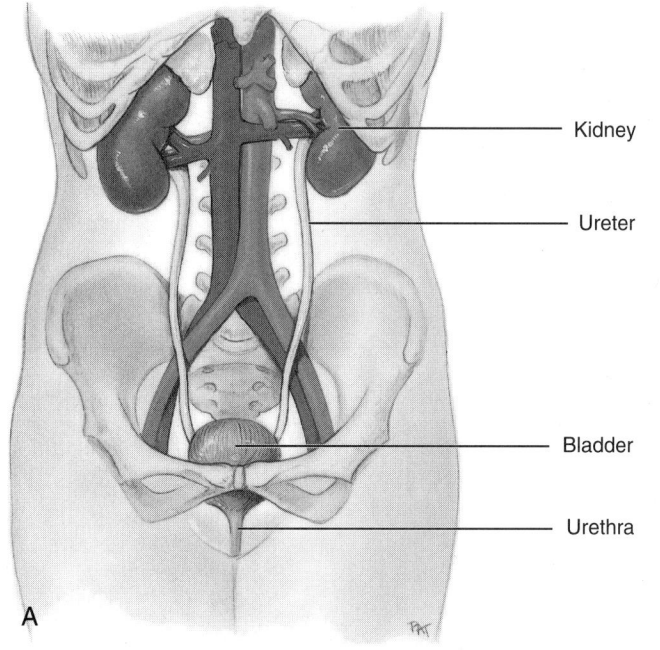

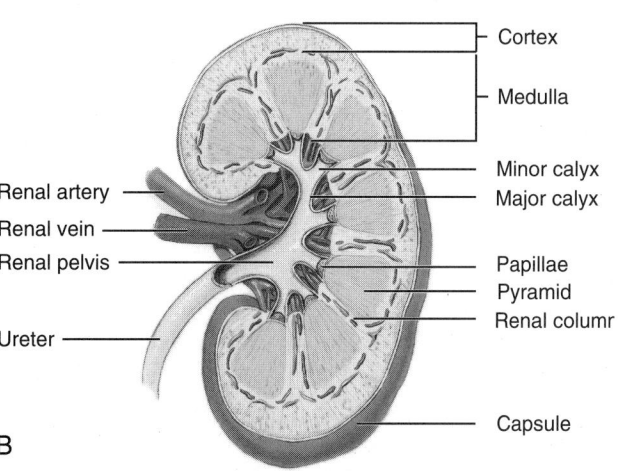

Figure 6-8
The urinary system (A). The kidney (B). (Reprinted with permission from Applegate, E. J. *The Anatomy and Physiology Learning System.* Philadelphia: W. B. Saunders, 1995, p 372.)

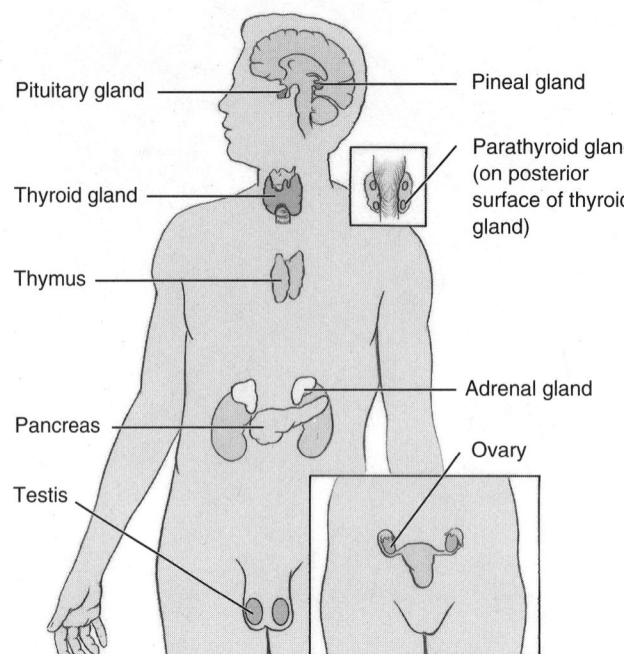

Figure 6-9
The endocrine system. (Reprinted with permission from Applegate, E. J. *The Anatomy and Physiology Learning System*. Philadelphia: W. B. Saunders, 1995, p 208.)

we often see the characteristic bulging eyes, strained appearance of the face, intense nervousness, and very rapid metabolic rate.

The other glands found in the endocrine system include the two pairs of *parathyroid* glands which are embedded in the thyroid and which control use of the calcium and phosphorus by the body; the two *adrenal* glands, located just above each kidney, which are responsible for controlling the release of energy in order for us to meet emergency situations and use of water and salt by the body; and the *islands of Langerhans,* which are small groups of cells located in the pancreas that secrete insulin. Insulin controls the rate of glucose metabolism and regulates blood glucose concentration. While the islands of Langerhans are located in the pancreas, they do not produce digestive juices and have no digestive function.

The sex glands are also part of the endocrine system. In the male, they are called the *testes,* and in the female, they are the *ovaries*. The testes produce hormones that regulate the production of sperm cells and the development of male secondary sex characteristics. The hormones produced by the ovaries regulate the development and activity of the female reproductive system. They are also responsible for the development of the female secondary sex characteristics and for body changes that occur during pregnancy.

Two glands, which are little understood by most scientists, are the *thymus* gland and the *pineal* gland.

The thymus gland is located in the chest just behind the sternum. It produces hormones that stimulate the production of lymph tissue and lymphocytes. It also functions in immune reactions. However, its function is not well understood. We know that it is present at birth, but begins to atrophy after about the age of 16. The function of the pineal gland, which is located in the cranial cavity at about the middle of the brain, is also little understood.

The disease most often asssociated with the endocrine system is *diabetes mellitus*, which results from a lack of insulin being produced in the islands of Langerhans. Without enough insulin, the body is unable to use glucose. Patients with diabetes must control the disease by regulating their glucose intake. They do this by taking additional insulin at regular intervals.

The Reproductive System

The reproductive system includes both the internal and the external reproductive organs. Its purpose is to produce reproductive cells, called ovum in the female and sperm in the male, which unite in a process called fertilization to produce a new human life.

The male reproductive system includes the scrotum, testes, penis, and prostate gland (Figure 6-10). The *scrotum* is a sac or pouch suspended between the thighs, which contains a pair of testes. The *testes*

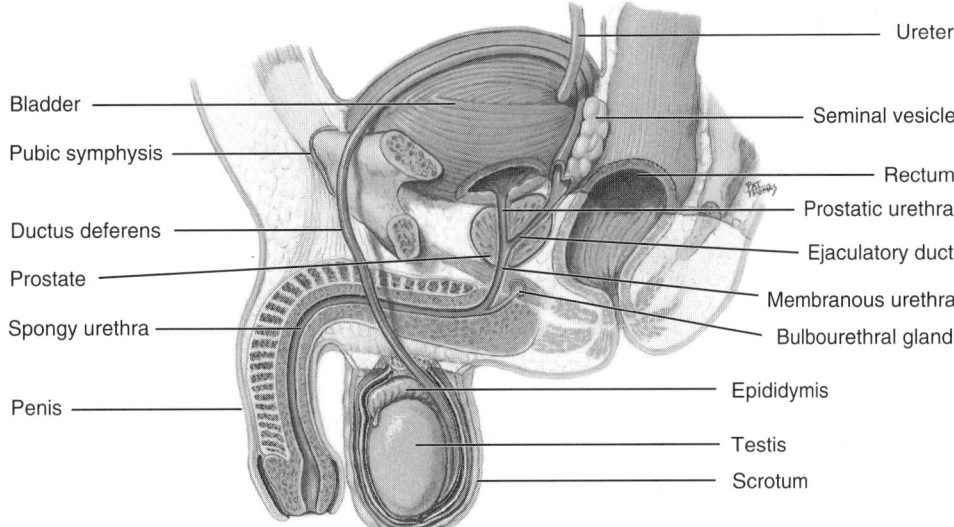

Figure 6-10
The male reproductive system. (Reprinted with permission from Applegate, E. J. *The Anatomy and Physiology Learning System*. Philadelphia: W. B. Saunders, 1995, p 392.)

produce sperm cells and male sex hormones. The *penis* is the male organ of reproduction and urination, which is located in front of the scrotum. The *prostate gland* surrounds the neck of the bladder and the urethra and produces a fluid that helps sperm cells keep their motility.

The female reproductive system includes the ovaries, fallopian tubes, uterus, vagina, and vulva (Figure 6-11). The *ovaries* produce egg cells, called ova or ovum, and female sex hormones. They are located in the upper part of the pelvic cavity. The *fallopian tubes* are ducts which lead from an area near the ovary to the uterus and carry the egg cells from the body cavity to the uterus. The *uterus* is the organ that contains and nourishes the embryo during pregnancy. The *vagina*, which is often referred to as the birth canal, is the muscular tube connecting the uterus with the exterior of the body. The *vulva* is the external part of the female reproductive system.

The union of the female's ovum with the male's sperm is called *fertilization*. This union usually occurs in the fallopian tubes, with the fertilized egg continuing its passage down through the tube until it reaches the uterus. Six-to-eight days after fertilization, the ovum becomes implanted in the uterine wall. The duration of pregnancy is about 40 weeks or 280 days. At the end of the pregnancy, hormones stimulate contractions of the uterine wall and birth occurs.

If an ovum is not fertilized by the sperm, a phenomenon known as *menstruation* occurs. It is the discharge of bloody fluid from the uterus at about 28-day intervals. Ova are produced at about the middle of this 28-day period. The uterine wall prepares for the implantation of a fertilized egg, but when fertilization does not occur, the lining of the uterine wall sloughs off and is discharged. This loss of uterine lining leaves some areas bleeding, which causes blood to be discharged from the uterus. The entire menstrual cycle is controlled by hormones.

There are many disorders of the reproductive system, but the most common include *dysmenorrhea*, which is painful menstruation; *leukorrhea*, which is a vaginal discharge; *orchitis*, an inflammatory condition of the male's testes, which is often due to trauma, mumps, or other infections; *salpingitis*, which is an inflammatory condition of the fallopian tubes; and *sterility*, a less common condition in which either the male or the female may be unable to reproduce.

Summary

In this chapter, we talked briefly about the basic structure and function of the human body. We discussed the fact that there are nine systems of the human body, and each of these systems has individual structures that function both independently and as part of a group in order to provide us with the necessary processes and functions required to live. We also talked about some of the more commonly seen disorders and medical conditions associated with each of these systems.

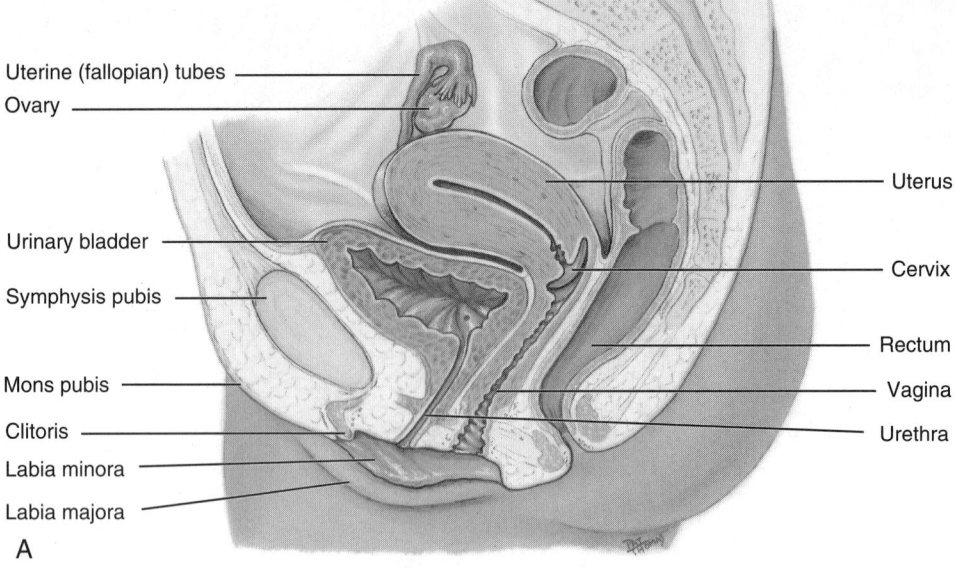

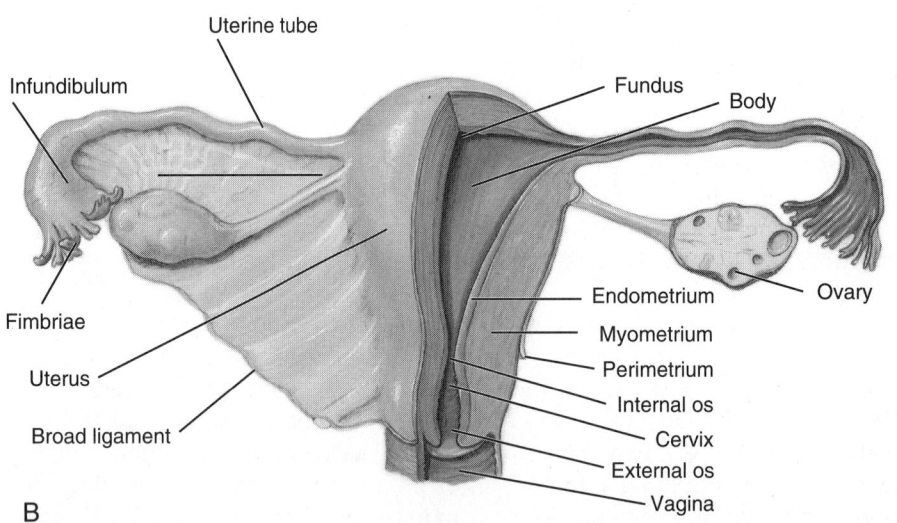

Figure 6-11
(A) The female reproductive system. (B) The uterus and uterine tubes. (Reprinted with permission from Applegate, E. J. *The Anatomy and Physiology Learning System*. Philadelphia: W. B. Saunders, 1995, pp 400, 403.)

Review Questions

1. Briefly explain the following diseases:

    a. Arthritis _____

    b. Dysentery _____

    c. Prostatism _____

2. Any abnormal dilation or ballooning of a blood vessel is called a(n) _____.

3. Structures which are responsible for holding bones together are called:

    a. tendons

    b. ligaments

    c. joints

4. The location at which two bones come together is called a:

    a. tendon

    b. ligament

    c. joint

Administrative Medical Services

5. List the nine systems of the human body:

   a. _____
   b. _____
   c. _____
   d. _____
   e. _____
   f. _____
   g. _____
   h. _____
   i. _____

6. Another term for a red blood cell is a:

   a. platelet
   b. leukocyte
   c. erythrocyte

Section II

Administrative Medical Assisting: Basic Concepts and Applications

7
Administrative Management and Communications

8
Admissions, Transfers, and Discharges

9
Filing Procedures and Medical Records Management

10
Data Abstracting

11
Insurance Coding and Indexing

12
Release of Information and the Law

13
Providing Quality Assurance in the Health Care Environment

14
Basic Computer Applications

Administrative Management and Communications

Performance Objectives

Upon completion of this chapter, you will be able to:

1. Discuss how to receive and direct patients, their family members, and visitors.
2. Explain how to communicate effectively using the telephone, a pager, and an intercom system.
3. Describe and be able to demonstrate how to record telephone messages.
4. Explain how to effectively receive mail and process outgoing mail in the health care setting.
5. Discuss the purpose of making duplicate copies in the health care setting.
6. Briefly explain how to prepare monthly statements.

Terms and Abbreviations

Beeper also referred to as a personal pocket pager; an electronic device which is carried by a person and which can be activated by a radio signal in order to produce a "beep" or buzzing sound.
Fax (facsimile) a machine which is designed to transmit information on a document over telephone lines to a terminal.
Intercom also referred to as an intercommunication system, which is used to send and receive messages, using a loudspeaker and a microphone at each location.
Paging a process by which a loudspeaker system is used to summon a person by name, number, or beeper to a telephone.

Screening mail the process of separating mail into specific categories.
Screening telephone calls the process of identifying a telephone caller before permitting him or her to speak to the person called, and then referring or handling the call in the correct manner.
Statements requests for payment for services rendered.
Transfer the process of switching a caller to another telephone number in a facility.

The patient's first contact with a health care facility almost always occurs in the reception area. The receptionist, whether he or she works in a private doctor's office, the business office of a large medical facility, or a single department within a hospital, must be prompt, friendly, tactful, and knowledgeable in dealing with people. In general, most people will be cooperative, keep their appointments, and pay their bills as obligated. In general, all patients prefer to have a relaxed and friendly relationship with the people they encounter in the health care facility. And such a relationship usually starts with the very first person they encounter.

Receiving and Dealing with Patients, Family Members, and Visitors

Health care workers who are assigned to work in the administrative or reception area of a clinic or private medical office, are generally responsible for reviewing the appointment sheet either at the beginning of each new day or prior to that day. This is important because this employee must be aware of which patients are expected in the office. If you are uncertain as to who is expected in your facility and, in some cases, unsure of the correct pronunciation of a particular name, you may want to consult with your coworkers. In fact, you may even want to write down a phonetic pronounciation in order to help you address the patient correctly.

If you are the person responsible for making initial contact with the patient, you may want to follow the guidelines listed below:

- If possible, always call the person by his or her last name, for example, "Mrs. Green" or "Mr. Jones."
- If a person who is not scheduled for an appointment suddenly appears at your facility, you must still be gracious and find out why the patient has arrived; for example, you might say "Good afternoon, Mr. Jones. How may I help you?"
- If a person makes a request without giving you his or her name, you should pleasantly ask, for example, "May I have your name please?"
- Never call the patient by his or her first name unless you are certain this familiarity is acceptable; older patients may expect to be addressed as "Mr." or "Mrs." by members of the office or department who are younger.
- It is generally acceptable to call children and teenagers by their first names.
- Whenever possible, you should always wear a name tag since this is considered a considerate gesture that allows new patients to learn your name quickly.
- If there must be a delay in an appointment, explain the reason to the patient as soon as possible and estimate the length of time which may be involved.

In addition to the guidelines listed above, you should try never to make a patient feel hurried. On the other hand, you cannot allow anyone to consume long amounts of time in needless conversation. It's important for you to learn to read behavior and be able to work with patients so that each individual patient looks forward to a visit in your office or department.

Dealing with Family Members and Visitors in the Hospital

If you are required to work in the administrative or reception area of your individual department, you may be required to provide information and assistance to family members and others visiting patients in the hospital. Try to make everyone you encounter feel that you are there to serve. Many relatives and friends are already worried about the illness of their loved ones, and may seem angry or upset. Your job is to be as courteous and polite as possible. Some family members may ask you questions or make demands that are not within the scope of your responsibilities. Listen to them politely. Some demands may be valid, while some may not be. Always refer medical inquiries to the patient's doctor, and try not to become involved in any private family matters or disputes.

In dealing with family members and visitors of patients in the hospital, it is important to respect the confidentiality of all information. All patient information is considered confidential. If a member of the patient's family or a friend persists in asking you questions pertaining to the patient, inform the person that the physician will discuss the patient's condition and progress with those entitled to know.

Communication Systems Used in the Health Care Environment

Telephone Techniques

The telephone is an essential and important tool of communication which is used in all health care settings. It, along with personal paging systems and remote mobile telephones, is considered to be the most rapid means of communication. All members of the health care delivery team must learn how to answer, transfer, and respond to telephone calls properly and accurately record and relay messages to the appropriate persons.

The most common type of telephone used in most health care facilities and physician's offices is the multibutton or multiline telephone (Figure 7-1). This type of telephone allows calls to come in and go out at the same time. A lighted button generally indicates that a line is in use.

When a call comes in on the multibutton system, the telephone will ring and the line will be indicated by a lighted button. When this occurs, you should depress the button, lift the receiver, and answer the call.

When answering the telephone, always identify the name of the health care facility or deparment, followed by your own name. For example, if you are working in the ECG Department, you might say, "ECG Department, Mrs. Jones speaking."

If you are required to place an outgoing call, plan what you want to say prior to placing the call. Think about what questions you are going to ask and what points you want to cover with the caller. Always have a pen or pencil and paper close by in case you

Administrative Medical Services

Figure 7-1
The multiline telephone system. (Reprinted with permission from LaFleur, M. W. and Starr, W. K. *Health Unit Coordinating,* 2nd ed. Philadelphia: W. B. Saunders, 1986, p 44.)

want to make notes. You may also need to decide at what time the person you are calling is most likely to be available. All of your calls, of course, should be made during business hours, and you should never use the business or department's telephone to make personal calls.

If you are using a multibutton or multiline telephone system, and you wish to place a call, punch an unlighted button, pick up the receiver, and punch in the number. Identify yourself and your department and the name of the health care facility as soon as you reach the desired number. Give the reason for your call. If you must ask the person called any questions, always make sure that they are asked in a tactful but direct manner. At the end of the call, make sure that you thank the person called for his or her time and respond to any questions. You should end the conversation with a simple "good-bye." Replace the telephone receiver quietly, and remember that the person who called is the one who should end the call.

Holding and Transferring Telephone Calls

The "hold" button is used to close off one call from all the other lines. To place a call on "hold," simply pick up the receiver and ask the caller if he or she will please hold. If the person cannot hold, offer to call back. When the caller agrees to hold, depress the key two to three seconds and then release it. A flashing light generally indicates that the line is on hold.

It is important to return to the caller often with a progress report, such as "I'm still looking for your report" or "Dr. Smith will be with you as soon as possible." Get the caller's attention when you return to the line by saying, "Thank you for holding." If you do not use a hold button, remember that the caller can hear everything you say.

If at all possible, try to help the caller. If you cannot, offer to transfer the call to someone who can. Tell the caller why you are transferring the call and to whom he or she should speak. In most cases, your facility will have a specific procedure for you to follow in order to transfer calls. If not, you can generally assist the caller by dialing the new number or by going through the facility's switchboard operator.

Telephone Etiquette

Whenever you are required to speak on the telephone, you should remember that the listener is the person who immediately forms an impression of you and your entire facility. A good telephone personality imparts friendliness, concern, sincerity, and helpfulness. The caller should be made to feel that the call is important to you.

Your telephone call will be much more effective if you also remember to use short, simple, business-like descriptive words, such as "yes" and "of course," instead of "ok" or "sure." You should also try to avoid the use of abbreviations or lengthy terms that the caller may not understand. Also, never use slang. Always try to be brief if the situation requires an explanation, and remember to use "please" and "thank you" during your conversation. When speaking on the telephone, use a normal conversational tone and, if at all possible, use the person's name during your discussion. Also, remember to speak directly into the mouthpiece so that your words will be clear and distinct.

Often, if you are answering your department's telephone or if you are working in the reception area of the private medical office, you may be required to take a message. If this is the case, always make sure you have a message pad and pen or pencil close at hand. Take notes as you receive information, and if you are not certain that you understand, ask to have the information repeated. If names are unusual, verify the spelling, and be sure that the telephone number and dates are correct. You may even want to repeat the entire message back to the caller in order to ensure its accuracy.

Information on a telephone message should always include the date and time of the call, as well as the name and telephone number of the caller. In some cases, it may also be necessary to record the caller's address. Above all, make sure you record the message accurately and note any action that the caller may require. Recording a message is generally completed by making sure that the person who recorded it also initials it at the bottom of the message form.

Screening Telephone Calls

Screening a telephone calls deals with initially identifying the caller before permitting him or her to speak to the person called. While this is a common practice in many health care facilities and private doctor's offices, many callers often find it irritating. If screening is practiced where you work, never say, "who's calling please?" Answering the caller with, "may I tell him or her who's calling, please?" tends to be less disturbing to the caller. Screening also involves handling the call correctly by relaying it to the person called or someone else or by obtaining information to enable the call to be returned. Acceptable phrases such as "she's not at her desk," or "he's out of the office for a few minutes," are often used during the screening process.

Paging Systems

Large medical centers and health care facilities often have a telephone switchboard or PBX system that employs an operator who is responsible for controlling the flow of telephone communications throughout the facility. Paging physicians, directors of departments, and emergency personnel is generally done through this system. *Paging* uses a loudspeaker system to summon a person by his or her name or number to the telephone.

At night, or during holidays, many hospitals and private physician's offices may also use a telephone answering service. In these cases, the telephone operator takes information from the caller and notifies the doctor, who in turn returns the patient's call.

Most physicians and other hospital personnel carry a small beeper. A *beeper* or small pocket pager is an electronic device that can be easily activated by a radio signal in order to produce a "beep" or buzzing sound. Each person has an individual call number which can be activated by the PBX operator or the telephone answering service, and the individual carrying the pager can then be alerted to telephone the hospital or office.

Intercom Systems

An *intercom* system is a type of communication system that allows patients and your co-workers or other staff members to communicate with one another from different areas within a health care facility (Figure 7-2). Each location in which the intercom is used must also have a loudspeaker and microphone. These systems may be used to allow a doctor's office to communicate with examining rooms or to connect examining rooms with offices. Most often, in the hospital setting, the intercom is located at the nurse's station and is almost always accompanied by a patient's call light system. The panel connects to a signal cord which has been attached to a wall panel near the head of the bed in the patients' rooms. The end of the cord has a button which the patient can use to flash a light on the panel. Once the signal has been activated by the patient from his or her room, the health unit coordinator or the person in charge of running the nurse's station should immediately respond to the signal by turning it off and answering the patient's request.

Intercoms, like telephones, are also considered a "first line" of communication between the patient and the health care professional. Therefore, it is im-

Figure 7-2
Intercom system.

portant that, whenever you are responding to the patient using the intercom system, you address the patient by his or her name. Once you have answered the patient's signal, it's just as important to provide the assistance requested, or relay the message to the appropriate person. Also, don't forget to cancel the call at the nurses' station once you have completed your communication.

Communicating by Fax Machines

A type of communication system which has recently become valuable within the health care setting because of its ability to send documents from place to place using the telephone lines is the fax machine. A *fax*, or *facsimilie machine*, is designed to copy a document placed in it and then transmit the information over telephone lines to a waiting terminal. A major advantage of the fax machine is that all types of records can be easily transmitted to physicians or to medical records rooms from any location.

Processing Written Communications within the Health Care Environment

If you are ever involved in the day-to-day administrative tasks of your department or your individual health care facility, you will soon discover that being part of the health information management team also includes processing mail and other types of written communication. Such tasks always require that you use great speed, accuracy, and efficiency. Processing mail involves both incoming and outgoing mail. No matter which department or facility you work in, you must remember that all correspondence from your facility or department must be grammatically correct and neat in appearance. All incoming mail must be conveyed to the addressee and dealt with in order of priority.

Processing Outgoing Mail

It is always important to establish some type of central collection point for processing all outgoing mail. All correspondence should be deposited there so it can be easily picked up by a postal worker for delivery to a post office.

The mail which you will be processing will include correspondence which generally falls under one of five categories. These include informational correspondence, referrals, reminders, account due statements, and interoffice mail. *Informational* correspondence is generally used to request or provide information to a patient regarding his or her care or treatment. *Referrals* are when one health care professional communicates to another a request that a second professional examine an individual patient. A *reminder* is a type of correspondence which can be used either to inform the patient that it is time for a periodic examination or provide the time and date of the patient's next scheduled appointment. *Account due* statements are a "friendly" reminder to a patient that his or her account is past due. *Interoffice mail* is a type of office memorandum and usually includes all memos and correspondence between members of a department or a health care facility.

Most health care facilities require that all mail be presorted into specific categories before the mail can be sent out. These categories generally include local mail, out-of-town mail, and metered mail. Bulk mail is also sorted according to zip codes. First-class and third-class mail is often separated before it is processed and mailed.

Postage Meters

Some health care facilities with a large volume of daily mail choose to use postage meters to save time and facilitate the processing of mail. When a postage meter is used, postage is paid for in advance and the machine set by the post office. The postage meter imprints postage on envelopes and records the amount used on a meter. The date on the imprinter must be changed daily. All mail that is metered is generally stacked separately, with addresses facing in the same direction. Five or more pieces of mail, according to U.S. Postal regulations, must be bundled together.

Types of Mail

The U.S. Postal Service offers health care facilities many different types of mail service. If you know that you will be required to process mail as part of your responsibilities, you should become familiar with the various classifications of mail and how to use the services offered by the postal service.

There are nine types of services offered by the postal service: first-, second-, third-, and fourth-class; special delivery mail, certified and registered mail, express mail, and mailgrams. *First-class mail* consists of letters, business reply mail, bills, and account statements. *Second-class mail* includes newspapers and published periodicals. *Third-class mail* includes all books, catalogs, and printed materials weighing less than one pound. *Fourth-class mail* is made up of printed matter and packages weighing more than one pound. *Special delivery mail*, which can be delivered seven days a week including holidays and weekends, may be purchased for all classes of mail for which a doctor or the facility wishes to ensure a rapid delivery from the post office.

When mail is sent either certified or registered, the sender wishes to go one step further to ensure the prompt and exact delivery of the mail. *Certified mail* offers the sender proof of both mailing and reception. The sender receives a receipt, and the receiving post office keeps a record of the delivery on a receipt signed by the addressee. *Registered mail* goes a little further by assigning a registration number to each item mailed. The item's value must be specified and the sender receives a receipt, while the receiver or addressee is required to sign on delivery.

The post office's fastest service is *express mail*. Special labels are provided for each item sent by express mail, and the item must be received by the post office before a specified time (usually before 5:00 p.m.) at the express window for delivery by 3:00 p.m. the following day.

A *mailgram* is a combination of a night telegram service and a postal service and is not often used, except in situations in which the facility must notify the addressee during "off hours" of regular mail service delivery.

Processing Incoming Mail

If you are working for a private physician, you will find that doctors usually require the administrative staff person to screen their mail and separate those items that require immediate attention. Any mail marked *personal* or *confidential* is usually given to the physician unopened, unless he or she has pre-designated another member of the staff to open such mail. Other mail must be sorted, opened, and directed to the appropriate person for action.

Opening, Removing, and Dating Mail

When mail is first received, it must be placed on the desk in a specific order for easy recognition and opening. The order for opening is (1) telegrams; (2) special delivery letters; (3) regular mail; (4) advertisements; and (5) magazines, newspapers, and other periodicals. If there are only a few letters, they may be opened with a hand letter opener. If there are many, you may want to use an electronic letter-opening machine. To use most electronic letter-opening machines, you must first turn the envelopes upside down lightly to get the top edges even, and then run them through the machine for opening.

To remove mail from an envelope, place the opened edge to the right. Remove the letter, unfold it, and carefully fasten any enclosures to it. If the enclosures mentioned in the letter are missing, write the word "NO" beside the enclosure notation in the margin of the letter. If a letter is opened by mistake, it should be resealed with tape, marked "Opened by Mistake," and initialed.

All mail that is to be opened must be checked to ensure that the name and return address of the sender are provided in the letter. If they are not, the envelope should be attached to the back of the letter rather than discarded. Other reasons to save an envelope include when an envelope is incorrectly addressed, when the return address in the letter appears to be different from that on the envelope, when no signature is found in the letter, when the letter date appears to be different from the date stamped on the envelop by the postal service, when an enclosure mentioned has not been included in the envelope, and when a letter contains a bid, an offer, or acceptance of a contract. It's always important to keep the envelope, since it could be used as legal evidence, should the need arise.

In addition to properly removing and categorizing mail, some health care facilities also require that all incoming mail be dated with the receipt date. If this is the case, a rubber stamp or dating machine is generally provided for stamping the date. If your facility requires this procedure, information included usually consists of the date, time, and other departmental information. If you are using a dating method, you must also remember never to allow the stamped date to cover other information.

Processing Incoming Payments

In some cases, you may find that payments for services rendered by your department or facility may arrive as part of your daily incoming mail. If this is the case, remember that all payments received should be handled immediately. Cash payments should be counted as soon as you receive them and then recorded so that the prompt credit is given. Checks should be endorsed immediately, using a "For Deposit Only" stamp, and then credited to the patient's account and held for routine deposits.

Office Machines Used in the Health Care Environment

In addition to computers and fax machines, there are two types of office machines which are used quite frequently in the health care environment. These are the office duplicating or copying machine and the office calculator. In many health care facilities, a fax machine may be used to duplicate documents going to distant points. However, when copies are needed locally, you may be required to use a photocopier to duplicate specific patient data or other types of office or departmental communications. A *photocopier* or *copying machine* is a device that makes a photographic reproduction of printed material. Copiers may have other capabilities, such as printing on both sides of a sheet of paper, sorting pages, printing overhead transparencies, and printing on various sizes and types of paper stock.

When more than one person or department is required to share patient data, you may be able to provide such information rapidly and accurately if you have access to a photocopier. All health care workers should be familiar with the operation of these machines so that they can troubleshoot minor disruptions in service.

Some health care facilities, generally those that choose to perform their own billing and insurance services, use an electronic calculator to process numbers. *Calculators* perform the basic mathematic functions, that is, addition, subtraction, multiplication, and division. Some calculators also have the ability to provide the user with a spreadsheet which can be needed for accounting services. If you are the person required to perform any of the billing services in your facility, it is in your best interests to learn how to use the calculator. However, you must also remember that all health information must be accurate. This means that, even if you do use the calculator to perform your mathematical equations, you must still check all your work. If your calculator has paper tape for recording, you can easily check the calculations by matching them against the printed hard copy found on the paper tape.

Summary

In this chapter, we discussed the importance of knowing that the patient's first contact with any health care facility, department, or private physician's office takes place in the reception area or over the telephone. We also noted that it is the responsibility of any person working in the administrative or clerical department to be prompt, friendly, tactful, and knowledgeable in dealing with people. Besides dealing with the public personally or by telephone, this person must also be able to communicate by intercom or using a paging system, and must be able to properly and accurately record messages for others.

In addition to talking about the telephone as a means of communication, we also discussed how the written word is a form of communication used by all health care facilities. This type of communication involves receiving and screening incoming mail, processing outgoing mail, recognizing types of mail and postal services, and using postage meters. Finally, we talked about other types of machines which may be used in the health care facility, noting that such machines generally included a fax machine, photocopier, and calculator.

Review Questions

1. Briefly explain the following communication systems used in a health care facility:
   a. beeper _____
   b. intercom _____
   c. paging _____
   d. fax _____
   e. telephone _____

2. Briefly explain the process of screening telephone calls.

3. Briefly explain the process of screening mail.

4. Define the following types of mail:
   a. first-class mail
   b. certified mail
   c. registered mail
   d. third-class mail
   e. special delivery mail

5. Briefly explain the process involved in holding and transferring telephone calls.

Admissions, Transfers, and Discharges

Performance Objectives

Upon completion of this chapter, you will be able to:
1. Briefly discuss the role of the administrative medical worker as it relates to admissions, transfers, and discharges.
2. Identify three admission forms used by most health care facilities, and list the information that is usually included in each form.
3. List steps involved in transferring patients.
4. Discuss three types of discharge situations that would require the use of forms.

Terms and Abbreviations

Admission agreement a form used by most medical facilities which sets forth the services that will be provided and requires an authorization signature for care to be rendered; in some cases it will also include a release of liability.

Admission registration of a patient in a medical office, clinic, or hospital.

Confidential keeping something private and undisclosed without the patient's permission.

Demography the study of human populations relating to vital statistics.

Discharge the process of completing all records and business as a patient leaves a health care facility.

Medical power of attorney the designation of a person to authorize or refuse care for the patient should that patient be unable to do so.

Responsible party a person who is legally liable for charges for care or the person who is the parent or guardian of a patient not legally able to contract for his or her own care.

Transfer the process of moving a patient from one unit to another or from one facility to another.

Depending on where you are employed as an administrative medical worker, you may be responsible for obtaining specific information and completing the appropriate forms once the physician has written an order for the patient to be admitted, transferred, or discharged from a medical facility. The exact information and the nature and content of the forms necessary will vary with the needs and desires of the facility for which they are being used. The purpose of this very important task will generally be to organize information and prepare permission forms necessary for the patient's treatment. As an efficient and professional member of the administrative medical team, your responsibility will be to reassure the patient that he or she will be cared for promptly, and that the information needed on the forms is a necessary step for admission and treatment. Therefore, it is very important that you always maintain a friendly and caring attitude and not just hand the forms to the patient or guardian. As silly as it may seem, you should also remember to smile.

The Admission Process

If you are preparing the patient's admission in the medical office, or if the patient's treatment only involves minor care, the admission form you use may be brief or may not even be necessary. It is important to become familiar with the exact forms and procedures used in your facility so you will know when it is or is not appropriate to use the admission form. Often this information can be found in a procedure or policy handbook in your facility.

Admission Forms

Admission forms are used for several reasons. They may be used to register information concerning the patient, acquire the legally necessary written permission to treat the patient, disclose information to designated persons, set forth the terms for the patient's care and treatment, and designate who is responsibile for the charges incurred in caring for the patient.

Two types of forms are generally completed for each patient prior to beginning treatment. These are the *patient demographic form*, or what some facilities refer to as the personal information form, and the *patient medical history form*. The demographic form usually provides the following patient information:

- Name, address, and telephone number;
- Date of birth, marital status, and social security number;
- Who the responsible party is and who should be notified in an emergency;
- Employer and employer's address and telephone number;
- Insurance coverage; and
- Permission forms that may be necessary to provide treatment and to release any information to specified persons or under specified conditions.

The patient's demographic form may be either a separate form, such as the one shown in Figure 8-1, or it may be a combined form (Figure 8-2), which includes the release from the physician's office for the care and treatment of the patient. A new patient information form showing the person and the insurance company responsible for charges incurred may also be used (Figure 8-3). It's important to understand that all information provided on these forms is considered confidential and, therefore, may not be disclosed to anyone who does not have a right to the information.

The second part of the admission form, which provides the physician with a detailed history of the patient's medical background as well as his or her present medical complaints, is the *patient medical problems and history form* (Figure 8-4). It may be used by nurses and physicians as part of the necessary information needed to evaluate the patient's physical condition and to choose the appropriate care and treatment. It is also the portion of the form that deals specifically with the physical signs and symptoms for which the patient is being seen. The following information is generally included in this part of the admission form:

- The current reason the patient is seeking care;
- Any medications currently being taken by the patient;
- Any allergies, including any known drug allergies, which the patient may have;
- A list of any previously diagnosed illnesses;
- A list of any previous surgeries; and
- A family medical history

The Admission Agreement

In hospitals or surgical day care centers, a more extensive admission form may be used that includes an *admission agreement form* (Figure 8-5). This may be part of the demographic information form or may be a completely separate form. When used, the admission agreement form provides specific information for the hospital or medical facility that sets forth the terms for the care and treatment of the patient. It generally includes the following:

- Information about medical services, including statements as to the relationship of the physicians to the hospital or institution.
- Information regarding nursing services that can normally be expected.
- Notice indications that care may be provided by or procedures observed by students if the institution is a teaching facility.
- Information about procedures for handling money and valuables.
- Statements of release of information to the patient's insurance company, third-party payers, medical audits, medical research, or other health care providers.
- Information regarding a designation of medical power of attorney, if one is needed.
- Agreements concerning payment of the patient's account.

The admission agreement form usually requires the signatures of the patient and a witness and must be dated at the time of completion. It is often a multiple-copy form, with one copy being provided to the patient and other copies being designated to other receivers or being placed in the patient's medical chart.

The Transfer Process

A transfer is not generally used in medical offices. However, it may be used in various critical and long-term care areas as the need arises. To provide the best

1 PATIENT HOSP.NO.(M.R.#) 987-654 | INFO STATUS | ACCOUNT NO. (BUS. OFF.) 123-456

2 PATIENT NAME LAST Andrews | FIRST Iver | MIDDLE S. | ADM. DATE MO. DAY YR. 8/7/XX | ADM. TIME 1345 | HOW BROUGHT TO HOSPITAL Amb.

3 PATIENT'S CURRENT ADDRESS STREET, P.O. BOX, APT. NO. 701 East Danish Lane | CITY Carpenterville, | STATE AZ | ZIP CODE 85013 | TELEPHONE NO. 246-XXXX

4 PATIENT'S PERMANENT ADDRESS STREET, P.O. BOX, APT. NO. Same | CITY | STATE | ZIP CODE | TELEPHONE NO.

5 SEX 1. MALE 2. FEMALE [1] | MARITAL STATUS 1. SINGLE 2. MARRIED 3. SEPARATED 4. DIVORCED 5. WIDOWED [5] | RACE 1. WHITE 2. BLACK 3. INDIAN 4. ORIENTAL 5. OTHER [1] | RELIGION 1. CATHOLIC 2. JEWISH 3. PROTESTANT 4. OTHER 5. LDS [3] | AREA OF RESIDENCE

6 BIRTHDATE 10/6/XX | AGE 69 | PLACE OF BIRTH N.J. | MAIDEN NAME | SOC. SEC. NO./MEDICARE NO. 151-18-XXXX

7 PATIENT'S OCCUPATION Retired | UNION & LOCAL NO. | PATIENT'S EMPLOYER | ADDRESS | TELEPHONE NO.

8 PREVIOUSLY TREATED HERE? ☒ YES ☐ NO | NAME USED Same | PREV. ADM. DATE MO. DA. YR. 2/3/XX | PREV. ADMISSION 1. INPATIENT 2. OUTPATIENT [1] | IF NEWBORN, MOTHER'S HOSP. NO.

9 UNIT A5 | ROOM NO. | ACCOM. CODES 1. PRI 2. SEMI 3. NURSERY 4. PREMIE 5. ICU 6. RCU 7. CCU 8. VIP [2] | ROOM RATE | PAY STATUS | CLASS OF ADMISSION 1. EMERGENCY 2. ELECTIVE 3. URGENT 4. OTHER [3]

10 ADMITTING DIAGNOSIS Cerebrovascular accident

11 PHYSICIAN NAME I.M. Human | PHYSICIAN NO. 432 | ADM. SERVICE Med. | INFORMATION OBTAINED FROM: daughter

12 SPOUSE OR NEAREST RELATIVE (NEXT OF KIN) Kay Ellis | RELATIONSHIP daugh | ADDRESS 301 West Restful Dr. Phoenix | TELEPHONE NO. 258-XXXX

13 SECOND RELATIVE OR FRIEND Marie Darrow | RELATIONSHIP sister | ADDRESS 12 Center St. Danstown, CA | TELEPHONE NO. 837-XXXX

14 RESPONSIBLE PARTY NAME Self | RELATIONSHIP | SOC. SEC. NO.

15 RESP. PARTY ADDRESS STREET P.O. BOX APT. NO. | CITY | STATE | ZIP CODE | TELEPHONE NO.

16 RESP. PARTY OCCUPATION | NO. YRS. IN THIS EMPLOY | RESP. PARTY EMPLOYER | ADDRESS | TELEPHONE NO.

17 LENGTH OF TIME IN ARIZ. 10 yrs. | 1. OWN HOME 2. RENT HOME [1] | TYPE OF HOME Single | BANK NAME & BRANCH Desert National - Camelhead | 1. SAVINGS 2. CHECKING [1]

18 CREDIT REFERENCES | 1. NAME Deep River S&L | ADDRESS 9832 N. LasVegas Pl. Phoenix | TELEPHONE NO. 943-XXXX

19 | 2. NAME Yucca Federal | ADDRESS 1903 W. Bottletree Ave. Phoenix | TELEPHONE NO. 246-XXXX

20 INDUSTRIAL INJURY | DATE: MO. DA. YR. | CLAIM NO. | EMPLOYER'S NAME AND ADDRESS AT TIME OF INJURY

21 BLUE CROSS | NAME OF PLAN | GROUP NO. | IDENTIFICATION | EFFECTIVE DATE MO. DA. YR. | CITY | STATE

22 CHAMPUS DATA | PATIENT'S ID NO. | CARD EFFECTIVE MO. DA. YR. | CARD EXPIRES ON MO. DA. YR. | PATIENT OR SPONSORS BRANCH OF SERVICE | SERVICE CARD NO.

23 | SPONSORS NAME | RANK-SERVICE NO. | DUTY STATION

24 OTHER INSURANCE (INC. BLOOD BANK & BLUE SHIELD) | INS. CO. NO. 1000 | COMPANY NAME Desert State | POLICY HOLDER NAME Iver Andrews | POLICY NO. A657483 | DATE ISSUED MO. DA. YR. 2/17/XX | CITY | STATE AZ

25 | INS. CO. NO. | COMPANY NAME | POLICY HOLDER NAME | POLICY NO. | DATE ISSUED MO. DA. YR. | CITY | STATE

26 NAME OF HEALTH FACILITY DISCHARGED FROM WITHIN LAST 60 DAYS None | ADDRESS

27 OTHER INFO. | V. A. ☐ | COORDINATION OF BENEFITS ☐ | INTERVIEWED BY G. Talker | TYPED BY W.K.S.

28 REMARKS:

Figure 8-1

Patient demographic form. (Reprinted with permission from LaFleur, M. W. and Starr W. K. *Health Unit Coordinating*, 2nd ed. Philadelphia: W. B. Saunders, 1986, p 482.)

care for a patient, it may be necessary to move a patient to another unit of the hospital or to another hospital that has the equipment or expertise needed by the patient. A *transfer* is the form used when moving a patient. Because procedures for transferring vary from one facility to another, the Policy and Procedure Manual of the specific facility is usually the best source of information on how to proceed in such situations.

Some hospitals do not use forms to transfer patients. When a patient is to be transferred to another unit, a call is usually made to the unit clerk. The clerk is told that the patient is coming. A note is then made in the patient's medical chart regarding the move, and the chart is sent with the patient to the new unit. Some facilities may use a form such as the one shown in Figure 8-6.

Administrative Medical Services

🏥 Huntington Hospital
Discharge Instructions

Discharge Date: _____ Time: _____

(ADDRESSOGRAPH)

Accompanied by: _____ Discharge Questionnaire Given: ❏ Yes ❏ No ❏ N.A.

Discharge to: ❏ Home ❏ Other _____ ❏ A.M.A.

Per: ❏ Car ❏ Ambulance ❏ Other: _____ **Via:** ❏ Wheelchair ❏ Ambulatory ❏ Other: _____

Valuables: ❏ None
- Dentures: ❏ Upper ❏ Partials ❏ Cane ❏ Glasses ❏ Electrical appliance: _____
- ❏ Lower ❏ Other ❏ Crutches ❏ Contacts ❏ Clothing
- ❏ Pre-admit drugs ❏ Walker ❏ Hearing aid ❏ Pressure mattress
- ❏ Valuables in admitting ❏ W/C ❏ Prosthesis: _____
- ❏ Other belongings: _____

Follow-up: M.D. office: _____
Dispensary: _____ Date: _____ Time: _____
Referrals: _____ Phone: _____

Special Instructions (activity, diet, bath, wound care, dressing, cast, equipment): _____

_____
_____
_____
_____
_____
_____
_____

Instructed by: _____

Medication Instructions: _____

_____
_____
_____
_____
_____
_____
_____
_____

Pharmacist: _____

I have received and understood these instructions. ❏ Received valuables

_____ _____ _____
Patient/Other   Relationship   Signature of HMH Staff

Figure 8-2
Combined form with release and care agreement. (Courtesy of Huntington Memorial Hospital, Pasadena, CA.)

Huntington Memorial Hospital
ADMISSION INFORMATION SHEET

DATE: _____

TIME: _____

(ADDRESSOGRAPH)

INFORMATION OBTAINED FROM: _____

ADMITTED FROM: ☐ Home ☐ ECF ☐ ER ☐ Doctor's Office ☐ Dispensary
☐ Other _____

VIA: ☐ Wheelchair ☐ Guerney ☐ Ambulance ☐ Ambulatory ☐ Carried

ACCOMPANIED BY: _____ Identification Band Checked: ☐

REASON FOR HOSPITALIZATION: _____

_____

_____

_____

_____

_____

_____

VALUABLES

☐ None

Dentures:	☐ Upper	☐ Partials	☐ Cane	☐ Glasses	☐ Electrical Appliance: _____
	☐ Lower	☐ Other	☐ Crutches	☐ Contacts	_____
	☐ Denture cup at bedside		☐ Walker	☐ Hearing Aid	
			☐ W/C	☐ Prosthesis: _____	

Check here if no other valuables: ☐

Other Valuables: Description   Disposition

_____ | _____

_____ | _____

_____ | _____

_____ | _____

ORIENTATION TO: ☐ Call Lights ☐ Bed ☐ Side Rails ☐ Bathroom ☐ Room Lights
☐ Phone ☐ Meals ☐ Visiting Hours ☐ TV & Educational Channels

☐ Unable to orient patient. Explain: _____

VITAL STATISTICS	Height	Weight	T	P	R	BP	
	☐ By Report	☐ Bed scale				R	L

ADMITTED BY: _____
Signature and Title

Figure 8-3
New patient information form. (Courtesy of Huntington Memorial Hospital, Pasadena, CA.)

Administrative Medical Services

Huntington Memorial Hospital
CASE HISTORY

(Addressograph Plate)

CHIEF COMPLAINTS

AGE

PRESENT ILLNESS

PERSONAL
   MEDS.
   KNOWN ALLERGIES
   HABITS

SYSTEMIC REVIEW
   SKIN
   HEENT
   CARDIAC
   RESP.
   G.I.
   G.U.
   GYN.
   MUSCULOSKELETAL
   NEUROPSYCHIATRIC

PAST HISTORY
   SURGICAL

   MEDICAL

FAMILY HISTORY

Figure 8-4
Medical history form. (Courtesy of Huntington Memorial Hospital, Pasadena, CA.)

1. **NURSING CARE:** This hospital provides only general nursing care unless the patient's physician orders more intensive nursing care. If the patient's condition is such as to need the service of a special duty nurse, it is agreed that such must be arranged by the patient or his/her legal representative. The hospital shall in no way be responsible for failure to provide the same and is hereby released from any and all liability arising from the fact that said patient is not provided with such additional care.

2. **MEDICAL AND SURGICAL CONSENT:** The patient is under the care and supervision of his/her attending physician, and it is the responsibility of the hospital and its nursing staff to carry out the instructions of such physician. The undersigned recognizes that all physicians and surgeons furnishing service to the patient, including the radiologist, pathologist, anesthesiologist, emergency room physician, and the like, are independent contractors and are not employees or agents of the hospital. The undersigned hereby consents to X-ray examination, laboratory procedures, anesthesia, emergency treatment, medical or surgical treatment, or hospital services rendered the patient under the general and special instructions of the physician.

INITIALS	DATE

3. **RELEASE OF INFORMATION:** To the extent necessary to determine liability for payment and to obtain reimbursement, ***the hospital or attending physicians may disclose portions of the patient's record,*** including his/her medical records, to any person or corporation which is or may be liable, for all or any portion of the hospital's charge, including but not limited to, insurance companies, health care service plans, or workers' compensation carriers.

   Special permission is needed to release this information where the patient is being treated for alcohol or drug abuse.

4. **PERSONAL VALUABLES:** It is understood and agreed that the hospital maintains a safe to protect the patient's personal property, money and valuables. The hospital shall not be liable for any loss or damage to the patient's personal property, money or valuables unless those items have been deposited within the hospital safe.

INITIALS	DATE

5. **SAFE ENVIRONMENT FOR PATIENT CARE:** Weapons or other dangerous objects, illegal drugs, and drugs not prescribed by the patient's physician are not permitted in the patient's room. The Medical Center's obligation to provide a safe environment for patient care must override the patient's right to privacy. The Medical Center reserves the right to search the patient and room and to confiscate such objects upon reasonable probable cause.

6. **FINANCIAL AGREEMENT:** The undersigned agrees, whether he/she signs as agent or patient, that in consideration of the services to be rendered to the patient, he/she hereby individually obligates himself/herself to pay the account of the hospital in accordance with the regular rates and terms of the hospital and/or as set forth by the terms of managed care contracts entered into by Saint Joseph Medical Center, and/or applicable Workers' Compensation regulations. Should the account be referred to an attorney for collection, the undersigned shall pay actual attorney's fees and collection expense. All delinquent accounts shall bear interest at the legal rate. Emergency room patients will be billed separately by the emergency room physician and all terms and conditions of this paragraph apply to emergency room bills.

7. **ASSIGNMENT OF INSURANCE BENEFITS:** The undersigned authorizes, whether he/she signs as agent or patient, direct payment to the hospital of any insurance benefits otherwise payable to the undersigned for this hospitalization at a rate not to exceed the hospital's regular charges. It is agreed that payment to the hospital, pursuant to this authorization, by an insurance company shall discharge said insurance company of any and all obligations under a policy to the extent of such payment. It is understood by the undersigned that he/she is financially responsible for charges not covered by this assignment. The terms and conditions above also apply to emergency room treatment which does not require hospital admission.

8. **MEDICARE INSURANCE BENEFITS AND EXCLUSIONS:** I certify that the information given by me in applying for payment under Title XVIII of the Social Security Act is correct. I authorize any holder of medical or other information about me to release to the Social Security Administration or its intermediaries or carriers any information needed for this or a related Medicare claim. I request that payment of authorized benefits be made on my behalf. Some services may not be covered by Medicare, such as the following: 1) Worker's Compensation, 2) Dental, 3) Cosmetic Surgery, 4) Custodial Care, 5) Personal Comfort items, and any service determined to be unnecessary or unreasonable by Medicare.

9. **HEALTH CARE SERVICE PLANS:** This hospital maintains a list of health care service plans with which it has contracted. A list of such plans is available upon request from the financial office. The hospital and/or emergency department has no contract, express or implied, with any plan that does not appear on the list. The undersigned agrees that he/she is individually obligated to pay the full cost of all services rendered to him/her by the hospital if he/she belongs to a plan which does not appear on the above mentioned list.

10. **CONSENT TO PHOTOGRAPH:** The taking of pictures of medical or surgical progress and the use of same for scientific, education, or research purposes is approved.

The undersigned certifies that he/she has read the foregoing, receiving a copy thereof, and is the patient, or is duly authorized by the patient as patient's general agent to execute the above and accept its terms.

TRANSLATION (If necessary) – I have accurately and completely read the foregoing document to the signator identified below in the patient's/patient representative's primary language. He/she understood all terms and conditions and acknowledged his/her agreement by signing this document in my presence.	PATIENT/PATIENT REPRESENTATIVE'S PRIMARY LANGUAGE: SIGNATURE OF TRANSLATOR

SIGNATURE OF PATIENT / PATIENT'S REPRESENTATIVE X	DATE	TIME	REASON PATIENT DID NOT SIGN
RELATIONSHIP TO PATIENT ☐ PATIENT ☐ SPOUSE ☐ PARENT ☐ CHILD ☐ SIBLING ☐ GUARDIAN ☐ OTHER (specify)	WITNESS		TITLE/DEPARTMENT

CONDITIONS OF TREATMENT AND/OR ADMISSION

DISTRIBUTION: ORIGINAL – Medical Records
CANARY – Business Office
PINK – Patient

60-0208-R0594/8560

ADM 02/22/95
ACCT 25173238

Admit Date

Figure 8-5
Admissions agreement form.

Administrative Medical Services　　　　　　　　　　　　　　　　　　　　　　　　　　　　　　　　71

(Use Typewriter or Ballpoint Pen — Press Firmly)

CONTINUING CARE TRANSFER INFORMATION

TO BE COMPLETED AND SIGNED BY NURSING SERVICE (Please attach a copy of the Nursing Care Plan)

PATIENT'S NAME　　Last　　First　　MI	DATE OF BIRTH	SEX	RELIGION	HEALTH INSURANCE CLAIM NUMBER

PATIENT'S ADDRESS (Street number, City, State and Zip Code)	ATTENDING PHYSICIAN　　Name　　Address

RELATIVE OR GUARDIAN　Name	Address	Phone Number

Name and Address of Facility Transferring FROM	Dates of Stay at Facility Transferring FROM	Facility Name and Address Transferring TO
	Admission　　Discharge	

PAYMENT SOURCE FOR CHARGES TO PATIENT:

☐ Self or Family　　☐ Private Insurance ID Number _____　　☐ Blue Cross/Blue Shield ID Number _____　　☐ Employer or Union
☐ Public Agency _____　　☐ Other (specify) _____

PATIENT EVALUATION:

SPEECH: ☐ Normal ☐ Impaired ☐ Unable to speak	HEARING: ☐ Normal ☐ Impaired ☐ Deaf	SIGHT: ☐ Normal ☐ Impaired ☐ Blind	MENTAL STATUS: ☐ Always Alert ☐ Occasionally Confused ☐ Always Confused	FEEDING: ☐ Independent ☐ Help with Feeding ☐ Cannot Feed Self
DRESSING: ☐ Independent ☐ Help with Dressing ☐ Cannot Dress Self	ELIMINATION: ☐ Independent ☐ Help to Bathroom ☐ Bedpan or Urinal ☐ Incontinent		BATHING: ☐ Independent ☐ Bathing with Help ☐ Bed Bath with Help ☐ Bed Bath	AMBULATORY STATUS: ☐ Independent ☐ Walks with Help ☐ Help from Bed to Chair ☐ Bed Bound

NURSING ASSESSMENT AND RECOMMENDATIONS:　　　　　　　　TREATMENTS:

APPLIANCES OR SUPPORTS:　　or check none ☐　　　　　Last Medication: _____/Dose: _____
　　　　　　　　　　　　　　　　　　　　　　　　　　　　Date: _____　　　　Time: _____

　　　　　　　　　　　　　　　　　　　　　　　　　　　Signature　　Title　　Date

TO BE COMPLETED AND SIGNED BY THE ATTENDING PHYSICIAN

ECF Admitting Diagnosis:　　　　　　　　　　　Please send a copy of the following records with patient:
　　　　　　　　　　　　　　　　　　　　　　　　☐ Summary Sheet (face sheet)
　　　　　　　　　　　　　　　　　　　　　　　　☐ Discharge Summary
　　　　　　　　　　　　　　　　　　　　　　　　☐ Physical Examination and History
　　　　　　　　　　　　　　　　　　　　　　　　☐ Consultation
　　　　　　　　　　　　　　　　　　　　　　　　☐ Other (specify) _____

Patient knows diagnosis: ☐ Yes ☐ No　　　　Transfer by: ☐ Ambulance ☐ Car ☐ Other (specify) _____

Surgical Procedures: (current admission)　　　Allergies: ☐ No ☐ Yes (specify) _____

　　　　　　　　　　　　　　　　　　　　　　　VDRL: ☐ Positive ☐ Negative

　　　　　　　　　　　　　　　　　　　　　　　Anticoagulant: ☐ Taking now ☐ Previously

Orders: Diet, medication and special therapy *(To be renewed in 48 hours)*　　Chest X-Ray Diagnosis: _____

　　　　　　　　　　　　　　　　　　　　　　　I will care for this patient after admission to new facility: ☐ Yes ☐ No

　　　　　　　　　　　　　　　　　　　　　　　Medication Regimen is stabilized: ☐ Yes ☐ No

　　　　　　　　　　　　　　　　　　　　　　　Anticipated length of stay for extended care _____ days

　　　　　　　　　　　　　　　　　　　　　　　Physician's Signature　　　　Date

If necessary, attach order sheet — The above constitutes valid temporary orders only if signed by a physician.

　　　　　　　　　　　　　　　　　　　　　　　Address　　　　Telephone Number

Figure 8-6
Patient transfer form. (Reprinted with permission from LaFleur, M. W. and Starr W. K. *Health Unit Coordinating,* 2nd ed. Philadelphia: W. B. Saunders, 1986, p 515.)

The Discharge Process

There are three ways a patient can leave the health care facility. The first is by the physician releasing the patient and providing him or her with the appropriate medical orders for the discharge. The second is when the patient signs him or herself out of the facility, usually against the advice of the physician. The third is when the patient dies. A *discharge* is the process of completing all the necessary records and business as a patient leaves the facility.

Discharging the Patient with Medical Records

Discharging a patient from the care of a medical facility can be handled in different ways. Some facilities use a checklist form that includes the steps to be taken by various staff persons and the patient. Other facilities simply note on the patient's chart that he or she was discharged to go home. The patient or his or her escort is sent to check out at the business office where arrangements are made for handling charges. Other medical facilities may use very detailed forms that include documentation of all steps in preparing the patient to leave the hospital. These forms may include the following information:

- The status of the patient's health upon discharge;
- Physician's orders that the patient received when he or she was discharged;
- Instructions given to the patient on specific limitations and care that might be needed after leaving the facility;
- Medications to be sent home with the patient at the time of discharge;
- Information pertaining to whether the patient or his or her representative was sent to the business office;
- Information pertaining to whether any clothing or valuables were returned to the patient; and
- The name of the person who escorted the patient to the exit at the time of discharge.

Discharge Against Medical Advice

In some cases, the patient may feel that he or she is not receiving the type of care that is needed, or perhaps the patient may believe that no improvement can be expected from further hospitalization. For whatever reason, the patient chooses to leave the hospital without the approval or written order of his or her doctor.

In cases in which the patient does choose to leave the hospital against the advice of his or her physician, the doctor must first be notified. Prior to the patient leaving, everything possible should be done to encourage the patient to remain and discuss the situation with the doctor. However, the patient cannot be restricted from leaving if he or she chooses. Doing so would violate the rights of the patient. In such a situation, the patient should be asked to sign a form. This form states that the patient is making this choice against the advice of the doctor and hospital staff, that the patient understands that it may constitute a health risk, and that neither the doctor nor the hospital can be held liable. The signature of the patient should be witnessed by two persons. Once the form has been signed by the patient, it should be added to the medical chart and kept as part of the patient's medical record.

Discharge of the Patient after Death

Patients who have a terminal illness, that is, a disease that will ultimately result in the patient's death, are of advanced age, or who have suffered a severe traumatic injury may die in the hospital. There are specific laws governing the procedures that are followed in these cases. In most states, a death certificate is completed by the doctor, followed by a report of the death being filed with the Bureau of Vital Statistics. Depending upon the circumstances at the time of the patient's death, an autopsy or examination of the body after death may be requested or required by law. The family or the patient's doctor may request the autopsy in order to determine the cause of death or for research purposes. Permission to perform the autopsy or the examination must be granted by the family, and a consent form should be signed by the next of kin, that is, the closest family member.

In cases in which the patient's death is due to sudden, violent, or unexplained circumstances, such as an accident or possible criminal action, the county medical examiner or coroner may be required to examine the body. The specific forms used in this situation may vary, but either the mortician or the coroner must sign before the body can be released and taken from the hospital. Often, these forms are handled by the pathology department of the hospital.

Summary

In this chaper, we discussed the various steps and forms used in the admission, transfer, and discharge process. We determined that the specific steps taken and the types of forms used for these procedures are usually detailed in the Policy and Procedure Manual for the hospital or medical care facility. Finally, we

Administrative Medical Services

talked about the various forms which may be used for admitting, transferring, and discharging patients, noting that such forms are extremely important and require great accuracy and care in completing to assure that the patient, those caring for him or her, and the facility are protected.

Review Questions

1. Briefly explain what an *admission agreement* is.

2. What does the term *responsible party* mean?

3. _____ pertains to keeping something private and undisclosed without the patient's permission.

4. The study of human population relating to vital statistics is called:
   a. geography
   b. demography
   c. geology

5. The process of moving a patient from one unit or facility to another is called _____.

6. A(n) _____ is the designation of a person to authorize or refuse care for the patient should that patient not be able to do so.

7. Give at least two examples of information that might be found on a patient's medical history form:
   a. _____
   b. _____

8. All patients who choose to leave the hospital without the doctor's written order must sign a(n) _____ form.

Filing Procedures and Medical Records Management

Performance Objectives

Upon completion of this chapter, you will be able to:

1. Explain the necessity for properly maintaining all patient information in a medical record.
2. Briefly discuss the different types of indexing systems used by most health care facilities.
3. Explain why accurate filing is essential to the efficient operation of a health care facility.
4. Discuss the basic methods involved in filing patient records.
5. Identify the standard rules for alphabetical filing, and be able to alphabetize files correctly.
6. Briefly explain the role of the Joint Commission on Accreditation of Healthcare Organizations as it relates to reviewing medical records.
7. Identify and briefly describe the methods used for properly assemblying medical charts.
8. Identify the responsibilities of special records clerks.
9. Identify the forms that would be contained in a patient's medical record.
10. Discuss the actions that would be taken by members of the medical records department relative to incomplete patient files and medical records.

Terms and Abbreviations

Accreditation   special recognition for meeting a standard.
Alphabetical filing system   a method in which patient records are filed by alphabetical last name sequence.
Chart back   a metal or plastic cover to which medical chart forms can be added.
Chronological order   arrangement of items or events according to their dates.
Coding   assignment of specific numbers by a classification system.
Color-coding filing system   a method by which alphabetical groupings are color-coded and number-coded on the patient record in order to made presorting and filing of charts easy.
Deficiencies   when used in connection with medical records, pertains to incomplete records.
File guide   a divider which is placed in a file to show the divisions of a particular filing system.

Index   anything that serves to guide or facilitate a reference.
JCAHO   the Joint Commission on Accreditation of Healthcare Organizations, which is a regulatory agency responsible for establishing the criteria and inspections of hospitals for accreditation.
Medical abstracts   summaries of medical data from patients charts.
Numerical filing system   a method in which each patient is assigned a specific number, making it possible for each record to be filed in numerical order; an alphabetical system is generally used to cross-reference the assigned number.
Outguide   an indicator which is placed to show that an item has been removed from a file.
Patient record   a permanent, legal document of a patient's illnesses, diagnoses, symptoms, treatment, and care; also referred to as a chart or medical record.

Records management the organization, implementation, and maintenance of a filing system for patient records.
Statistical data facts that are tabulated numerically.
Tagging the process of placing a name and/or number on the medical chart.

All the information concerning a specific patient's illnesses, diagnoses, symptoms, treatment, and care is compiled in a permanent, legal document commonly referred to as the *patient's medical record*. The patient record is generally kept in a color-coded manila folder or some type of hard chart back. To conduct daily transactions, a health care facility needs to have an efficient system for maintaining patient records so that they can be stored, retained, and retrieved when necessary. In time, all patient information will be maintained in a computer networking system. At the present time, many health care facilities already maintain basic patient information in a computer database. This makes it a great deal easier to access specific information as it may be needed by the physician and other members of the health care staff.

Indexing Medical Information

In order for hospitals and other medical facilities to be able to use and access specific patient information, computerized indexes are maintained. These indexes serve as valuable reference tools for information concerning the patient and the health care facility. They are important sources of information and play a vital role in patient care management, research, facility planning, and financial analysis. The amount and type of data maintained in an index greatly depend upon the specific needs of the individual health care facility.

Many hospitals maintain a *Master Patient Index*, a *Physician's Index*, and a *Disease and Operative Index*. The demand for information related to patient health care management requires facilities to maintain indexes that manage and retrieve information efficiently.

The Master Patient Index, or MPI, is responsible for keeping tack of all patients who have ever been admitted or treated by the health care facility. This index is almost always initiated by the admitting department, which is responsible for registering and assigning new patient numbers and retrieving the medical record number of patients who have been previously treated. Information is also gathered from nursing units. This information is available to all departments and usually includes the patient's full name and address, identification number or social security number, sex and date of birth, mother's maiden name, race or ethnicity, admission and discharge dates, and attending physician.

The Physician's Index is used to provide every member of the medical staff with an accurate record of the patients treated in a given health care facility. To maintain the confidentiality of these records, each physician is assigned a specific code number which appears in the index. The Physician's Index includes the doctor's identification number, diagnostic codes, the patient's sex and age, and the length of stay or date the patient was discharged.

The Disease and Operative Index is available in all health care facilities to provide specific data regarding complete medical and statistical reports. In addition to patient information, it includes the name of the attending physician or surgeon, the final result of hospitalization, the length of a specific patient's hospital stay, and the identification of any associated diseases or operations that occurred during the patient's stay in the hospital. This information can be used by the medical staff for any number of purposes, including research, determination of admission trends, identification of specific disease occurrences, facility use, and infection control studies.

Filing and Records Management

The process by which all medical records involving a single patient are organized and maintained in a health care facility is referred to as *records management*. Because no single system will meet the needs of every health care facility, management of patient records must be individualized to meet the specific needs of each facility. Every health care facility is required by law to maintain patient records in a systematic manner so that records can be easily and quickly retrieved with the highest degree of efficiency and accuracy. Most facilities use some type of filing system for maintaining patient records. The system chosen by the individual facility should allow for adequate expansion and provide security for confidential patient information. Whatever the system used, accurate filing is the key to a good system and is considered the most important aspect of filing.

Filing Systems

There are four basic systems of filing used by almost every health care facility and medical office. These include the *alphabetical* filing system, the *numercial* filing system, the *color-coded* filing system, and the *subject matter* filing system.

The alphabetical filing system is the oldest and most frequently used system of keeping files. It involves maintaining patient records by filing them alphabetically by the patient's last name. This system works well for facilities managing fewer than 10,000 records. Unfortunately, however, it is more subject to human error than other methods, and expansion of files generally requires the shifting of charts in order to redistribute the space available for filing.

When patient confidentiality is especially important, the numerical filing system is considered most effective. Under this system, each patient is assigned a number. The number is entered on the patient's record index card, which is filed by last name alphabetically. When the patient comes to the facility, the record index card is checked for the patient number, and the patient record is then retrieved by that number. Numbering can be done by *unit* or *serial assignment*.

In unit numbering, the patient is assigned a specific number on his or her initial visit to the facility, and then keeps that same number for all other visits. This provides a single record for each patient. With serial numbering, however, the patient receives a new number each time he or she comes to the health care facility. A new patient record is established with each visit, and the patient will have as many different files as visits.

Although the numerical filing system requires a two-step process for record retrieval, it does allow for unlimited expansion without moving files. Patient record numbers may be kept in the computerized Master Patient Index rather than a card file and can easily be retrieved by typing in the patient's name.

For health care facilities that have large daily transactions, the color-coded filing system allows for the greatest efficiency for retrieving many files, while providing a color-coded system for ease in presorting and filing a large volume of patient records. However, initially, it is a more difficult system to learn. Because alphabetical groups are color- and number-coded, a misplaced file stands out by breaking the color pattern of the files. The color code is applied using the second letter of the patient's last name. Most second letters are vowels, but the entire alphabet is included using the following arrangement: *A–D is green, E–H is red, I–M is blue, N–Q is yellow, and R–Z is purple*. Color coding of charts can also be used with other filing systems to indicate a specific type of insurance, diagnosis, geographical or demographical location, and so forth.

Certain special files used by a health care facility may need to be filed by a specific subject. Purchase orders, invoices, payroll information, and inventory control sheets, for example, are usually filed in this manner. The files are divided into broad categories, and each category is subdivided into specific subjects. Subjects are generally filed by date, with the most recent item in front. This method of filing is almost always used by the facility's business office.

Filing Rules

When many people have access to files, each person must use the same rules in order to ensure the proper organization and maintenance of each and every file. A haphazard filing system causes a lot of confusion, and time is spent unproductively when records cannot be retrieved quickly and efficiently. In most cases, basic filing rules for the health care facility are usually recorded in a procedure manual and taught to each employee who will have access to patient records. These rules generally include the following:

- The main filing unit is the patient's last name. When last names are different, letters are considered from left to right and filed in alphabetical order. For example:

  Jones, David   *Adams, James*
  Thomas, Brian   *Aikins, Cynthia*
  White, Denise   *Anderson, William*

- When last names are identical, the second filing unit, the patient's first name, is considered and filed in alphabetical order. For example:

  Smith, Andrew   *Green, Carol*
  Smith, Jason   *Green, Cathy*
  Smith, Steven   *Green, Constance*

- When the main and second filing units are the same, the order is determined by the middle name. For example:

  Jones, Carol Ann
  Jones, Carol Elizabeth
  Jones, Carol Jean

- When all letters of a unit are identical, the shorter unit name is filed first, that is, "nothing" before "something".

  *Stilson, Anne_*
  Stilson, Annemarie

- When an initial is used as a filing unit, it precedes all names in the same unit beginning with the same letter and is considered as a separate unit. For example:

    Garcia, A.C.
    Garcia, Allison M.
    Garcia, Annmarie
    White, J.J.
    White, J. Jason
    White, John J.

- An abbreviated first or middle name is always treated as if it were spelled out. For example:

    Raulerson, Eliz. C.
    Raulerson, Elizabeth K.
    Raulerson, Elizabeth Marie

- A prefix is combined with the word following it to form a single index unit. The space is ignored for filing purposes. For example:

    McMartin, Carla   MacHenry, Dennis
    Mcmartin, Carol   McDermitt, Mary

    An exception to this rule is when St. is used before a last name. It is considered SAINT for filing purposes. Therefore *St. Mark* would be filed as though it were *Saintmark*.

- A hyphenated name is always considered as a single filing unit. For example:

    Thomas-Smith, Nina
    Thomastone, Kathryn

- Each word in a compound personal name is treated as a separate filing unit. For example, *Jeannette Green Smith* and *Denise Smith Wilson* would be filed as:

    Smith, Jeannette Green
    Wilson, Denise Smith

- The names of married women are always indexed as written. In offices where "head of household" is used for billing purposes, a cross-reference should be made on the husband's chart of the wife's name or patient number. For example:

    James, Eileen M. Mrs.
    James, Eileen M. Ms.

- When religious or royal titles are followed by a single name, the title becomes the main filing index and the name is second. For example:

    Father James
    Princess Grace

- Degrees, titles, and seniority following names are generally ignored in filing, although *Jr.* may be filed before *Sr.* or the address may be used to determine the filing order.

- When the full names of two or more files are identical, they should be arranged according to the alphabetic order of the city address or by street address if the cities are also identical. For example:

    James, Connie   Atlanta, 3345 Freedom Way
    James, Connie   Augusta, 1066 Olympic Highway
    James, Connie   Marietta, 327 Hotsprings Lane

- Organization names are always indexed in the order in which they are written. If they are abbreviated, they are filed as though they were written in full. For example:

    AMA (American Medical Association)
    American Red Cross
    University of California, Los Angeles
    University of California, Santa Barbara

Storing Files and Medical Records

Many office equipment companies have developed various types of convenient filing systems for storing patient records and files. The vertical filing systems which many health care facilities use consist of stacks of shelves or drawers for filing patient folders. Each folder has a tab that indicates a patient number, name, or date.

Placing the patient's name and number on the chart tab is called tagging. There are three different methods of *tagging*—horizontal, lateral, and vertical. The type of tagging method used depends not only on the filing method used, but also on how the charts are stored. The vertical file is ideal for charts that have been tagged horizontally. The lateral file cabinet and shelf filing on stationary shelves built on the wall are most often used for charts that have been tagged vertically or laterally.

Letters of the alphabet or a series of numbers are separated by the placement of dividers called *file guides*. They are usually larger than the file folders, are made of a durable material, and can be lettered or numbered as needed.

An *outguide* is an indicator put in place to show that a specific item has been removed from a file. These are made of a durable material and may also be color-coded. An outguide can be a folder in which additions to the record may be added while the original is gone. All outguides indicate what information was removed.

Files should be kept for an indefinite period as a

protection against potential malpractice lawsuits. However, files may be moved to an inactive file area after two to three years. Large medical facilities generally store their inactive files on microfiche or microfilm to condense the stored information.

Medical Record Assembly and Analysis

Medical record assembly must meet certain criteria established by national and local regulating agencies such as the *Joint Commission on Accreditation of Healthcare Organizations (JCAHO)*, *Medicare* and *Medicaid*, and other state and federal organizations. As a legal document, the patient's medical chart must be complete, placed in the appropriate chronological order, and signed by the attending physician. It is the duty first of the administrative medical worker and ultimately of the members of the medical records department to ensure that these requirements are met.

The Role of JCAHO

Medical documents must meet specific regulatory criteria. The Joint Commission on Accreditation of Healthcare Organizations is a national regulatory agency that has established criteria that all participating hospitals agreed to meet. Every three years the Joint Commission is responsible for inspecting these participating institutions for accreditation. This agency reviews all departments, including medical records. Failure to comply with any of the regulations set forth by the Commission could result in a hospital losing its accreditation.

In 1993, the Commission developed their *standards for medical records*. According to these standards, every hospital or institution accredited by JCAHO must:

- Maintain medical records that are documented accurately and in a timely manner, are readily accessible, and permit prompt retrieval of information, including statistical data.
- Make sure that all medical records contain sufficient information to identify the patient, support the diagnosis, justify the treatment, and document the course and results accurately.
- Implement a procedure for maintaining the confidentiality and security of all medical records, as well as keep them current and up-to-date, authenticated, legible, and complete.
- Implement a procedure for maintaining the medical records department by providing it with the direction, staffing, and facilities necessary to perform all required functions.

Assembly and Analysis of Medical Records

Once a patient has been discharged, and before his or her medical chart is sent down to the medical records department, the unit secretary is responsible for removing all the chart forms from their holder and placing them in a specific order, with the most recent documentation on top. This action increases the efficiency of the medical records personnel in handling patient files.

When a patient's chart is received in the medical records department, clerks check on how the chart is assembled and whether its parts are in the appropriate chronological order. Members of the medical records department can detect and track *deficiencies* much more readily when the charts are arranged by date. Papers that the physician needs to complete and sign may be placed on top of the file. This arrangement ultimately provides the physician with greater access to the records.

There are several different methods of chart assembly, each of which depends greatly on the type of chart received by the department. Files from the nursing units are assembled in chronological order and may be placed in the format shown in Figure 9-1. Other specialty medical records may be assembled in a different fashion, such as the one shown in the example of the chart for a Day Surgery, Emergency Room, or Outpatient department, in Figure 9-2.

Analysis and Qualitative Check of the Medical Record

A *discharge analysis clerk* is a special position established in the medical records department of some hospitals. The responsibilities of the person in this position require that he or she examine each chart for deficiencies in information, required reports, and signatures, and then enter this information into a computer. Discharge analysis clerks use special checklists to assist them when they are reviewing records. These checklists serve as standards which are necessary for detecting deficiencies in repeated reviews. A sample Inpatient Record Analysis Checklist or deficiency sheet is shown in Figure 9-3.

In facilities that do not employ a designated charge analysis clerk, various record clerks in the health information section are responsible for assemblying records, analyzing and checking them for deficiencies, and entering the information into the computer. From completed medical charts, medical abstracts are created. A *medical abstract* is a summary of the data from patient charts. These abstracts provide the hospital with a database that can generate statistical data that has been tabulated numeri-

Chart Sections
Routine Nursing Unit

1. **Patient Data**
   Summary Sheet
   Insurance Date Sheet
   Diagnosis/Procedure Summary
   AMA Release
   Release of Remains
   Autopsy Permit

2. **Physician Documentation**
   History and Physical
   Specialty H&P (oral, podiatry, etc.)
   Progress Notes
   Consultations
   Emergency Room Encounter
   Physician's Orders

3. **Consents**
   Consent to Admit and Treat
   Other Consents

4. **Surgical**
   Operative Consent
   Anesthesia Record
   Pre/Post-Anesthesia Notes
   Report of Operation
   Tissue Report
   Intraoperative Record
   Post-Anesthesia Record
   Pre/Post-Operative Checklists

5. **Diagnostic Test**
   X-ray Reports/Nuclear Medicine
   ECG/Echocardiography
   Stress Test
   EEG
   EMG
   Pulmonary Function
   Investigation of Suspected Blood
     Transfusion Reaction
   Lab Reports

6. **Paramedical Progress Notes**
   Dietary Notes
   Home Health Notes
   Rehabilitation Notes (P.T., O.T., etc.)
   Respiratory Therapy Notes
   Social Services Notes
   Speech Therapy Notes

7. **Nursing Documentation**
   CPR Record
   Clinical Record
   Medication Record
   Fractional Diabetic Record
   Insulin Report
   Patient Education Record
   Patient Problem List
   Med/Surg Admission Assessment
   24-Hour Nursing Record
   Discharge Summary/Home Care Instructions

Figure 9-1
Assembled chart from nursing unit.

cally. Hospitals can use this statistical information to answer questions, plan for staffing needs, encourage cost containment, develop reports, improve patient care, justify expenditures, and promote the particular health care facility.

When all records are complete and no deficiencies exist, medical record clerks can code each individual chart. *Coding* is assigning specific numbers by a classification system. From these codes, procedures and diagnoses can be identified.

Chart Forms

A *hospital chart* is a written record that includes all the relevant information about the patient from the time of his or her admission until the time the patient is discharged from the hospital. All chart forms have two things in common: the patient's name and the name of the hospital. Usually, the hospital's name is stamped on the chart form with an imprinter card or filled in on the computer form.

The nature of the information included in a medical record demands that it remain totally *confidential*. This means that the information is private and can only be reviewed by authorized persons. Patients must feel free to discuss the most personal details about their illness to enable the doctor to make an accurate diagnosis. All medical workers, including the members of the administrative medical services staff and the medical records department, are obliged to respect this confidentiality by never revealing any information about patients. Confidentiality is not only a moral and ethical obligation but also a legal right of all patients.

The information contained in a patient's medical

Chart Sections
Specialty Nursing Units

Day Surgery List
Summary Sheet
Patient Data Sheet
Diagnosis/Procedure Summary
Medical History and Physical
Specialty H&P (oral, podiatry, etc.)
Physician Progress Notes
Physician's Orders
Discharge Prescriptions
Consent to Admit and Treat
Refusal to Permit Blood Transfusion
Consent to Operate
Anesthesia Record
Pre/Post-Anesthesia Notes
Operative Report
Surgery Flowsheet p.1
Surgery Flowsheet p. 2
Recovery Room Record
X-ray Reports
ECG
Lab Reports
Paramedical Progress Notes (P.T., etc.)
Patient Education Record
Preoperative Nursing Care Record
Nursing Discharge Summary
Day Surgery Post-Op Evaluation Form

Emergency Room List
Summary Sheet
Patient Data Sheet
Physician Encounter
Physician Dictation
Follow-up Instructions
Nurse's Notes (triage first)
Ambulance/Paramedic Notes
Consent Forms
X-ray Reports
ECG
Lab Reports

Outpatient List
Summary Sheet
Patient Data Sheet
Consent Forms
X-ray Reports
ECG
Lab Reports

Figure 9-2
Assembled chart for specialty nursing unit.

record depends greatly on the institution and its particular needs. Medical and dental offices, clinics, long-term care facilities, and hospitals are all responsible for maintaining accurate patient records. All members of the medical records department, therefore, must become familiar with the different chart forms used in their particular facility.

The following is a list of some of the forms more frequently used by most medical and health care facilities:

- *Patient Information Form.* This form usually includes the patient's name, address, telephone number, age, date of birth, sex, marital status, religious preference, insurance number and responsible party, and attending physician.
- *History and Physical Form.* This form includes information relevent to the patient's medical history, results of a physical examination, the current diagnosis, and the prognosis as it relates to the patient's current complaints, signs, and symptoms.
- *Nursing Notes.* These include a record of the daily care, treatments, vital signs, diet, medications, and pertinent observations made by the members of the nursing staff.
- *Doctor's Notes/Progress Notes.* These include all notations made by the patient's physician regarding his or her daily progress or lack thereof.
- *Physician's Orders.* The doctor uses this form, which must be signed, to write orders necessary to direct all diagnoses and treatments for the patient.
- *Graphic Sheet.* This form is used to record the patient's vital signs (blood pressure, temperature, pulse, and respiration), and may be a valuable tool to both the physician and the nursing staff in providing important information regarding variations in the vital signs.

Administrative Medical Services

Inpatient Record Analysis Checklist

Name _____ Pt. No. _____ Date _____
Physician _____ Dr. No. _____ Pt. No. _____

	Complete	Signed
Final Diagnosis	_____	_____
Sign on Summary	_____	_____
History & Physical	_____	_____
Obstetrics Summary Sheet	_____	_____
Prenatal Record	_____	_____
Newborn Record	_____	_____
Progress Notes	_____	_____
Discharge Note/Condition	_____	_____
Discharge Summary	_____	_____
Discharge Order	_____	_____
Consultation	_____	_____
Operative Record	_____	_____
Emergency Room Record	_____	_____
Labor & Delivery Record	_____	_____
Signed Orders	_____	_____
Special Reports	_____	_____
Blood Transfusion Notes	_____	_____
Pre-Op Note	_____	_____
Post-Op Note	_____	_____
Autopsy (Provisional)	_____	_____
Autopsy (Final)	_____	_____
Patient Destination	_____	_____

St. Mary's Community Hospital

Figure 9-3
Inpatient record analysis checklist.

- *Laboratory Reports.* These forms provide the physician with the results of all laboratory tests performed on the patient.
- *Special Forms.* These forms provide the physician with the results of any special treatments or diagnostic procedures performed on the patient.

Importance of Providing Complete Data

Physicians on the staff of a hospital have a specified period in which to complete and sign their patients' charts. Charts that have been flagged as deficient are placed in a special location within the department to give physicians easy access and encourage them to complete their charts.

If charts are not completed and signed within the specified period, a deficiency system goes into effect in the medical records department. This system generates letters to the doctors involved, notifying them of their delinquent status. Failure to complete charts in a timely manner can result in the physician's hospital privileges being suspended.

Summary

In this chapter, we discussed the importance and necessity of properly maintaining all patient information in medical records. We discussed the different types of indexing systems, explained why accuracy in filing is essential, and discussed the basic methods involved in filing patient records. We also talked about the various methods used to assemble medical charts properly, and identified the types of forms contained in the patient's medical record chart.

Review Questions

1. Define the following terms:
   a. tagging _____
   b. indexing _____
   c. filing _____
   d. coding _____

2. Briefly explain the following types of filing systems:
   a. alphabetical filing system _____
   b. numerical filing system _____

3. When is a file guide used?

4. When is an outguide used?

5. What information is provided for under statistical data?

6. Briefly explain the importance of records management in the health care environment.

10

Data Abstracting

Performance Objectives

Upon completion of this chapter, you will be able to:

1. Explain the purpose and process involved in abstracting data.
2. Briefly explain and demonstrate how to calculate the length of stay and average stay of a hospital patient.
3. Explain the process involved in determining a hospital's patient census for a specific day.
4. Describe the functions of registers.
5. Discuss how to complete a birth certificate properly.
6. Describe the various records maintained for patient deaths and fetal deaths.

Terms and Abbreviations

Census the number of patients present in a health care facility at any given time.
Data abstracting compiling pertinent information from a patient's record.
Pluralty a term referring to the order of delivery in a multiple or plural birth.
Register a chronological list of data maintained by a health care facility to reference basic information such as births, deaths, and fetal deaths.
Registrar the person who is responsible for keeping lists and statistics.
Vital statistics data that pertains to births, deaths, and fetal deaths.

Patient records are considered the primary source of data used in compiling medical care statistics that are required for state licensure of health care facilities participating in various insurance programs, including Medicare and Medicaid.

Abstracting Data

The term *abstracting* refes to the process of compiling pertinent information from a patient's medical record. While some information can be collected upon the patient's admission, and then indexed, most patient data are compiled after the patient has been discharged. All data collection is facilitated by a *data entry source document*, which is a list of the data that must be obtained from the patient's record. Once the data have been properly abstracted, they are either entered directly by the medical records department or sent to data processing for computer entry.

Discharge information is entered and the results kept for a number of reasons such as to determine the number of patients per medical staff unit, maintain admission and discharge dates, facilitate infection control, and record births, deaths, consultations,

83

autopsies, and so forth. This information is needed for an accurate record of the daily transactions of the health care facility. From this basic information, calculations can be made to determine the average length of stay, average census, and other pertinent information that can be used to justify staffing, equipment use, and other budgetary concerns and may be important for medical research.

Determing the Average Length of Stay

The average length of stay for each patient admitted into a health care facility can be calculated by subtracting the day of admission from the day of discharge. For example, if Nancy Smith was admitted to the hospital on March 16 and was discharged on March 20, then her length of stay would be 4 days $(20 - 16 = 4)$.

The average length of stay can be calculated by dividing the length of stay days for all discharged patients by the total number of discharged patients for a particular period. For example, in February, the total number of patients discharged was 2,173 and the total length of stay days of the discharged patients was 5,396. The average length of stay for February can be calculated as follows:

$$\frac{5396}{2173} = 2.5 \text{ days}$$

Determining Patient Census

Patient census is determined by calculating the number of inpatients present at any given time, with calculations beginning at midnight. Patient census is compiled at the same time each day by following the steps outlined below:

1. Determine the number of patients in the hospital at midnight on March 10 (n = 230).
2. Add the number of patients admitted on March 11 (n = 18).
3. Subtract the number of patients discharged on March 11, including deaths (n = 12).
4. The inpatient census for March 11 can be determined as follows: $230 + 18 = 248 - 12 = 236$.

Vital Statistics and Health Care

All states are responsible for keeping vital statistics for their individual state. For example, the Bureau of Vital Statistics of the California Department of Health is responsible for vital statistics for the state of California. *Vital statistics* are the data pertaining to live births, deaths, and fetal deaths. The Bureau of Vital Statistics for the State of California is responsible for collecting, recording, transcribing, compiling, and preserving these data for the state of California.

Hospitals are required by law to maintain birth, death, and fetal death registers. These registers are records which provide an official chronological listing of patient information and may be computerized or maintained in a book with bound pages. The hospital register is completed by the hospital staff prior to filing certificates with the local register.

For the purpose of registering births, deaths, and fetal deaths, all states are divided into counties or districts. The justice of the peace or the county clerk is responsible for the register and for securing records for each individual county or district. When a certificate is filed, the local registrar keeps a record on file and forwards the original to the state registrar for permanent reference. The state registrar sends a copy to the National Center for Health Statistics, which, in turn, sends information and data to the World Health Organization. Data obtained from the registration of vital statistics can be compiled for the state, nation, and world, and can be used for medical statistical research.

Processing and Filing Birth Certificates

A *live birth* is officially defined as the complete expulsion or extraction from the mother of a product of conception, irrespective of the duration of pregnancy, which, after such separation, breathes or shows any other evidence of life such as beating of the heart, pulsation of the umbilical cord, or definite movement of voluntary muscles, whether or not the umbilical cord has been cut or the placenta is attached. When a live birth has occurred, a certificate of birth must be filed according to the laws governing the individual state in which the birth took place. Certificate of Birth forms are provided by the individual state's Department of Health, Bureau of Vital Statistics. The medical records department of most hospitals is responsible for preparing the birth certificate and filing it in a timely manner. Births outside the hospital must be registered by someone in attendance during the birth, and all births must be reported and filed within five days. This period, however, can be delayed, so long as the proper applications and documentation have been provided to the appropriate government agency.

Birth certificates have many functions. They are needed to establish a person's citizenship or parentage, to obtain a passport, to show the age of a child for admission to public school, to register to vote, to get a driver's license, and to obtain social security benefits. They must be either typed or written in durable black ink, and all information must be complete and accurate. If it is impossible to secure particular information, the word *unknown* may be entered. To be considered a legal document, a birth certificate must not have any strikeovers, mark-throughs, erasures, or illegible entries.

The upper portion of the birth certificate is the legal portion that contains information required for identification of the individual. Information that must be entered includes the child's name, date of birth and sex; the county, city or town, and place of birth; the name of the hospital and whether it was within city limits; an indication of whether it was a single or multiple birth with the order of birth recorded; the father's name, date of birth, and birthplace; the mother's maiden name, date of birth, birthplace; and the residence and the race of both parents (if Hispanic, the specific country of origin); the exact time of birth; the name, address, and qualification of the attending physician; the registrar's signature; the date and file number; the signatures and social security numbers of the parents; and whether the parents want a social security number for their child. The last name of the child does not have to be the same as either parent, and the parents may provide any name they desire so long as it fits in the space.

All remaining questions on the certificate relate to medical and health information. This information can be useful in studying trends in childbearing, child spacing, infant mortality, congenital anomalies, obstetrical procedures, and medical factors in pregnancy.

Processing and Filing Fetal Death Certificates

A *fetal death* is officially defined as death prior to the complete expulsion or extraction from the mother of a product of conception, irrespective of the duration of pregnancy. The death is determined by the fact that, after such separation, the fetus does not breathe or show any other evidence of life such as beating of the heart, pulsation of the umbilical cord, or definite movement of voluntary muscles. All fetal deaths must be recorded in a fetal death register that is maintained by each local health care facility housing obstetrical facilities. The hospital register is completed by the hospital staff before the certificate is filed with the local registrar. When a certificate is filed, the local registrar keeps a record on file and forwards the original to the state registrar.

When a fetal death occurs, great care must be taken in completing the certificate of death. This certificate is very similar to the certificate of birth in format, with the upper portion containing information required for identification. The remaining sections relate to medical and health information. If parents do not provide a first or middle name, the item is left blank. As with the birth certificate, the last name does not have to be the same as either parent, and parents may provide any name they desire so long as it fits in the space provided.

Processing Death Records

The hospital in which a death occurred is responsible for maintaining an accurate death and fetal death registry, and for providing information concerning and surrounding the patient's death. Deaths occurring in the presence of an attending physician must have the death certificate signed by that physician. Deaths without medical attendance must be referred to the medical examiner or justice of the peace for future investigation and signature. The funeral director responsible for making disposition of the dead human body is responsible for filing the death certificate. All deaths must be reported and filed within 10 days, and all fetal deaths must be filed within five days.

Information obtained from the death certificate may be used for studying future mortality rates and statistics, accidents and related injuries, violent deaths, deaths due to disease, and to process insurance claims and other survivor benefits.

Summary

In this chapter, we discussed the function and process involved in abstracting data from the patient's medical record. We determined that information which is abstracted must be completely and accurately compiled so that it could be used to facilitate the operation of the health care facility. We also talked about the purpose and means of determining the average length of stay and patient census within a health care facility. Finally, we discussed the maintenance of records regarding the births, deaths, and fetal deaths taking place within a hospital, noting that each state is required to follow the correct guidelines for filing live birth and death certificates with their local registrar.

Review Questions

1. Define the following terms:

   a. census _____

   b. vital statistics _____

   c. register _____

   d. abstracting _____

   e. registrar _____

   f. pluralty _____

2. Briefly explain the process and calculation involved for determining a patient's average length of stay in the hospital.

3. Briefly explain the process and calculation involved in determining patient census in a hospital.

4. Who is responsible for filing a fetal death certificate?

Insurance Coding and Indexing

Performance Objectives

Upon completion of this chapter, you will be able to:

1. Identify and briefly discuss various types of insurance programs available for health care.
2. Discuss reference manuals available for coding medical insurance claims.
3. Describe the importance of accuracy as it relates to coding an insurance claim form.
4. Explain and be able to demonstrate how to produce a rough draft copy of an insurance claim form.

Terms and Abbreviations

Alphanumeric code a code with letters and numbers and sometimes other symbols, such as punctuation marks.

CHAMPUS abbreviation for Civilian Health and Medical Program for the Uniformed Services, which is a military program responsible for providing medical care for dependents of the military and retired military personnel.

CHAMPVA abbreviation for Civilian Health and Medical Program for the Veterans Administration, which is a government insurance program that covers family members of veterans with 100 percent service-connected disability, or the surviving spouse and children of a veteran who died as a result of a service-connected disability.

CPT abbreviation for Current Procedural Terminology codes, a coding system published by the American Medical Association and used to communicate procedures performed by a physician.

Deductible the amount of money a patient or responsible party must pay before the insurance company will start paying.

FECA abbreviation for Federal Employee Compensation Act, which is a program that reimburses federal employees for medical expenses related to accidents or injuries which occurred on-the-job.

HCFA abbreviation for Health Care Financing Administration, which is an administrative agency within the U.S. Department of Health and Human Services that is responsible for handling all issues related to Medicare and Medicaid.

ICD codes abbreviation for International Classification of Diseases codes, an internationally recognized coding system for all known diseases, that was originally designed by the World Health Organization of the United Nations.

Medicaid a state government program designed to help indigent or poor people by providing medical benefits.

Medicare a federal government program designed to help elderly and disabled people who have paid into the Social Security system by providing medical expenses, the deductibles and coverage associated with which change annually.

Medicare Part A the part of Medicare that pays a portion of an inpatient's hospital care or daily skilled care received in a skilled nursing facility, home health care, and hospice care.

Medicare Part B the part of Medicare that pays a portion of a patient's physician and outpatient services.

SNF abbreviation for skilled nursing facility.

Subscriber the person who holds the insurance policy; may also be called the insured.

Workers' compensation medical payments made by employers when an employee has a work-related illness or injury.

According to most health insurance providers, approximately 80 percent of the money paid to a physician's office or to a health care facility comes in the form of insurance payments from private insurance companies or government insurance programs. Since the costs of health care continue to rise on an annual basis, it is mandatory that physicians, hospitals, and clinics be paid on a timely basis. Therefore, as a member of the administrative medical services team responsible for working with health insurance agencies, you have a very important role in gathering the data necessary to file claims accurately for monies owed, as well as in grouping illnesses which allow for indexing.

Careers in Insurance Coding

While Section IV of this text will deal specifically with concepts and applications involved with working as an Informational and Insurance Coder, you should know that, with the advent of coming changes in the national health care system, accurate insurance coding clerks will be in great demand. Persons specializing in working with health insurance should be able to perform with a high degree of accuracy and attention to detail. A knowledge of medical terminology is a definite asset, and computer keyboarding skills are a major advantage. If you are really ambitious, you may choose to enroll in any one of a number of vocational schools or colleges throughout the country and become a Certified Coding Specialist.

Health Insurance Programs

There are two major programs that exist exclusively to pay for health care. These two systems include one which is financed by the individual and one which is financed by state and federal government funds. Approximately 2,000 different private health care programs are available, including all types of commercial plans, preferred provider organizations (PPOs), and health maintenance organizations (HMOs).

Comprehensive health care is provided by the federal government for large groups of qualified people. The groups that fall under this category include members of the uniformed services, retired military personnel, and their dependents through CHAMPUS and CHAMPVA, the elderly through Medicare, which is administered by the Social Security Administration, the disabled and indigent through Medicaid, which is also administered by the Social Security System, and those with job-related injuries or illnesses, with payment administered by FECA/Black Lung or individual workers' compensation insurance companies. All funds for payment through health insurance coverage are financed through general taxation or, in the case of the Social Security System, from a trust fund of contributions paid by employees, employers, and self-employed workers.

When a service or a procedure is performed by a health care provider, that individual expects to be paid. To obtain insurance compensation, either the patient or the provider must submit a completed form detailing the specific services which have been provided. *Codes* are used to standardize and simplify these forms. The code numbers, which are used to identify individual physicians; diseases being treated; and procedures, services, and supplies provided to a patient, are obtained from publications called *manuals*. This system is referred to as a *coding system*.

Since payment is made according to the code that has been assigned for a diagnosis and procedure, it is impertative that those working as insurance coding clerks be accurate in choosing the correct code. An incorrect code can result in a form being returned to the doctor's office or the health care facility or, in some cases, a reduced amount being paid.

Understanding and Using Coding Systems

When a doctor examines a patient in his or her office or at the hospital, a procedure has been performed for which the physician expects to be paid, by the patient, a private insurance company, or a government health care agency. The regular sequence of events is that physician or health facility will first identify the procedures performed for the patient. For uniformity in reporting procedures and treatments, the American Medical Association has developed and published a coding system called *Current Procedural Terminology* (CPT) codes. These procedures and their codes are generally marked on a superbill, a charge slip, or the medical record.

The CPT manual is used for determining the correct five-digit code. It is divided into several sections, including *Introduction*, *Medicine*, *Anesthesia*, *Surgery* (by individual body systems), *Radiology*, *Nuclear Medicine*, *Diagnostic Ultrasound*, *Pathology and Laboratory*, and *Appendix*. When your responsibilities include insurance coding, it is extremely important that you code the procedure accurately because insurance companies and government agencies will pay according to the coding used.

Administrative Medical Services

Identifying CPT Codes

If you are working with insurance codes, there are three steps you must follow to insure correct coding. These include the following:

- Referring to the alphabetical index at the end of the CPT manual to locate the procedure.
- Noting the five-digit code which has been provided for the procedure.
- Locating the correct numerical code in the manual for the exact description of the procedure.

To gain an understanding of how procedures are assigned a specific code number, let's look at the two examples provided in Figure 11-1. In Figure 11-1A, the codes represented are those which would be found under *Medicine*, while the codes shown in figure 11-1B, are codes found under *Surgery: Procedures for the Integumentary System*.

Coding under the International Classification of Diseases

Several years ago, the World Health Organization (WHO) of the United Nations was entrusted with the responsibility for devising a list of all known diseases. Once this list was developed, WHO assigned code numbers to the various diseases. Ultimately, the *International Classification of Diseases*, or *ICD* codes, were developed and are now used internationally.

ICD codes are identified in three bound volumes, known as ICD books. *Volume I* is a tabular list of diseases listed as three-digit numbers with sometimes a two-digit extension, such as 003.22 *Salmonella pneumonia*. *Volume II* of the ICD code book is an alphabetic index of diseases. An insurance clerk would first look in Volume II to find the alphabetic name, and then use the number given to refer to Volume I. It's important to remember that you should never code from Volume II alone. It should be used only as a reference. Coding from Volume I will provide you with much more extensive information.

Volume III of the ICD code books is both a tabular and an alphabetic index of procedures used for the purpose of billing hospital patients, and it coordinates with the Health Care Procedural Coding System (HCPCS) coding used for all government claims.

Coding under the Health Care Procedural Coding System

In 1983, HCPCS, which is pronounced "hicpics," was created by Medicare to standardize procedural coding for all Medicare and Medicaid claims. It is an alphanumeric coding system which was developed by the federal Health Care Financing Administration (HCFA), and designed to be used as a supplement to the CPT codes for filing insurance claims for Medicare, CHAMPUS, and Medicaid. These codes relay information about nonphysician procedures, services, and specific supplies.

There are three levels of coding found in the HCPCS system. *Level I* uses the CPT codes from the *Current Procedural Terminology* manual to code the services and procedures provided by physicians. The five-digit code represents Level I, for example, 90040, *brief office visit*. *Level II* codes are published and updated on an annual basis by the Health Care Financing Administration, and are used to code nonphysician procedures, services, and supplies. Level II codes are represented by one alphabetic character followed by four numbers, for example, A0040, *ambulance service, air, helicopter service*. *Level III* codes may be assigned by the Medicare office responsible for a specific geographic region. Local codes have an alphabetic letter, "S" through "Z," and four digits, and are provided by the regional Medicare office as they

90000	office and other outpatient medical service, NEW PATIENT; *brief* service	
90010	*limited* service	
90015	*intermediate* service	
90017	*extended* service	
90020	*comprehensive* service	

Figure 11-1A
Codes found under medicine.

10000	incision and drainage of infection or noninfected sebaceous cyst; *one* lesion	
10001	*second* lesion	
10002	*more than two* lesions	
10003	incision and drainage of infected or noninfected epithelial inclusion cyst (sebaceous cyst) with complete removal of sac and treatment of cavity	

Figure 11-1B
Codes found under surgery: procedures for the integumentary system.

are updated. Local codes have the highest priority and should always be used first.

Processing Coding Changes and Charges

As a member of the administrative medical services staff, you will be responsible for staying current with the different coding procedures, government rules and regulations, and newly assigned coding numbers. Updates are printed every year for each of the coding manuals, and if you are assigned to process insurance claims, you must make sure that you are using the very latest edition of the coding manuals. If an update, addendum, or appendix is printed, it should be placed immediately in the appropriate manual to insure you are using the current codes.

Each year, Medicare changes the amounts for deductible charges and pricing for procedures that can be charged by the physician who is a Medicare provider. Medicaid also changes its rules and regulations often. Medical office workers who are responsible for handling Medicare and Medicaid are also responsible for noting any changes and charging the patients for their deductible part accordingly. In addition, you must also bill the government for the Medicare/Medicaid-approved amount.

Processing Insurance Claim Forms

In order for the physician's office or the health care facility to receive payment from an insurance company or government insurance system, an insurance claim form must be properly filled out. The form most widely used by private insurance companies and government insurance agencies is FORM HCFA-1500 (Figure 11-2). It is commonly referred to as the "universal insurance claim form," because it can easily be adapted for almost every private and government insurance agency. If you are responsible for filling out this form, you must remember that there are two types of information required. The first includes all routine information regarding the patient, his or her subscriber, and the insurance company. The second type involves all the medical information covering any health-related issues which is provided by the patient's physician.

The most accurate way to obtain information regarding health insurance is to ask the patient or the person insured for his or her insurance identification card and make a copy of both the front and the back of the card. This card contains essential information required to process the claim, and usually includes the person's social security number and group number. In addition, the insurance company's name, address, telephone number, and service representative are usually printed on the back of the identification card.

If you are responsible for processing Medicare claims, you should also make a copy of the patient's or subscriber's identification card so that verification can be made concerning coverage under Parts A and B. Part B is currently optional for all patient's receiving Medicare benefits.

Filling Out the Insurance Claim Form

If you plan on receiving payment on behalf of your physician or the health care facility, especially in a timely manner, it's extremely important that the claim form be accurately filled out. Any errors can result either in not receiving the payment or in waiting an undue time for payment. Therefore, when processing the HCFA-1500 insurance claim form, you must respond to each and every section of the form in a specific way.

At the top of the form, you must first check the applicable program block. If the patient has private or employer's group insurance, check *Other* and write the name of the insurance company in the top right hand corner. After you have checked the appropriate box or written in the name of the company, complete the form by following the instructions provided in Appendix C.

Upon completion of the insurance claim form, make sure you double-check it for accuracy and completeness. Check for any omissions of information and for any math errors. Once you have made copies for your billing records, send the completed form to the appropriate insurance company.

Summary

In this chapter, we discussed the process and skills involved in coding procedures and treatments performed by the physician or health care provider. We talked about how to use the Current Procedural Terminology manual for coding descriptions of procedures; the International Classification of Diseases, for coding a diagnosis or disease; and the Health Care Procedural Coding System (HCPCS), for coding non-physician procedures, services, and supplies for Medicare and Medicaid. Finally, we discussed the purpose of health insurance claim forms, including how to properly code and fill out the HCFA-1500, which is a universal health insurance claim form used by most private and government insurance agencies.

PLEASE DO NOT STAPLE IN THIS AREA

PICA **HEALTH INSURANCE CLAIM FORM** PICA

CARRIER

1. MEDICARE (Medicare #) | MEDICAID (Medicaid #) | CHAMPUS (Sponsor's SSN) | CHAMPVA (VA File #) | GROUP HEALTH PLAN (SSN or ID) | FECA BLK LUNG (SSN) | OTHER (ID)

1a. INSURED'S I.D. NUMBER (FOR PROGRAM IN ITEM 1)

2. PATIENT'S NAME (Last Name, First Name, Middle Initial)

3. PATIENT'S BIRTH DATE MM DD YY SEX M F

4. INSURED'S NAME (Last Name, First Name, Middle Initial)

5. PATIENT'S ADDRESS (No., Street)

6. PATIENT RELATIONSHIP TO INSURED Self Spouse Child Other

7. INSURED'S ADDRESS (No., Street)

CITY STATE

8. PATIENT STATUS Single Married Other

CITY STATE

ZIP CODE TELEPHONE (Include Area Code) ( )

Employed Full-Time Student Part-Time Student

ZIP CODE TELEPHONE (INCLUDE AREA CODE) ( )

9. OTHER INSURED'S NAME (Last Name, First Name, Middle Initial)

10. IS PATIENT'S CONDITION RELATED TO:

11. INSURED'S POLICY GROUP OR FECA NUMBER

a. OTHER INSURED'S POLICY OR GROUP NUMBER

a. EMPLOYMENT? (CURRENT OR PREVIOUS) YES NO

a. INSURED'S DATE OF BIRTH MM DD YY SEX M F

b. OTHER INSURED'S DATE OF BIRTH MM DD YY SEX M F

b. AUTO ACCIDENT? PLACE (State) YES NO

b. EMPLOYER'S NAME OR SCHOOL NAME

c. EMPLOYER'S NAME OR SCHOOL NAME

c. OTHER ACCIDENT? YES NO

c. INSURANCE PLAN NAME OR PROGRAM NAME

d. INSURANCE PLAN NAME OR PROGRAM NAME

10d. RESERVED FOR LOCAL USE

d. IS THERE ANOTHER HEALTH BENEFIT PLAN? YES NO *If yes*, return to and complete item 9 a-d.

READ BACK OF FORM BEFORE COMPLETING & SIGNING THIS FORM.

12. PATIENT'S OR AUTHORIZED PERSON'S SIGNATURE I authorize the release of any medical or other information necessary to process this claim. I also request payment of government benefits either to myself or to the party who accepts assignment below.

SIGNED DATE

13. INSURED'S OR AUTHORIZED PERSON'S SIGNATURE I authorize payment of medical benefits to the undersigned physician or supplier for services described below.

SIGNED

PATIENT AND INSURED INFORMATION

14. DATE OF CURRENT: MM DD YY ILLNESS (First symptom) OR INJURY (Accident) OR PREGNANCY(LMP)

15. IF PATIENT HAS HAD SAME OR SIMILAR ILLNESS. GIVE FIRST DATE MM DD YY

16. DATES PATIENT UNABLE TO WORK IN CURRENT OCCUPATION FROM MM DD YY TO MM DD YY

17. NAME OF REFERRING PHYSICIAN OR OTHER SOURCE

17a. I.D. NUMBER OF REFERRING PHYSICIAN

18. HOSPITALIZATION DATES RELATED TO CURRENT SERVICES FROM MM DD YY TO MM DD YY

19. RESERVED FOR LOCAL USE

20. OUTSIDE LAB? YES NO $ CHARGES

21. DIAGNOSIS OR NATURE OF ILLNESS OR INJURY. (RELATE ITEMS 1,2,3 OR 4 TO ITEM 24E BY LINE)

1. ___ . __ 3. ___ . __

2. ___ . __ 4. ___ . __

22. MEDICAID RESUBMISSION CODE ORIGINAL REF. NO.

23. PRIOR AUTHORIZATION NUMBER

24.	A DATE(S) OF SERVICE From MM DD YY To MM DD YY	B Place of Service	C Type of Service	D PROCEDURES, SERVICES, OR SUPPLIES (Explain Unusual Circumstances) CPT/HCPCS MODIFIER	E DIAGNOSIS CODE	F $ CHARGES	G DAYS OR UNITS	H EPSDT Family Plan	I EMG	J COB	K RESERVED FOR LOCAL USE
1											
2											
3											
4											
5											
6											

25. FEDERAL TAX I.D. NUMBER SSN EIN

26. PATIENT'S ACCOUNT NO.

27. ACCEPT ASSIGNMENT? (For govt. claims, see back) YES NO

28. TOTAL CHARGE $

29. AMOUNT PAID $

30. BALANCE DUE $

31. SIGNATURE OF PHYSICIAN OR SUPPLIER INCLUDING DEGREES OR CREDENTIALS (I certify that the statements on the reverse apply to this bill and are made a part thereof.)

SIGNED DATE

32. NAME AND ADDRESS OF FACILITY WHERE SERVICES WERE RENDERED (If other than home or office)

33. PHYSICIAN'S, SUPPLIER'S BILLING NAME, ADDRESS, ZIP CODE & PHONE #

PIN# GRP#

PHYSICIAN OR SUPPLIER INFORMATION

FOLD HERE / USE ENVELOPE NO. 1500E

(APPROVED BY AMA COUNCIL ON MEDICAL SERVICE 8/88) ***PLEASE PRINT OR TYPE*** APPROVED OMB-0938-0008 FORM HCFA-1500 (12-90), FORM RRB-1500, APPROVED OMB-1215-0055 FORM OWCP-1500, APPROVED OMB-0720-0001 (CHAMPUS)

Figure 11-2
HCFA-1500 insurance claim form.

Review Questions

1. Define the following abbreviations:
   a. HCFA _____
   b. CPT _____
   c. CHAMPVA _____
   d. CHAMPUS _____
   e. FECA _____
   f. SNF _____

2. Briefly explain the difference between *Medicare Part A* and *Medicare Part B*.

3. Briefly explain the difference between the following programs:
   a. Medicare
   b. Medicaid

4. Who is covered for injury and illness under a workers' compensation program?

5. What does the term *subscriber* mean?

Release of Information and the Law

Performance Objectives

Upon completion of this chapter, you will be able to:

1. Identify the patient's rights and discuss the importance of upholding these rights.
2. Describe the steps involved in assuring the patient's rights.
3. Discuss the legal requirements for a person to sign a consent.

Terms and Abbreviations

Informed consent the process by which consent for a medical procedure is obtained by giving a person a full explanation of what he or she is giving permission for, and identifying what the risks, alternatives, and probable outcomes of the procedures are before the person signs the consent form.

Invasion of privacy an unauthorized release of information concerning a patient or unnecessary exposure of a patient.

Liability the condition of being subject to legal responsibility.

Litigation any legal action that takes place in a court; a lawsuit.

Negligence failure to perform legal duties that result in damage or injury to another.

Standard of care action that would normally be taken in a given situation by persons of equivalent training.

Tort a wrongful act arising from a breach of duty that is not a crime but is subject to litigation.

As a professional member of the administrative medical services team, it is important that you become familiar with your job description and the duties for which you are responsible. Usually, your job description will be part of the policy or procedure manual found in each department of the facility in which you are employed. If your employer has not shown you such a listing, you will need to ask for it as soon as possible. Under the law, you can be held responsible for any and all tasks listed in your job description, whether or not you have been specifically trained to perform them. If any item listed in your job description is unfamiliar to you, you have a responsibility to ask for training during your initial employment probationary or training period.

Patient Privacy and the Law

Early in the history of patient care, all matters were kept secret, and the physician and his or her employees could not be held responsible for any results of their care or neglect. Since the 1970s, however, when the American Hospital Association developed a document known as the Patient's Bill of Rights, the law has become quite specific with regard to the physician and the health care worker being responsible for their own actions and the results.

Several of the rights identified in the Patient's Bill of Rights concern the maintenance of privacy. Keeping all information regarding the care and treatment of the patient confidential is a major responsibility of all health care workers. No information, including whether a person is a patient, can be released to anyone who does not have a right to that information. To have that right, a person must have direct responsibility for the care of the patient, and the information requested must be related to the person's area of responsibility.

The patient can allow the release of information by signing a form that specifies what information and under what conditions it is to be released to whom. For instance, insurance companies need to have certain information regarding diagnosis and related issues in order to pay a patient's charges for care. The release is usually a part of the patient's admission form.

If you are asked for any information regarding your patient, you must be sure that the person to whom you release it is the agent for the company specified on the third-party release of information form. You should not hesitate to require proper identification prior to making any records available. You will also need to become familiar with the specific forms used by your facility and the information covered by those forms.

During your employment, you may encounter situations in which you are required to provide the patient with privacy while completing admission and history forms. An area should be provided where patients cannot easily see information about another patient. Care should also be taken to prevent any verbal communication where other patients may overhear information they are not entitled to hear. If a patient has questions about the forms that he or she must sign, be sure you escort them to a private area and then determine who can best answer the questions.

If you are ever questioned by anyone outside your facility regarding the care or treatment of a specific patient, it is better to simply state that information cannot be discussed without written permission from the patient. In this case, it is better to err by not releasing information, and perhaps making that person unhappy, than to expose the privacy of a patient and later find out the person to whom you were speaking was not entitled to the information. Remember that *invasion of privacy* can occur anytime there is an unauthorized release of information regarding the patient's care or treatment, or the patient suffers any unnecessary bodily exposure while under the care of a health care provider.

Informed Consent and Liability

One right the patient expects is that of informed consent. *Informed consent* means obtaining the patient's consent by providing him or her with information concerning any procedures, surgeries, or treatments that are planned for his or her care. This information must include what each procedure entails, as well as what the expected outcome of the procedure might be, and any risk factors which may be involved. It must also include provision of any possible alternatives to the suggested procedure. It is the responsibility of the physician, or in some cases the nursing staff, to provide the patient with this information. That this information was made available to the patient must be documented in a consent form and then be made part of the legal record of the patient's care prior to the performance of such procedures. In many health care facilities, this consent form is often part of the admission agreement.

Liability

To be *liable* for something means to be responsible for one's own acts. *Negligence* is failure to perform the legal duties assigned to you, as part of your job, when such failure could result in damage or injury to another person. You can be held liable for failing to perform any job duties in a way that another person of equivalent training would have done. In health care, this is referred to as *standard of care*. In court or when litigation occurs, the standard of care is usually established by the testimony of expert witnesses, that is, persons employed in a job similar to yours or who have had similar training.

If you are deemed liable for an act that is covered in your job description, your employer can also be held responsible. If, however, you perform an act that is not one of your responsibilities, and someone is injured or harmed by that act, you alone will be held liable and responsible.

If you have acted in a negligent manner and the action is brought into a civil court, you will be held liable and charged with a tort. A *tort* is a wrongful act that arises from any breach of duty that is not considered a crime but is subject to litigation. If grievous damage occurs, such as the death of a patient due to your negligence, you could be charged with a crime and tried in a criminal court.

The Medical Record and Liability

One important aspect of the liability involved in managing health information is the maintenance of all medical records according to the law. These records are considered legal documents and, as such, must meet some basic requirements. They must be readable and written in blue or black ink, with all errors corrected in a specific way. No erasures or use of liquid paper is allowed, and if an error is made it is to be corrected by placing one single line through

Administrative Medical Services

the error and writing in the correction. The error must also be initialed by the person making the correction, and in some cases the correction must be dated. Physicians and nurses are required to correct their own errors, since members of the medical records department are only allowed to change an error on a typewritten report that they are typing.

Patient Consent Forms and Documentation

There are legal rules that dictate who can and cannot legally sign admission and consent forms. In certain cases, the patient may be deemed "not responsible to consent." In such a situation, a responsible person must be designated by law as the patient's *next of kin*, a term referring to the closest relative who can legally sign a consent form. In most states, persons who are deemed unable to provide their own consent include minors, that is, anyone under the legal age for adulthood in that state, anyone under the influence of a mind-altering drug or medication, such as alcohol or narcotics; anyone who is not of stable mind, such as those suffering from senility or a psychotic disease; and anyone who is unconscious or in a coma.

Being a Witness

Occasionally, you may be asked to witness the signature of someone signing a consent or release of information form. In doing so, you are simply saying that you saw the person who was named on the document sign the form. If the person is not someone you know, do not hesitate to ask to see some form of picture identification that states the person's name. You may later be called upon to verify this fact. Remember, your signature means only that you witnessed the signature, not that any information given is correct or factual.

Medical Records as Legal Document

All medical records, charts, and consent and release of information forms are legal documents that are the property of the health care facility in which they originated. Patients have the right to see their records, but must do so with their physician present and should never be given copies of their records without the doctor's permission. This avoids any possibility of misunderstanding comments or test results.

The patient's medical chart legally belongs to the physician if it is compiled in his or her office or to the health care facility when compiled there. A patient pays for the physician's advice based upon the information provided in the chart. A copy of a patient's chart is sometimes sent to a consulting doctor, but it still remains the property of the physician or facility that prepared it.

If a chart or document is to be destroyed, it should be burned or shredded to preserve the patient's confidentiality. It must never be placed in a wastebasket where it might be read by an unauthorized person.

Summary

In this chapter, we discussed the importance of protecting the patient's rights, particularly as they relate to the release of information and consents for care and treatment. We noted that, by being aware of these rights, you will be better able to avoid uncomfortable or threatening situations that may arise through failure to provide for the best care of the patient and for his or her legal and ethical rights.

Review Questions

1. _____ pertains to action that is normally taken in a given situation by persons of equal training.

2. The process of being subject to legal responsibility is called:
   a. responsibility
   b. accountability
   c. liability

3. A wrongful act arising from a breach of duty that is not considered a crime, but is subject to litigation, is called a:
   a. liability
   b. tort
   c. battery

3. Define the following terms:
   a. litigation _____
   b. lawsuit _____

4. What does *informed consent* mean?

5. What does *malpractice* mean?

6. What does *negligence* mean?

Providing Quality Assurance in the Health Care Environment

Performance Objectives

Upon completion of this chapter, you will be able to:
1. Discuss the history of evaluation and accreditation in the health care industry.
2. Identify and describe the four parts of quality assurance in the health care setting.
3. Explain quality assurance as it relates to being an ongoing process.

Terms and Abbreviations

Audit   an older method used to evaluate professionals and professional standards by members of the same profession.
Credentialing   the process of giving permission to a medical staff member to provide patient care in a designated health care institution.
Diagnosis   the scientific method used to determine the cause of a person's illness.
DRG   the abbreviation for Diagnostically Related Groups, a system by which illnesses and conditions are grouped into categories according to diagnosis.
Discharge planning   the process of planning care established for a patient following his or her discharge from a health care facility.
JCAHO   the abbreviation for Joint Commission on Accreditation of Healthcare Organizations, which is a formal group responsible for evaluating and accrediting health care institutions relative to the quality of care provided as measured according to minimum standards.
LOS   abbreviation for Length of Stay, which refers to the number of days a patient is confined to a hospital, SNF, or other health care facility.
Malpractice   a term used to describe incorrect or negligent treatment of a patient by persons in the health care industry.
Monitoring   the formal process of observation to detect any current or potential problems in the quality of care.
Peer review   a survey of the conduct of professionals or professional standards by members of the same profession.
Quality assessment   a system of monitoring and evaluating the level or quality of health care provided in a health care facility.
Quality assurance   the overall continuing activities and efforts of a health care institution to assess the quality of patient care.
Risk management   activities and policies that a health care institution adopts to provide an environment that is as safe as possible for patients, staff, and visitors.
Standards   established criteria for safe quality patient care.
UR   abbreviation for Utilization Review, a process which is required in order to determine whether health care is provided in the most effective and efficient manner possible.

Evaluating health care is not a new concept. In fact, as long ago as the 1800s, hospital size was evaluated and compared relative to the number of patients surviving surgery. According to the history and evolution of health care facilities throughout the United States, in 1918 only about 12.9 percent of hospitals were accredited, compared with the 95 percent that were accredited in 1956.

Administrative Medical Services

The *Joint Commission on Accreditation of Healthcare Organizations*, or *JCAHO*, was the first formal group entrusted with the responsibility for evaluating the quality of medical care provided by health care institutions. In 1952, they developed specific *standards* which would be used to assist hospitals in reviewing the quality of patient care in their facility. These standards, which are still in effect today, were considered the established criteria for providing safe, quality care. The *audit*, which grew out of these standards, became the acceptable procedure for evaluating professionals and professional standards by members of the same profession.

Since early efforts in quality assurance looked only at a certain number of patient records within individual deparments, by 1979, JCAHO was forced to change its evaluation system. Today, hospitals coordinate their quality care activities for an entire facility through committees or teams, rather than through fragmented or individual reviews.

Since 1985, standards of quality and evaluation have been included throughout all JCAHO manuals. Quality assurance efforts have evolved into a system of continuous monitoring to improve both the quality and the outcome of patient care. The Joint Commission on Accreditation of Healthcare Organizations is not the only group to effect quality assurance efforts. Medicare also requires rigid adherence by hospitals and other agencies to quality assurance programs and, ultimately, to more efficient patient care.

Understanding Quality Assurance

Quality assurance is a broad term that is generally used to refer to the overall activities and efforts which have been made by an institution to assess the quality of patient care at an individual health care facility. *Patient care quality*, as it has been defined by JCAHO, is the degree to which patient care services increase the probability of desired patient outcomes, and thus reduce the probability of undesired outcomes, given the current knowledge.

No health care facility or institution can ever be perfect. Problems will continually develop, be solved, and new problems will be identified. Health care providers will always seek to improve the level of care they provide. It is because of this premise that assessment of the quality of patient care is an ongoing concern.

Assessing the quality of patient care within a health care facility involves four separate components: *utilization review*, *quality assessment*, *credentialing*, and *risk management*. Many institutions are able to integrate these components into a coordinated effort called quality assurance.

Utilization Review

Utilization review, or *UR*, is a process used to determine whether health care is being provided in the most effective and efficient manner possible. Utilization review teams are responsible for determining, based on the documentation found in the patient's medical record, whether a patient's condition warrants hospitalization, if so, how long a hospitalization period is justified, or whether an alternate care plan would be more suitable. Private insurance companies, as well as government plans such as Medicare and Medicaid, now require UR for reimbursement.

The utilization review process involves reviewing the medical record in detail. It may involve discussions with the patient's doctor to clarify the diagnosis or cause of illness, as well as the treatment plan during the patient's hospitalization. Patients are often discharged from the hospital even though skilled nursing care or therapy may still be necessary. For example, a patient with an indwelling urinary catheter or one who may require frequent blood testing may be well enough to be discharged from the hospital, but at the same time may require home health care or an outpatient center for such services. Such determinations can be made through the process of *discharge planning*, which is the plan of care established for a patient following his or her discharge from the hospital. Health care providers have to be certain that the patient's needs can be adequately met in the least costly environment.

As a result of the utilization review process, it may be determined that the patient requires the use of the facility's social services department. Services provided by this department are sometimes used before and after the patient's discharge. For example, an elderly patient's diagnosis and treatment plan may indicate that he or she is ready to be discharged from the acute care facility. Because the patient lives alone, a social services referral may be indicated to ensure that the patient can function alone at home until a full recovery has been achieved. The patient may need referrals for home meal delivery or home health care monitoring before he or she is discharged from the hospital.

In the 1980s, the government determined that a new system needed to be instituted and put into place for reimbursing health care facilities based upon a patient's illness and medical condition. This method was based upon *diagnostically related groups*, or *DRGs*, rather than on a patient's *length of stay*. The DRG process groups illnesses and conditions into specific categories according to diagnosis and expected duration of treatment. The hospital can then be reimbursed according to this schedule, regardless of what the cost of treatment is at the time of the patient's discharge. If the actual treatment costs more than the allotted amount, the hospital must pay the difference. If the treatment costs less, the hospital is allowed to keep the difference. The success of DRG payment plans greatly depends upon the diligence of the UR team in monitoring the overuse and underuse of health care services. Diagnostically related groups and UR work closely together to keep down the escalating costs of health care.

Another important aspect of the utilization review process is that it provides protection of the patient against extended, unnecessary hospital services, while assuring that a patient receives hosital care when that level of care is needed. It also protects the hospital from expensive and nonreimbursed care. Protection of both the patient and the health care facility can only be achieved through the ongoing process of utilization review from the time the patient is admitted until such time as he or she has been discharged.

Quality Assessment

Quality assessment, the third component of quality assurance, is a system of monitoring and evaluating the level of health care quality provided by a health care facility. It is also a requirement for accreditation by JCAHO in order to ensure that a minimum standard of care is provided by an institution.

Clinical services, such as those utilized by the nursing and laboratory departments and the medical staff, which is made up of physicians, must show evidence of ongoing monitoring and evaluation of patient care. Each department within a health care facility is required to state the objectives of that specific department and identify problem areas. Actions taken to resolve or improve those problems must be documented. Once a problem has been identified, an important part of quality assessment is to communicate the findings and action plan to the staff involved to ensure that the problem does not recur. This communication may be conveyed through staff meetings, written memos and documentation, the bulletin board, or through electronic messages. More complicated problems, however, may require staff inservice or training.

Credentialing

Credentialing is another component of quality assurance that deals with giving specific permission to a medical staff member to provide patient care in a designated health care facility. Physicians are required to follow the staff regulations, as well as the policies and regulations issued by their individual state medical board. To ensure that only qualified staff are given permission to provide care in the facility, most institutions keep detailed records on all their staff members. These records usually include a history of previous practice, including any prior malpractice claims or dismissals from staff privileges, a record of the member's education, licensure information with the date of expiration and history of any revocation or suspension of license, if any occurred, the narcotics number issued to the practitioner, any information regarding insurance/provider participation, any notices of fraud regarding Medicare and Medicaid, and the date of the commencement and expiration of any designated clinical privileges. Access to this information is usually strictly limited since it is sensitive and confidential information that could be misused by others. Medical records related to physician credentialing usually fall under the responsibility of the facility's medical records department.

All physicians who practice in a health care facility are required to participate in peer review activities which also assist in the overall quality assurance process. *Peer review* is a survey of the conduct of professionals or professional standards by members of the same profession. If a physician does not comply with regulations or commits a nonethical or illegal act, such as Medicare fraud, the institution has the right to revoke or suspend the physician's hospital privileges. Ultimately, the physician could lose his or her license to practice medicine.

Risk Management

The fourth and final component of a quality assurance program involves a process known as *risk management*, in which an institution implements certain activities and adopts policies that will provide an environment that is as safe as possible for patients, staff, and visitors. Risk management focuses on individual incidents involving the potential for injury and possible lawsuits or claims.

By studying the potential risk of injury and claims, the health care facility can often eliminate the underlying reasons for the original problem. It may be an isolated incident, such as a patient falling out of bed that results in a fractured hip. Conversely, careful study may show that there have been many such incidents over the past year and that they all occurred on one specific floor of the hospital. Having studied this case and made several determinations regarding these incidents, the risk management team can then provide education and inservice for the employees on the hospital floor involved concerning

Administrative Medical Services

side rails and frequent patient monitoring for safety precautions. This action could help prevent similar accidents, saving the hospital from potential claims of negligence and its patients from possible injury. The cost of health care should be lower due to fewer malpractice and negligence cases.

Quality Assurance and the Medical Record

In order for quality assurance to be successful or even attempted, it is imperative that the patient's medical record be carefully documented from the time the patient is first admitted to the hospital through the time he or she is discharged. All health care providers who are involved in making notes on a medical record should realize that, according to the law, unless an action, treatment, or precaution is properly and accurately documented, it was not done. The efforts toward quality assurance always depend on accurate, timely documentation in the record by all members of the staff.

Summary

In this chapter, we discussed how the evaluation and accreditation of health care is a necessary part of providing patient care, noting that such a concept is not new to the medical field but has, in fact, undergone major evolutionary changes during the past two decades. During our discussion, we discovered that this process is referred to as quality assurance and determined that all health care providers should be concerned with it in their individual facilities to assure that all patients receive the highest level of care and treatment. Finally, we discussed the goals of a good quality assurance program, noting that the ultimate goal of quality assurance is to provide effective and appropriate care for patients through efficient methods within an appropriate period of time.

Review Questions

1. Define the following terms:
    a. audit _____
    b. peer review _____
    c. monitoring _____
    d. risk management _____
    e. credentialing _____

2. Briefly explain the difference between *quality assessment* and *quality assurance*.

3. What do the following abbreviations mean?
    a. LOS
    b. JCAHO
    c. DRG
    d. UR

4. What is the term used to describe the scientific method for determining the cause of a person's illness?

5. What is the term used to describe the established criteria for safe quality patient care?

6. The process of planning care for a patient following his or her discharge from a health care facility is called _____.

14

Basic Computer Applications

Performance Objectives

Upon completion of this chapter, you will be able to:
1. Identify and discuss ways in which computer technology has increased productivity in the health care environment.
2. Identify and describe different types of computers and their main components.
3. Describe the relationship between initial data entry and ways in which data can be used to generate forms used in a hospital or physician's office setting.
4. Identify medical forms that can be generated on the computer.

Terms and Abbreviations

Database   a computer application program that allows entered data to be arranged and used as needed.
Display   the screen used for the temporary viewing of data stored in a computer.
Floppy disks   small disks, either 5 1/4 inches or 3 1/2 inches square, which are used to store limited amounts of information.
Font   the size and style of type used in printing; refers to the letters, numbers, or characters in a data processing system.
Formating   the process involved in preparing a new disk for storing data.
Hard copy   the printed paper copy of data retrieved from a computer.
Hard disk   hardware that is built into the computer by the manufacturer which is used to store programs that require large amounts of memory.
Hardware   the physical equipment of the computer, such as the central processing unit, keyboard, disk drives, printers, and mouse drivers.
Interface   connecting another device or machine to a computer.
Mainframe computers   large, sophisticated computers that are capable of storing massive amounts of data and that can be used by a hospital to store medical records.
Memory   the place where data is stored in the computer's main unit.
Menu   a computer display showing a list of items from which a selection is made to direct the computer to some action.

Modem   a device which is used for converting electronic signals from one form to another and which can transmit information from one computer to another by using telephone lines.
Network   computers that are connected in order to communicate with one another.
Offline   information stored on disks and/or terminals not in direct contact with the computer.
Online   information stored on disks and/or terminals immediately accessible to a computer.
Peripheral device   hardware that allows computers to communicate or adds an auxillary function.
PC   abbreviation for personal computer, which is a computer that can easily fit on a desktop and is widely used in individual offices, businesses, schools, and homes.
Security   protection of files and information by passwords, codes, and locks.
Software   the application programs that have been programmed to do a specific task, such as word processing, medical office management, graphics, or spreadsheet and database management.
Spreadsheet   data arranged in a row and column format, as in the case of budgets, expenditures, and statistics.
Word processing   an application program that is used for formatting text into an attractive, readable form, and which can be used for reports, letters, discharge summaries, and medical histories.

Administrative Medical Services

Because of rapidly changing technology, computers are used much more often in the everyday operations of today's health care environments. Computers aid us in performing routine tasks such as tracking patient data and printing out charges. The medical receptionist in a clinic, for example, finds the computer an indispensable tool for keeping track of patients' appointments, billing, and medical records.

Most computers used in the hospital setting are connected to one another and can communicate easily through this *network*. The health unit coordinator, for example, can send and retrieve vital medical information regarding the patient using the computer. Linking departments by computers results in greater efficiency for all health care providers. Nursing stations, the pharmacy, the admissions desk, and the medical records department can share information quickly and accurately. Hospitals can also be tied into networks with other facilities by use of a modem. This means that information about patients, such as an ECG or a blood test, can easily be sent through telephone lines to experts for quick interpretation and diagnosis.

Results of specific medical tests can be determined by specialized computer programs, and the *hard copy,* that is the printed paper copy of the data that has been retrieved from the computer, can be easily and efficiently filed in the patient's medical record for future reference.

Computers have many different uses in health care. A medical transcriptionist, for example, can format a physician's dictation through the use of a computer in order to produce a final copy that can later be placed in the patient's medical record. The computer can also be used to compile and build information for a database that can provide an institution with valuable statistical documentation in medical research. The information can also be retrieved from the database and arranged in lists as needed. And numerical information can be arranged in a row and column format called a *spreadsheet.*

Defining Computers and Their Components

A *computer* is an electronic device that can be used to create, organize, store, and retrieve information called data. Computers come in many different sizes and have various uses throughout the health care industry. One type is called a *mainframe computer,* which is a large, sophisticated computer capable of storing massive amounts of data that is generally found in large hospitals, clinics, and research facilities.

With the development of smaller, less expensive models, computers have appeared in private physician's offices, dental offices, chiropractic offices, and even in veterinary offices and pet hospitals.

Computer System Components

A computer or data processing system consists of three basic parts: hardware, software, and the auxillary equipment needed.

Hardware consists of the mechanical equipment needed to make the computer work. It includes the central processing unit (CPU), the keyboard, a screen or display terminal, and a printer that is attached and necessary to produce hard copy (Figure 14-1). Components may all be separated or combined in one or more units depending on the manufacturer. The CPU must have a disk drive and a memory. The *memory* is the place where data is stored in the computer's main unit. The keyboard resembles that of a typewriter but with additional keys. The *screen* or *display terminal* looks very much like a television set and is used for the temporary viewing of data stored in the computer. The *window* is an area of data viewed on the screen at any given time.

A printer may be located close to or at a distance from the other three hardware components. It may be a dot matrix or a laser printer. The size and style of the printed material is determined when the person operating the computer selects a *font* from the menu or list of items appearing in the display window. A "letter-quality" printer produces hard copy that has the appearance of being typed on an electronic typewriter.

Peripheral devices may also be added to your data processing system. They are considered hardware and are designed to connect computers to each other, thus providing the computer with the ability to communicate and perform auxiliary functions. A *modem* is a type of peripheral device that can be used to convert electronic signals from one form to another. A computer can be connected to a telephone by a modem for transmission of data along telephone lines to another computer. When a modem is added, it is said to interface with the computer. *Interface* means that the modem is linked to a computer and is under its control. Peripherals are separated from the computer but are connected by cables or wire. A printer may be considered a peripheral device.

The term *software* is used to identify the programs, disks, and operation manuals. Software also includes the applications programs for a specific task, such as word processing, medical office management, graphics, spreadsheets, and database man-

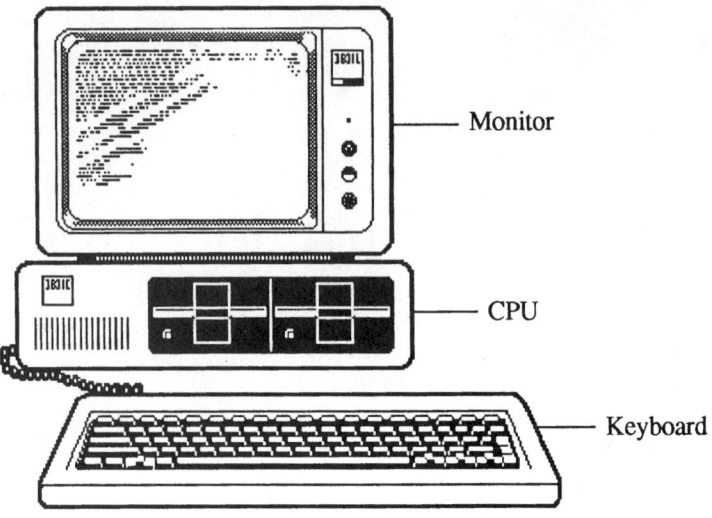

Figure 14-1
The computer and its components. (Reprinted with permission from Bonewit-West, K. *Computer Concepts and Applications for the Medical Office.* **Philadelphia: W. B. Saunders, 1993, p. 13.)**

agement. Modifications can be made to suit the user's needs. Disks may be fixed, also known as a hard drive, and floppy. *Fixed* or *hard disks,* are built directly into the computer by the manufacturer and are used to store programs that require large amounts of memory. *Floppy disks* are small disks, either 5 1/4 inches or 3 1/2 inches square, that can store limited amounts of information. Before a disk can receive information, it must be prepared. We refer to this prepartion as *formatting* the disk. A manual with instructions for the operation of a computer is usually provided by the manufacturer.

Auxillary equipment is not part of the computer but is necessary to operate the system efficiently. The auxillary item may be a device or consumable supplies such as zigzag-folded computer paper, insurance forms, address labels, letterhead stationery, and plain paper. Stands, modular furniture, storage cases, and covers are all considered auxillary equipment. If disks are added for extra storage capacity offline, they are considered supplementary even when brought online. Cables to connect the hardware are also considered auxillary equipment.

Computer Programs and Applications

Day-to-day use of the computer in a specific medical or health care setting requires training to use the software that has been installed on the machine. In some cases, the physician's office purchases a software package enabling the office to make appointments, send reminders, do billing, and keep track of financial records and activities. Hospitals, on the other hand, might purchase a program that would link all departments into one system for medical records, financial activity, inventory control, and specialized needs. A research facility might have a program written to produce the type of documentation and testing needed for research. With all the different uses and applications the computer can perform, you can see how important it is for the computer operator to be trained to perform the specific tasks applicable to an individual program. The health care employer may provide training for a specific medical program, while individual companies that sell medical software packages are generally expected to provide training, support, and online help for those who have purchased their products.

Students, such as yourself, who may be interested in designing and writing medical software programs, will need to complete course work in programming language, computer math, and computer science. Computer programmers work with mathematics and logic in order to be able to break down clerical procedures and make them compatible with the capabilities of the computer.

Computer Applications

The ability of a computer to perform a specific task depends upon the purpose for which the software program was written. Word processing, for example, is an applications program that is used for formatting text into an attractive, easily read form. The examples of computer applications that are identified in this chapter are used primarily in the physician's office, and the software that is being discussed is designed to manage the routine tasks performed by the administrative medical services specialist, the medical receptionist, the insurance and coding clerk, and the medical office assistant.

A typical day in the physician's office might include the computer operator or office assistant generating specific forms that would include an appointment calendar or scheduler, a new patient data entry

Administrative Medical Services

form, an invoice or bill, an insurance claim, reminders to the doctor, checks, daysheets, bank deposit reports and deposit slips, and correspondence or letters. There are many software applications already in use and available to the physician's office that enable the administrative office specialist to generate much of the information and data needed by simply inserting a programmed disk into the computer and turning the system on.

Computer Problems and Resolutions

While the implementation of computer technology has provided the health care industry with a much greater sense of security in terms of easing the employee's workload and generating more accurate and sophisticated documents, the use of computers also presents us with problems which must be recognized and resolved. *Errors* will appear in the system and may be undetected over time. Incorrect coding and careless processing of health information can be frustrating to the worker and can also result in the total loss of access to a patient's record. Corrections must be made as soon as an error is discovered.

One major disadvantage to using the computer is *downtime,* or the time when problems develop within the system requiring that it be closed down for servicing. Arrangement for prompt servicing is the best solution to this problem.

Other problems with computer usage may be electrical in nature. Static electricity and electrical surges, for example, can destroy stored information. By implementing specific safety measures, such as duplicating records and storing them in a safe place or installing antisurge equipment, you can avoid having to recreate documents because of the sudden loss or destruction of previously stored data.

One major area of concern in health information management, particularly as it relates to computer technology, deals specifically with the confidentiality of patient medical records. All patients expect their records to be kept in strictest confidence. Therefore, all precautions must be carefully thought out and observed when it comes to the management of these records in the computer system. Terminals should be placed in areas away from accidental viewing by unauthorized persons. Special security systems should also be employed that protect files and information with the use of passwords, codes, and locks.

Summary

In this chapter, we discussed the function of computer technology, specifically as it relates to the health care industry. We determined that the computer is an electronic device consisting of three basic parts: hardware, software, and the auxillary equipment that accompanies these parts. We also talked about how computers are widely used in a medical setting for recording and processing data, noting that the use of a computer in a specific setting or department requries training to use the software installed on the machine. Finally, we talked about computer programs and software applications, determining that a software program can be written to generate forms for use in a physician's office, network with other computers in a hospital, or prepare data for medical research.

Review Questions

1. Define the following terms:

    a. database _____

    b. menu _____

    c. memory _____

    d. hard disk _____

    e. modem _____

    f. floppy disk _____

2. Briefly explain the difference between *hardware* and *software*.

3. An applications program used for formatting text into an attractive readable form is called:

    a. spreadsheet

    b. database

    c. word processing

4. Briefly explain what a *network* is.

5. What is the difference between being *online* and being *offline*?

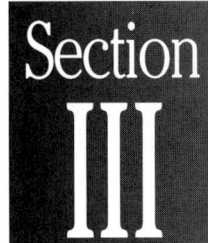

Section III

Medical Transcription: Basic Concepts and Applications

15
Introduction to Medical Transcription

16
Working as a Medical Transcriptionist

17
Sentence Structure and Letter Writing

18
Transcribing Medical Reports and Records

Introduction to Medical Transcription

Performance Objectives

Upon completion of this chapter, you will be able to:

1. Describe the role of the medical transcriptionist.
2. Identify the training and certification available to the medical transcriptionist.
3. Discuss the role medical terminology and anatomy and physiology play in performing tasks required of the medical transcriptionist.
4. Distinguish the differences between medical reports and medical records.

Terms and Abbreviations

AAMT abbreviation for the *American Association of Medical Transcriptionists*.
AMA abbreviation for the *American Medical Association*.
CMT abbreviation for *Certified Medical Transcriptionist*.
CRT a display terminal, similar to a television screen, which is attached to word processing equipment.

HMO abbreviation for *Health Maintenance Organization*.
PPO abbreviation for *Preferred Provider Organization*.
Transcription the process used to convert recorded dictation to a hard copy document.

One of the most interesting and highly paid professions available to the administrative medical services worker is in the field of medical transcription. Graduates of this type of training program enjoy a sense of independence and experience the strong feeling they are providing a needed service to the patient and medical community. While many transcriptionists attend formal education and training programs, others come from prior positions held in the health care industry.

If you choose to become a medical transcriptionist, you should be the type of person who likes to work independently. You should also be self-reliant, intuitive, and, most important, dependable. You must have a fair amount of intelligence, have an inquisitive mind, and possess the ability to provide others with accurate and professional work (Figure 15-1).

The Role of the Medical Transcriptionist

As a professional member of the administrative medical services team, your role as a medical transcriptionist is to provide others with the very best, most accurate, and most attractive medical reports, records, and data related to patient care and treatment. Since all medical records are considered legal documents, you must also be knowledgeable in medical law and ethics, medical terminology, and basic anatomy and physiology, since all of these play a key role in transcribing spoken and written records into professionally typed and legible printed documents. While the medical transcriptionist rarely has contact with patients, he or she must have the ability to provide the caregiver with a document that describes what the patient is experiencing and the signs and symptoms

Figure 15-1
The medical transcriptionist.

he or she is exhibiting to the physician. In other words, as a medical transcriptionist, your role is to take the spoken or recorded information provided by the caregiver and incorporate this information into accurate medical records and reports that can be put into the patient's medical chart or used as legal documents in a court of law.

Staying within the Law

All medical transcriptionists, like other administrative and clinical health care workers, are required to follow a specific code of ethics established for this profession. For medical transcriptionists, the code of ethics established by the American Association of Medical Transcriptionists (AAMT) is the standard.

All medical transcriptionists have a responsibility to protect and guard medical records and reports with which they come in contact. You must remember that these medical records are the property of the facility or physician with whom they originated. No matter who requests a copy of these medical records, you are never allowed to release a copy or other information regarding a patient without the proper written consent of the patient and permission of the owner of the records.

Training and Career Opportunities

One of the greatest opportunities you will have as a medical transcriptionist is the ability to work, not only in almost any medical environment you choose, but also to work as an independent contractor. Positions for the transcriptionist can be found in health maintainence organizations (HMOs), preferred provider organizations (PPOs), private and public health care facilities and medical centers, skilled nursing facilities, and even in private health insurance companies and law offices. If you have the entrepreneurial spirit, and you choose to work independently, you may work part- or full-time in your home, as an independent contractor or consultant in a medical facility, or in your own office or business facility.

Training to become a medical transcriber can be obtained through any number of ways. If you have no prior medical training, you may attend a vocational training school or community college. These programs can range anywhere from six months to two years, depending upon whether you choose to be certified. If you are already working in the health care field and would like to transfer to the medical transcription department, this can usually be accomplished by taking part in an on-the-job training program, which can generally be arranged between yourself, your supervisor, and the medical transcription department.

If you plan to excel in the field of medical transcription, it will be in your best interest to join the American Association of Medical Transcription (AAMT). This is a professional organization that sets the standard for all professional medical transcriptionists. Once you have been employed as a transcriptionist for three years, you may choose to become a certified medical transcriptionist (CMT) through the AAMT. If you are interested in this, you may contact the AAMT at P.O. Box 6187, Modesto, CA 95355.

Administrative Medical Services

Summary

In this chapter, we discussed the exciting role of the medical transcriptionist. We talked about this occupation as one of the most stimulating and lucrative careers available to the administrative medical services worker, and described the many opportunities and training available to the health care worker interested in this field. Finally, we talked about ethics and professionalism as they relate to the medical transcriptionist, noting that, as a professional member of the administrative medical services team, the medical transcriptionist has a responsibility to maintain the confidentiality and ethical standards expected of him or her with regard to medical records and reports.

Review Questions

1. Define the following abbreviations:
   a. CRT_____
   b. PPO_____
   c. CMT_____
   d. AMA_____
   e. AAMT_____

2. Briefly explain the function of a *Health Maintenance Organization*.

3. Briefly discuss the role of the medical transcriptionist.

4. What is the name of the organization responsible for establishing the standards of professionalism for medical transcriptionists?

5. Identify at least three ways in which a medical transcriptionist may be trained:
   a. _____
   b. _____
   c. _____

6. Identify at least five locations in which a medical transcriptionist may be employed:
   a. _____
   b. _____
   c. _____
   d. _____
   e. _____

16

Working as a Medical Transcriptionist

Performance Objectives

Upon completion of this chapter, you will be able to:

1. Identify and discuss the various types of equipment used by the medical transcriptionist.
2. Discuss the role grammer plays in medical transcription and be able to identify the various parts of speech, punctuation, and numbers used in transcription.

Terms and Abbreviations

Cardinal number any number used in simple counting or in answer to "how many."
Cassette a magnetic tape that is wound on two reels and encased in a plastic or metal container; used to mount or insert into a playback or recording device.
CRT a display terminal that is very similar to a television screen and is attached to word processing equipment.
Dictation the process of taking the spoken or recorded word and transferring it onto a tape cassette.
Hard copy any written, typed, or printed document.
Hardware the physical parts of a computer system.
Menu a list of previously stored items displayed on a CRT screen from which the transcriptionist can choose the word processing document to be edited.
Modem a device that is used to convert data into signals for telephone transmission.
Ordinal number term used to describe or indicate the order or succession of items in the same class.
Program any software used in the computer.
Software programs or instruction used to support a piece of computer equipment.
Word processing a software program used to produce printed communications; a "method" or "process" by which words can be produced.

While the field of medical transcription can offer you a great deal of flexibility and high earning potential, these benefits do not come easily but require intense involvement in your training, hard work in your job, and commitment to the challenge of understanding the many complexities and components involved in this profession. Such accomplishments can only be achieved if you are willing to spend the time and energy necessary to learn how to use computer, dictation, and transcription equipment; and the fundamentals of grammar, punctuation, and basic sentence structure, all of which are required if you are interested in pursuing a career as a medical transcriptionist.

Basic Medical Transcription Equipment

Before you are ready to tackle the use of basic medical transcription and dictation equipment, it's important to have a basic understanding of some of the more frequently used pieces of equipment required of some transcriptionists. These generally include manual, electric, and electronic typewriters (Figure 16-1). If you were lucky enough to take your mother's advice and learn how to type in your middle or senior high school classes, you will find this part of your job a lot less stressful and, in some cases, mundane.

Since the advent of highly electronic and sophis-

Administrative Medical Services

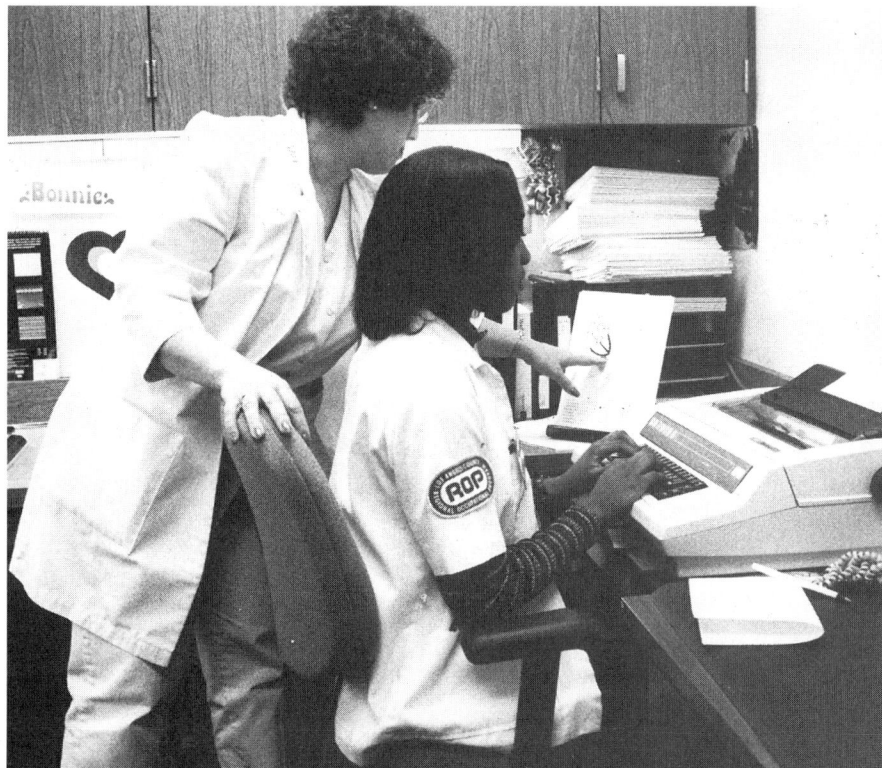

Figure 16-1
Electric and electronic typewriters.

ticated word processing systems, we rarely see manual typewriters in today's health care environment. However, if you are employed in a small private medical office, or just starting off in your own business as a transcriptionist, you may be more likely to use this basic piece of equipment. Manual typewriters are quite capable of performing all the necessary functions required of the medical transcriptionist. There are two major disadvantages to using these machines. The first is that all of the functions must be performed manually, at a much slower rate than with their electric and electronic counterparts. The other major disadvantage is that maintenance and repair of these machines are much more tedious and, in some cases, more expensive than electric and electronic typewriters.

Many health care facilities and medical transcription entrepreneurs purchase electric and electronic typewriters, and use them routinely in many administrative functions. The major difference between the electric machines and the electronic typewriters is that many of the electronic systems have the capability of storing information, and thus act as "pseudo" computing systems. For individuals and facilities that are unable to spend thousands of dollars to purchase a personal computer or network computing system, the electric and electronic typewriters will serve the purpose for a much lower cost. As a matter of fact, if you are planning to start your own medical transcription business, you may want to purchase an electronic typewriter initially before you go to the expense of purchasing a personal computer system. Once you have mastered the electric or electronic typewriter, and feel you are able to expand your business, you may then consider buying a computer system.

Using Medical Transcription Equipment

Medical transcription involves processing, storing, and communicating information from one machine to another. While the use of typewriters is acceptable for performing these tasks, the best available technology for accomplishing these goals is transcription equipment (Figure 16-2).

A medical transcription "set-up" usually consists of four components: a recording or dictation device, a foot pedal, earphones, and a means of printing out a *hard copy* of the information that has been transcribed, such as a personal computer or a typewriter. Most large health care facilities use desk top transcription systems, while many smaller offices purchase hand-held or portable systems, such as microcassette recorders and dictation machines.

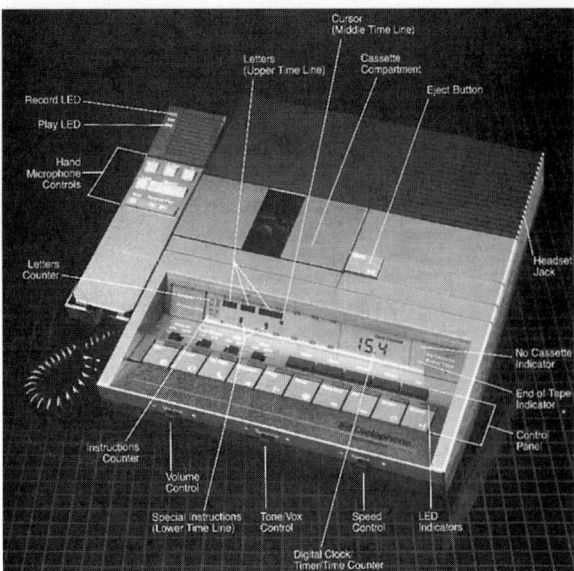

Figure 16-2
Types of transcription systems. (Reprinted with permission from Diehl, M. O. and Fordney, M. T. *Medical Typing and Techniques and Procedures*, 3rd ed. Philadelphia: W. B. Saunders, 1991, p. 34.)

Putting Your Equipment Together with Your Knowledge

It makes little difference what type of equipment you will be using if you do not possess the knowledge and understanding required to perform the basic tasks involved in medical transcription. Before you can even begin to transcribe the highly technical data and information used in most medical records and reports, you must first gain an understanding of the basic tools necessary to create these documents. Such tools involve the mastery of grammar, punctuation, and sentence structure. Once you have mastered these basic skills, you will be ready to put your equipment together with your knowledge and begin your career as a medical transcriptionist.

Understanding and Using Basic Grammar

To be competent in medical transcription, that is, to be able to create accurate documents, records, and medical reports, you must first be proficient in the use of the written word. To accomplish this, you must gain a basic understanding of grammar.

Grammar involves use of the parts of speech. These individual components include *nouns, pronouns, adjectives, verbs, adverbs, prepositions, conjunctions,* and *interjections*.

Nouns and Pronouns

A noun is a common or proper name of a person, place, or thing, while a pronoun takes the place of a noun. For example, the word *typewriter* is a common noun. If you refer to the typewriter as a *Brother Typewriter*, it is a proper noun because Brother is the name of the manufacturer of the typewriter and is capitalized.

Pronouns may be either *personal pronouns* or *relative pronouns*. Personal pronouns include first person pronouns, such as those used when a person is referring to him- or herself. Some examples of first person personal pronouns are *ours, us, my, I, mine*, or *we*; second person pronouns include *yours* or *you*; and third person pronouns include *her, she, his, he, they, theirs,* and *them*.

Relative pronouns are often used to replace nouns in adjectival clauses. For example, if you wanted to refer to a book that was purchased from a bookstore about the evolution of the health care industry, you might write the following: "The book *that* I purchased from the bookstore was about the evolution of the health care industry." The word "that" is the relative pronoun used to refer back to the noun book.

Adjectives, Articles, Verbs, and Adverbs

Adjectives are words used to describe a noun or pronoun. For example: *desktop* transcription machine, *new* system, or *clogged* arteries. In these examples, desktop states the type of transcription machine, new states the type of system, and clogged states the type of arteries.

The most often used adjectives are *the, a,* and *an*. These are called articles and are frequently used to define a specific situation. For example: *The* medical transcriptionist performs the job well; *an* x-ray report was generated by the transcriptionist; *a* supervisor helped me finish my work.

A verb is a word that can be used either to express an action, such as run, walk, or drive, or to "link" or join two words, such as was, be, been, or is.

Adverbs are words used to modify a verb, adjective, or another adverb. Some examples of adverbs used to modify verbs include *soon, too, always,* and *very*.

Prepositions, Conjunctions, and Interjections

A preposition is a word that links a noun or pronoun with another word or phrase in the specific sentence. For example, "the nurse *walked into* the patient's room." In this case, the word *into* was used to connect or show a relationship between the nurse and the fact that he or she went into the patient's

Administrative Medical Services

room. Some of the more frequently used prepositions include *about, along, among, during, into, like, over, under, upon, with,* and *without.*

In some instances, a preposition may include more than one word. They are referred to as *compound prepositions*, and often include such phrases as *prior to, in addition to, because of,* and *instead of.*

One of the easiest parts of speech to recognize is the *conjunction*, which is most often used to join words or groups of words.

There are three basic forms of conjunctions: *coordinating, correlating,* and *subordinating*. A coordinating conjunction is used to join equal parts of a sentence. Examples are *and, but,* or *so*. Correlating conjunctions are always used in pairs and include *not only, but also,* and *either, or*. Finally, subordinating conjunctions are used to introduce a subordinate part of a sentence or an adverbial clause that cannot stand alone, for example, *after, because, though,* or *while.*

You will also need to become familiar with interjections. An *interjection* is often used to express emotion and, therefore, possesses no grammatical association or relationship to any other word in a sentence. They are frequently accompanied by an exclamation mark or a comma. Examples of interjections include such words as *well, wow!* and *no!*

Punctuation and Capitalization

The proper use of punctuation and capitalization is a necessary part of sentence structure, and, thus, is a vital component in developing and creating accurate documents, records, and reports.

End Marks

Every statement and sentence must end with a punctuation mark, such as a period, question mark, or exclamation point. Any sentence that tells or informs ends with a *period*. For example: I have chosen a career in medical transcription. Any sentence that asks a question must end with a *question mark*. For example: Would you like to attend a training program in medical transcription? And any sentence that expresses extreme enthusiasm or emotion always ends with an *exclamation mark*. For example: I must go to school to become a medical transcriptionist!

Commas, Semicolons, and Colons

Unless you know how to use these three punctuation marks, you may find yourself completely overwhelmed with when or how commas, semicolons, and colons are used in a sentence.

Commas may be used in dates, for example, March 15, 1995; in addresses, such as Philadelphia, Pennsylvania; to separate elements in a series, such as Jane wants to become a medical transcriptionist, an admitting clerk, and a health unit coordinator; and to separate introductory elements, such as Yes, I do want to become a medical transcriptionist.

Semicolons are used to link independent clauses. For example: Only one person enrolled in the medical transcription class; the class was put off for a week. A semicolon can also be used between items in a series that already contains several commas. For example: There are medical transcription training programs located throughout the United States, for example, Los Angeles, California; Philadelphia, Pennsylvania; Miami, Florida.

A *colon* is used whenever you want something to stand out, before a formal statement or quotation is used, between the hour and the minutes in stating the time, and after the greeting or salutation in a letter. Following are examples:

1. The medical transcriptionist uses four pieces of equipment in his or her work: a transcribing machine, a foot pedal, earphones, and a computer.
2. The instructor said: it's time to take your test.
3. 7:30 p.m.
4. Dear Sir:

Underlining, Hyphenating, and Using Dashes

Even though underlining, hyphens, and dashes appear to be related, they have very different uses and are in no way related to or associated with one another.

When a word is underlined, it is generally because the person writing the word wants it to stand out. These words include titles of books and periodicals; works of art, such as paintings and statues; and names of specific planes, trains, and other forms of transportation.

Hyphens are used to divide a word at the end of a line, to join compound words, and when an adjective precedes the noun it modifies. Whenever you use a hyphen to divide a word at the end of a line, you must make sure you follow the proper rules of grammar. These rules include the following:

- Always divide the word between its syllables.
- To hypenate a word containing double consonants, try to divide it between the consonants.
- If a word has a prefix or suffix, try to divide the word between the prefix and the root or the root and the suffix.

- Never divide words which are already hyphenated.

The purpose of using a dash is to provide the reader with a blunt or abrupt break in the thought you are trying to convey. If you are using a dash in typed copy, you must make sure that you follow correct usage of this punctuation mark. All dashes, when typed, must be typed twice (--), with no spaces before or after the dash marks.

Apostrophes, Quotation Marks, and Parentheses

An apostrophe is used to show ownership or to combine two words. It can only be used with a noun or a pronoun. For example: the medical *transcriptionist's* machine is in perfect working order. When the apostrophe is used to combine two words, the apostrophe replaces part of the second word. Examples of this usage include *don't* (do not), *isn't* (is not), *it's* (it is), *I've* (I have), *you're* (you are), and *haven't* (have not).

Quotation marks are most frequently used when you are quoting someone's exact words, and they are always used at the beginning and at the end of the quote. If you are using quotation marks, you must follow some basic rules of punctuation, including:

- Always begin a direct quote with a capital letter.
- If you must divide a quotation, always use two sets of quotation marks, and never capitalize the second part of the quote.
- Always set off the direct quote from the rest of the sentence with a comma, question mark, or exclamation point.

When we want to add a comment or explanation that is not necessary to the sense of the sentence, we use parentheses to enclose the word or phrase. The parenthetical material can be placed within the sentence or, if it belongs to the sentence as a whole, the parenthetical material is placed outside the sentence.

Understanding and Using Capitalization

Part of using correct grammar is knowing when and when not to capitalize a word. Being able to disintinguish these differences involves a basic understanding of the rules of capitalization. These rules include:

- Only capitalize the first word of a sentence, regardless of what type of sentence it is. For example: *T*his is the best textbook I've ever used.
- Capitalize all proper nouns and proper adjectives of persons and titles. For example: *W*illiam *B*enjamin *R*aulerson, *J*r.
- Capitalize all geographical names, including towns, cities, countries, states, counties, townships, continents, islands, bodies of water, mountains, streets, parks, and specific parts or sections of a country. For example: *N*orth *H*ollywood, *C*alifornia; *P*acific *O*cean; *M*t. *S*t. *H*elens; *U*nited *S*tates of *A*merica.
- Capitalize the names of all government agencies and institutions. For example: *F*ood and *D*rug *A*dministration; *H*ouse of *R*epresentatives.
- Capitalize the names of all historical and special events. For example: *E*aster; *I*ndependence *D*ay.
- Capitalize the names of all organizations and businesses. For example: *A*merican *A*ssociation of *M*edical *T*ranscriptionists; *H*untington *M*emorial *H*ospital.
- Unless otherwise told, always capitalize the first word in every line of poetry.
- Always capitalize the pronoun *"I"* and the interjection *"O."*

Transcribing and Understanding the Use of Numbers

As you begin to use your medical transcription skills typing and transcribing important documents and medical reports, you will soon find it necessary to include numbers in many of the materials you transcribe from cassette tapes to hard copy. In order to do so, you should have a basic understanding of the various types of numbers which may be used, as well as when and how these numbers should be implemented into your documents.

Numbers are most often used in medical reports to identify specific numbered pages, as part of abbreviations, to express age, to identify the day of a particular month and year, as part of an address, in decimal fractions, as a roman numeral (e.g., John Smith, III), and in describing a particular piece of equipment or medical supply. In order to make it easier for you to understand when and how numbers are used in medical reports, simply follow the basic rules outlined below:

- If the medical report you are transcribing requires specific lists of information, be consistent with your numbering, use Arabic numerals, capitalize the first word of each numbered item, and number either vertically or horizontally: for example, *1.* Head trauma; *2.* Broken arm; *3.* Internal bleeding.
- When using symbols or abbreviations, only use Arabic numerals and remember to leave a space between the numeral and the abbreviation, except when you are typing numbers with multi-

ple symbols or identifying measurements, in which case, you may leave one space between the two: for example, Take *1* aspirin p.r.n.
- If you are required to write a plus or minus sign with a number, always use figures and symbols: for example, *Protein −2.*
- Always use figures with metric abbreviations: for example, *10 cc; 1000 ml; 20 G.*
- If you are transcribing two or more symbols and they are used together, always remember either to spell them out in full or use their abbreviated forms: for example, The patient had a temperature of *99.8* degrees F. When ill, a patient's temperature can reach a temperature of *99.8* degrees Fahrenheit.
- When transcribing electrocardiogram reports, use combinations of Arabic and Roman numerals and abbreviations for the different leads. For example, the patient's chest leads are *V1, V2, V3, V4, V5,* and *V6.*
- When transcribing addresses, spell out numbers one through ten for street numbers; for house or post office box numbers, spell out only the number "one"; for other numbers, use numerals. For example:
    Two Independence Square West
    631 Walnut Street
    P.O. Box *Five*
    Route *4*
    5311 South 10th Street
    P.O. Box *1029*
    13659 Victory Blvd.
- Use numerals whenever you are identifying someone's age unless the age is expressed indefinately. For example: The woman is *45* years old; she looks like she's in her *forties.*
- If transcribing a date, always remember to spell out the day of the month and year, use cardinal numbers to express the date after the month, and use ordinal numbers to express the date before the month. For example:

    People play practical jokes on April *1st.*
    Her birthday is April *1, 1995.*
    Her birthday was April *1.*
- When transcribing lab tests, medications, and body temperatures in decimal numerals, always remember to place a zero before a decimal point that is not a whole number.
- When transcribing reports that require the use of Roman numerals, always remember to follow the rules below:
    1. Only use uppercase Roman numerals (I, IV) after a person's name.
    2. Roman numerals may be used to describe a specialty (Para II). Arabic numerals may also be used.
    3. You may use a combination of Arabic and Roman numerals to describe a technique, factor, phase, class, or stage.
- When describing certain specific types of equipment, always transcribe them using Arabic numbers. For example: The patient was given an injection of *75* mg of Demeral with a *1 1/2*-inch *21*-gauge needle.

Summary

In this chapter, we learned about and identified the two most important areas for a beginning medical transcriptionist: equipment and grammar. We discussed the various components of the transcriber's equipment, including the use of manual and electric and electronic typewriters, computer systems, and the actual transcribing machine. After we completed our discussion of the equipment, we discussed how this equipment is used in beginning medical transcription, that is, how to use proper grammar and punctuation in the transcription process. As part of this discussion, we identified the various parts of speech, punctuation, and capitalization, and the use of numbers.

Review Questions

1. Define the following terms:
    a. modem _____
    b. menu _____
    c. program _____
    d. hardware _____
    e. software _____

2. What is the name of the method used to provide printed communications?

3. Define the following:
    a. ordinal numbers _____
    b. cardinal numbers _____

4. The process of taking the spoken or recorded word and then transferring it onto a tape cassette is called _____.

5. List at least three types of equipment which may be used to transcribe documents:

   a. _____

   b. _____

   c. _____

6. Define the following terms:

   a. noun _____

   b. pronoun _____

   c. adjective _____

   d. verb _____

   e. preposition _____

7. Pronouns may be identified as either _____ or _____.

8. Give at least three examples of adjectives:

   a. _____

   b. _____

   c. _____

9. Briefly define a *conjunction*.

10. Briefly discuss when the following may be used: semicolon, colon, comma.

11. Briefly discuss the different uses of underlining, hyphenation, and dashes.

12. Give at least two examples of when Roman numerals might be used:

    a. _____

    b. _____

Sentence Structure and Letter Writing

Performance Objectives

Upon completion of this chapter, you will be able to:

1. Distinguish between simple and complete sentence structure.
2. Demonstrate an understanding of sentence structure by being able to compose brief and accurate sentences.
3. Describe the purpose of using proofreading skills in letter writing and be able to identify and use common proofreading symbols in sentences and letters.
4. Discuss the four basic business letter styles and briefly explain the criteria for each.
5. Demonstrate an understanding of letter writing skills by composing various types of medical correspondence.

Terms and Abbreviations

Complement a word or group of words used to complete the meaning begun by the subject and predicate in a sentence.
Predicate a word or words used to tell something about a person or thing within a sentence.

Subject term used to describe a noun, noun phrase, or noun substitute in a sentence about which something is said.

Now that you have a fundamental understanding of the tools of basic grammar, it will be important for you to learn how to use these tools to develop clear and concise sentences. Once you can do this, you will be ready to take the necessary steps involved in letter writing, and eventually be able to put all your skills together to transcribe medical reports and other important documents used in the health care system.

Understanding Basic Sentence Structure

Four basic types of sentences are used in the English language. These include *declarative* sentences, which are used to make a statement; *imperative* sentences, which are a command or a request; *interrogative* sentences, which ask a question; and *exclamatory* sentences, which express a strong feeling. Sentences may also be described according to their complexity.

For example, a *simple* sentence is one which has one main clause and no subordinate clauses, thereby making it very easy to understand. A *compound* sentence, on the other hand, is more involved because it provides the reader with two separate statements or independent, coordinate clauses which are often related to one another. A *complex* sentence, that is, one which consists of a main clause and one or more subordinate clauses, is often more involved and lengthy, because it can be either compound or complex.

All sentences, whether they are simple, compound, or complex, must include a *subject* and a *predicate*. A subject is the person or thing that the sentence is about. It may be either simple or complete. For example, "Betsy is going to the store" is a simple sentence and *Betsy* is the simple subject. By adding another word that describes, identifies, or explains more about Betsy, the subject becomes complete. For example, Betsy, Ben's *mother*, is going to the store. The fact that we have now described Betsy as Ben's mother makes the subject complete.

All sentences must inform the reader of some action or state of being about the person or thing who is the subject of the sentence, and a predicate is used to complete such a task. All predicates must contain a verb. For example, in the phrase Betsy *drives*, the verb drives is the simple predicate of a simple sentence. If we say Betsy *drives Ben to the store*, we have explained, or modified, the action. The sentence now contains a complete predicate.

When we design a sentence, we must also be concerned with how we communicate a specific meaning of the sentence. To do this, we must use a specific word or group of words called a sentence complement. For example, if we say Betsy drives *well*, the word well is used to complement how Betsy drives.

Sentence Fragmentation

When speaking, we often pay little or no attention to how we make a statement or complete a sentence. In speech this tends to make little difference. However, if we do this when writing a sentence, it makes the sentence incomplete. The result is called a *fragmented* sentence. While fragmented sentences are generally accepted in oral presentations, once written they appear incomplete and incomprehensible.

Proofreading and Sentence Structure

The skills involved in learning how to properly design sentences, and ultimately how to complete letters and transcribe medical documentation, also involve learning how to present an accurate and professional-looking document. We refer to the process involved in completing a mailable, accurate, and highly professional-looking document as *proofreading*. To complete such a task, you must be free of all distractions, concentrate, and read through your document at least two times.

Basic Proofreading Marks

There are standard proofreading marks and symbols that are basic to all modes of transcribing, and which are acceptable and used throughout the United States. Figure 17-1 has a list of these symbols, their function, and how they are used throughout text. You should try to learn these symbols, since they will ultimately make you more aware of the importance of accuracy and professionalism when creating documents.

Letter Writing Skills and the Medical Transcriptionist

Now that you have learned the basic components of grammar, sentence structure, and proofreading skills you are ready to put all these skills together and write letters.

In health care professions, communication and accuracy are two of the most important aspects of good letter writing. During your career, you may be required to write or transcribe any number of letters, including informal communications and memoranda, announcements and invitations, promotional pieces, and various forms of official letters and business communications. The most important of these letters for a medical transcriptionist will have to do with official documentation, consultation, and business communication.

Unless you understand the importance of appearance and using the correct format, wording, grammar, punctuation, as well as the use of abbreviations and proofreading marks and symbols in your correspondence, it makes little difference what type of letter or type of communication you are required to design or transcribe. The bottom line is that all of these points must be considered when you write or transcribe a letter or other type of communication.

Spacing and Letter Styles

Before we can discuss the basic style of letters which have been adopted by most health care facilities, it is important for you to understand the concept of spacing and the individual parts of the letter.

Unless otherwise indicated, all letters contain the following parts:

- *Date line* This is typed three lines below the letterhead or on the fourth line.
- *Inside address* The inside address should be typed at least three to 12 lines below the date line, according to the length of the letter, and should consist of two or more lines, with all of the lines spelled out.
- *Salutation or Greeting* The salutation or greeting line should be placed two lines below the inside address or attention line, if one is used.
- *Body* The body of the letter should be placed at least two lines below the subject line, if one is used. If no subject line is used, the body should begin two lines below the salutation or greeting. There should also be two lines between each paragraph.
- *Complimentary closing* The complimentary closing begins two lines below the body of the letter.

Administrative Medical Services

Symbol	Function	Usage in Completed Text
.... or stet	Ignore correction	Physician and therapist
ℯ	Delete	Doctor and nurse
/ or —	Delete and add	doctor and nᵤrse
(/)	Delete and close up	Dr. Smith and his nurse
	Close up:	
⌒	Horizontal	ECG Technician and doct or
( )	Vertical	ECG Technician and doctor / ECG Technician and assistant
?	Verify accuracy	The procedure is set for 7:30 a.m. (?)
⟲	Move text	Dr. (James) and Nurse Jones
sp	Spell out word(s)	Third St. sp / sp 3rd Street
— or ital	Italicize or underline	The procedure is being done here / The procedure is being done here
⌐ or #	New paragraph	# Ms. Smith is scheduled to work on Monday
	Spacing:	
	Vertical:	
DS>	Double	DS> Ms. Smith is not going to work on Monday after all.
SS>	Single	You can't be tired and still complete your work accurately and professionally. SS> Physician and technician
#	Horizontal	
	Transpose:	
∽	Words	You must attention pay to your work
∼	Letters	You must pay attention ot your work
	Case:	
/ or lc	Lowercase	Dr. JoNes and Ms. Smith
cap	Uppercase	Dr. jones and Ms. Smith
≡	Capitalize	Dr. Jones and Ms. smith
	Insert:	
⊙	Period	Dr. Smith is going to do the surgery
⋀	Comma	Today tomorrow and Sunday
⌄	Apostrophe	The ECG Technicians schedule
⌃	Hyphen	Late afternoon procedure
⌃	Dash	Dr. Jones needs you immediately
" "	Quotation marks	Ms. Smith said I'm really in pain
⋀	Letter or number	supplie and equipment

Figure 17-1
Basic proofreading marks

- *Typewritten signature and title* The name of the person sending the letter should be typed four lines below the complimentary close or the company name. If the letter appears to be very short, you may place the typed name on the sixth, seventh, or eighth line below the company name or complimentary closing. The title of the person sending the letter should be typed one line below his or her typed name.
- *Reference initials* Reference initials are typed two lines below the signature title. If your letter uses two sets of initials, the first is the author of the document and the second is the person typing the document.
- *Enclosure* If you are sending material(s) along with your letter, you must type the word "enclosure" two lines below the reference initials flush with the left margin. If you are sending more than one enclosure, you must indicate the number by enclosing it in a parentheses.

In some cases, you may be required to add other information to your letter. This generally includes an attention or subject line, a copy notation or mailing notation, and a second page heading. If this is the case, you may follow the information provided below:

- *Attention line* If you are using an attention line, it must be typed two lines below the inside address.
- *Subject line* A subject line may be typed flush left, indented, or centered in the middle of the page. It should be placed two lines below the saluation, may or may not be typed in all capital letters, and is often followed by the word *Subject*.
- *Copy notation* No professional letter or piece of documentation is ever sent out without a copy being kept for your records. Whenever copies are prepared for the information of others, the notation *"cc"* is followed by the name of the person the copy is being sent to.
- *Mailing notation* Whenever you are required to follow any special mailing procedure for your letter or documentation, it should be typed between the date and the inside address and is always placed flush with the left margin.
- *Second page heading* The second page heading always includes the name of the person to whom the letter is being addressed, the page number, and the date, all of which should be typed on the seventh line from the top of the page.

A letter style is defined as the way in which we format or set up a page of communication. In both health care and business, there are four basic letter styles which are most often used. These include *block* style, *modified block* style, *modified block style with indentations and mixed punctuation*, and *simplified* style.

Block style is most often used in the health care setting because it is simple, gives a good appearance, and can often be quite fast to produce (Figure 17-2). In producing a block style letter, the most important thing to remember is that everything begins at the left side of the paper. In a *full* block style letter, both the left side of the paper and the right side of the paper are flush to the margins. In a *semi-block* letter, the right side is not flush with the margin.

In a modified block style letter, everything except the date, complimentary closing, and writer's signature and title are flush left (Figure 17-3). Generally, the date is typed in the center of the page, whereas the closing, signature, and title are typed in the center of the document.

A document that uses a modified block style with indentions and mixed punctuation generally includes a subject line. In this type of style, the date is typed in the center of the page, the subject is indented five spaces from the left, as are the paragraphs, and the complimentary closing and signature and title lines are typed mid-center to right of the document (Figure 17-4).

The newest type of letter style, which has recently been adopted by many health care facilities, is the simplified style. In this style, the salutation and complimentary closing are both eliminated, and the subject line, which replaces the salutation, is typed on the third line below the inside address in all capital letters. The sender's name and his or her title are also typed in all capitals, three lines below the body of the letter (Figure 17-5).

Typing the Envelope

Typing the envelope, in which your letter or documentation is sent, is just as important as how you actually set up your letter. In fact, with all the exact requirements of postal service, unless you set up your envelope in the precise manner required by law, your local post office may not even deliver or pick up your correspondence.

There are two basic types of envelopes which are used for most professional and business corespondence. The first type is called a #6 envelope (Figure 17-6). The #6 measures 6 1/2 by 3 5/8 inches. The second type of envelope frequently used by health care facilities is a #10 envelope. It measures 9 1/2 by 4 1/8 inches (Figure 17-7).

SUPERIOR ORTHOPEDIC SERVICES
1234 Anywhere Street
My Place, CA 91207

March 20, 1995

Robert Green, M.D.
Green Sports Medicine Clinic
10761 Riverdale St.
Toluca Lake, CA 91602

Dear Dr. Green:

Thank you for seeing my patient, Robert Jones, in consultation. As I am sure you can see from my report, Mr. Jones seems to have some serious orthopedic problems with his back and extremities.

My plan for Mr. Jones is to do a complete orthopedic work-up and then, if necessary, schedule him for surgery.

Should you have any additional comments or further information regarding Mr. Jones' condition, I would greatly appreciate your input in this patient's treatment.

Sincerely,

James Smith, M.D.
Superior Orthopedic Services

rw:js

Encl.

cc:

Figure 17-2
Block style.

SUPERIOR ORTHOPEDIC SERVICES
1234 Anywhere Street
My Place, CA 91207

March 20, 1995

Robert Green, M.D.
Green Sports Medicine Clinic
10761 Riverdale St.
Toluca Lake, CA 91602

Dear Dr. Green:

SUBJECT: Robert Jones

Thank you for seeing my patient, Robert Jones, in consultation. As I am sure you can see from my report, Mr. Jones seems to have some serious orthopedic problems with his back and extremities.

My plan for Mr. Jones is to do a complete orthopedic work-up and then, if necessary, schedule him for surgery.

Should you have any additional comments or further information regarding Mr. Jones' condition, I would greatly appreciate your input in this patient's treatment.

                                                  Sincerely,

                                                  James Smith, M.D.
                                                  Superior Orthopedic Services

rw:js

Encl.

cc:

Figure 17-3
Modified block.

Administrative Medical Services

SUPERIOR ORTHOPEDIC SERVICES
1234 Anywhere Street
My Place, CA 91207

March 20, 1995

Robert Green, M.D.
Green Sports Medicine Clinic
10761 Riverdale St.
Toluca Lake, CA 91602

Dear Dr. Green:

    SUBJECT: Robert Jones

    Thank you for seeing my patient, Robert Jones, in consultation. As I am sure you can see from my report, Mr. Jones seems to have some serious orthopedic problems with his back and extremities.

    My plan for Mr. Jones is to do a complete orthopedic work-up and then, if necessary, schedule him for surgery.

    Should you have any additional comments or further information regarding Mr. Jones' condition, I would greatly appreciate your input in this patient's treatment.

                                            Sincerely

                                        James Smith, M.D.
                                        Superior Orthopedic Services

rw:js

Figure 17-4
Modified block with indentions and mixed punctuation.

SUPERIOR ORTHOPEDIC SERVICES
1234 Anywhere Street
My Place, CA 91207

March 20, 1995

Robert Green, M.D.
Green Sports Medicine Clinic
10761 Riverdale St.
Toluca Lake, CA 91602

ROBERT JONES

Thank you for seeing my patient, Robert Jones, in consultation. As I am sure you can see from my report, Mr. Jones seems to have some serious orthopedic problems with his back and extremities.

My plan for Mr. Jones is to do a complete orthopedic work-up and then, if necessary, schedule him for surgery.

Should you have any additional comments or further information regarding Mr. Jones' condition, I would greatly appreciate your input in this patient's treatment.

James Smith, M.D.
Superior Orthopedic Services

rw

Figure 17-5
Simplified style.

(2 lines)
SUPERIOR ORTHOPEDIC SERVICES
1234 Anywhere Street
My Place, CA 91207
(indent 3 lines)

                                                SPECIAL DELIVERY
                                                (5 spaces from edge)

(2 1/2 inches from edge)   Robert Green, M.D.
                                    Green Sports Medicine Clinic
                                    10761 Riverdale St.
                                    Toluca Lake, CA 91602

Figure 17-6
A #6 envelope.

(2 lines)
SUPERIOR ORTHOPEDIC SERVICES
1234 Anywhere Street
My Place, CA 91207
(indent 3 lines)

(4 inches from the edge)

(12 to 14 lines from the top)

Robert Green M.D.
Green Sports Medicine Clinic
10761 Riverdale St.
Toluca Lake, CA 91602

Figure 17-7
A #10 envelope.

Summary

In this chapter, we discussed three very important aspects of working as a medical transcriptionist. These included sentence structure, proofreading skills, and letter writing composition. As part of our discussion, we identified the different types of sentences, the types of marks and symbols used in proofreading, and the exact way of setting up letters and envelopes used in the professional health care and business fields.

Review Questions

1. Define the following terms:
   a. subject _____
   b. predicate _____
   c. complement _____

2. List the four different types of sentences which are used in the English language:
   a. _____
   b. _____
   c. _____
   d. _____

3. Briefly define a *fragmented sentence.*

4. Briefly define the purpose of *proofreading.*

5. Provide the proofreading symbols for the following functions:
   a. new paragraph _____
   b. delete _____
   c. single spacing _____
   d. move text _____
   e. spell out word _____
   f. comma _____
   g. verify accuracy _____
   h. quotation marks _____

6. List the four types of letter styles most frequently used for communication in the health care environment:
   a. _____
   b. _____
   c. _____
   d. _____

7. Define the number of spaces which would be found following the items listed below:
   a. inside address _____
   b. salutation _____
   c. body _____
   d. complimentary close _____

Transcribing Medical Reports and Records

Performance Objectives

Upon completion of this chapter, you will be able to:

1. Identify the most frequently used reference materials available to the medical transcriber for use in transcribing medical reports and records.
2. Identify and explain the function of various types of transcription reports.
3. Describe and identify how to properly format and type various medical reports, including a patient history and physical report, a radiology report, and a pathology report.
4. Discuss the steps required in properly correcting a medical report.

Terms and Abbreviations

Pathology the medical specialty that examines all fluid and tissue specimens removed during a surgical procedure.
PDR abbreviation for Physician's Desk Reference, reference book used to identify the various characteristics of drugs and medications.

The main function of the medical transcriptionist is to provide the physician or the health care facility with an accurate and asthetic-looking medical report, record, or document, which has been dictated or written out by the professional health care provider and which can be used as a legal document in a court of law. While there are many types of documents that the transcriptionist may be required to transcribe, those most frequently transcribed include the patient history and physical report, physician's progress notes, and pathology, radiology, and laboratory reports.

Using Reference Materials

Since many of the reports and records you will be transcribing will include terms and words that are foreign to you, you should spend some time reviewing medical terminology, medical symbols, and abbreviations before beginning to transcribe these documents. You should also start getting acquainted with some of the more commonly used medical reference books, such as the (PDR) *Physician's Desk Reference*, *American Drug Index*, and medical dictionaries, such as *Taber's Cyclopedic Dictionary* and *Dorland's Illustrated Medical Dictionary* (Figure 18-1). These medical reference tools are considered the "bibles" of all medical transcriptionists. You may also want to have a general dictionary, such as *Webster's Collegiate*, close by to check your spelling and punctuation of nonmedical words. Remember, there's no shame in using a dictionary or reference book to complete your work; the shame is in *NOT* using these tools, and turning out unprofessional, inaccurate work.

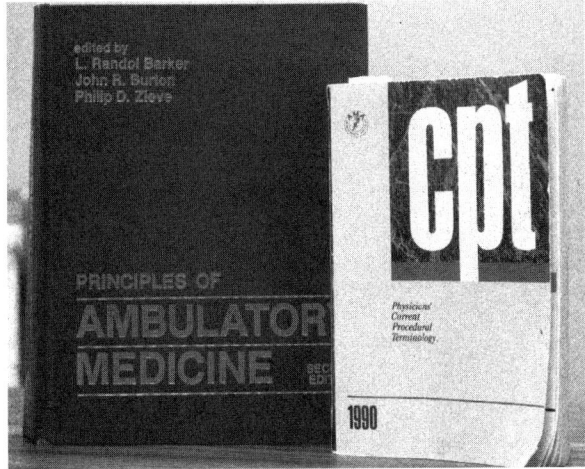 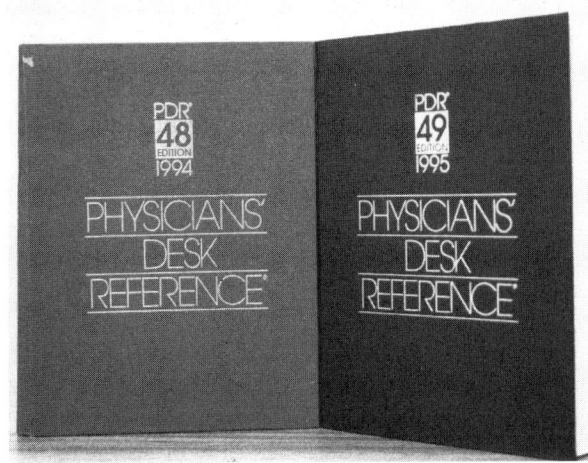

Figure 18-1
Reference materials used in medical transcribing.

Transcribing Patient History and Physical Reports

Two documents that are included in all patient medical records or charts are the patient's *history* and *physical* reports. Regardless of what type of format you use for these reports, the information contained in them is generally the same.

The physician or health facility uses the history and physical reports to provide the practitioner, and ultimately any other interested party, with a description of the patient's prior and present medical history. It is also a useful evaluation tool for describing the physical or objective findings of the various parts of the patient's body. The history is usually taken by the physician or his or her assistant on the patient's first visit to the office or hospital. It generally includes information regarding the patient's present illness, including its onset and duration; medical and surgical history; and allergies. The history also provides the physician with documentation regarding the patient's family history.

The history and physical report is generally transcribed as one document. It may be typed in a *full block* style, a *modified block* style, an *indented block* style, or a *run-on* format. The most frequently used style is the full block style (Figure 18-2). The patient's medical history evaluation usually includes the following:

- Chief complaint;
- Past medical history;
- Surgical history;
- Medication history;
- A list of any known allergies;
- A list of the patient's habits, including any use of alcohol, drugs, or cigarettes;
- A description of the patient's family and social history; and
- Review and brief discussion of the patient's body systems, including the patient's head, eyes, ears, nose, mouth, throat; cardiorespiratory, digestive, and genital-urinary systems; and any neuropsychiatric pathology if relevant to the patient's medical history.

Findings related to the patient's physical condition generally follow the information presented related to the patient's medical history. It usually includes data and a brief discussion related to the following:

- Vital signs;
- Head, eyes, ears, nose, and mouth;
- Neck;
- Chest, breasts, and lungs;
- Heart;
- Abdomen;
- Pelvis and rectum;
- Extremities;
- Neurologic signs; and
- Final impression.

Transcribing Pathology and Radiology Reports

The purpose of a pathology report is to provide the physician and/or the health care facility or other interested parties, with a description of the size, structure, pathologic physiology, and other pertinent information related to a specific tissue or fluid specimen. Once removed from the patient, these specimens are generally sent to a pathology laboratory for the pathologist to examine and confirm a preoperative

6/9/81

HISTORY AND PHYSICAL

CHIEF COMPLAINT: was admitted by Dr.   the evening of 6/9/81 following an episode of severe back pain in the admitting office while at work in the hospital. She was seen by Dr.   early that morning and complained of her back bothering her and had a pinpoint pain in the left side of her incision in the low back and pain radiating down the leg. She was given ROBOXIN by Dr.   r and told to use baths during the day and try to aleviate the poain. I do not think that she was at work but I am not certain and she was infact informed on a phone call earlier in the evening that she should in fact try to finish her job since she was there and then go home and resume the same regimen. Apparently she called Dr.   later in the evening and was still having pain and was admitted to the hospital by him with a complaint of low back pain.

PAST MEDICAL HISTORY: This is a 4'11" female who has had multiple problems throughout her life with weight control and at the present time is on a weight control weight reduction program and has lost some 19 to 20 pounds in the last 3-4 weeks. She has had a history of weight loss in the past but generally has been fobese most of her life.

Her second medical problem is that she has had recurrent right lower quadrant pains for many years and in the early months of 1980 finally underwent an exploratory laparotomy and removal of the uterus and the adnexa with no further pain.

SURGICAL HISTORY: In 1978 patient underwent an exploratory laparotomyu for hemorrhagic cyst and then in January of 1980 she underwent a hysterectomy and removal of the adnexa. Following that in July of 1980 the patient had an acute onset of low back pain following helping a patient remove herself from the toilet. At that time the pain radiated down the left leg and the patient was admitted to the hospital and in fact had a ruptured lumbar disc. Subsequent to that the patient came back with excruciating pain 6 months later and underwent surgery on 12/8/80 and in fact again had another portion of a ruptured disc present. Since that time the patient has had difficulty in finding a job because she has beenout on leave of absence and finally has gotten back to work at HPMC in the admitting office and appears to be enjoying the job and has had several episodes of intermittent low back pain possibly due to tear of adhesions in the area and has continually received support to try to continue her work.

MEDICATIONS: The patient is presently on a weight-reduction program with Dr.   and is having no food intake but only a substitute thing and she is in chronic ketosis.
CONTINUED----------------------
HISTORY AND PHYSICAL

PATIENT NAME

PATIENT NUMBER   16574751A

SIGNED

Figure 18-2
History and physical report using full block style. (Courtesy of Huntington Memorial Hospital, Pasadena, CA.)

6/9/81

HISTORY AND PHYSICAL

Page 2

ALLERGIES: There are none known.

HABITS: The patient smokes very heavily.

ALCOHOL: She drinks in moderation.

FAMILY HISTORY: The father recently passed away in 3/81 of a sudden heart attack with no previous knowledge of heart disease. The mother's history is unknown at this time but appears to be from the patient in relatively good health.

SOCIAL HISTORY: The patient is single. She lives in her own apartment. She works as an LVN. She likes to write and she has no permanent relations and no history of children prior to her hysterectomy.

REVIEW OF SYSTEMS:

HEAD: Patient has occasional headaches. No dizziness or syncope.

EYES: Vision is good.

EARS: hearing is good.

NOSE: There is no history of polyps or epistaxis.

MOUTH: She has her own teeth.

THROAT: There is no history of significant sore throat or tonsillitis.

CARDIORESPIRATORY: Patient denies pneumonia, tuberculosis. She has been a heavy smoker and she does have a chronic cough. There has been no heart condition noted.

GI: The patient has had relatively negative GI history except for her obesity.

GU: Patient has now post hysterectomy and has had episodes of anuria during hospitalizations but nothing has ever been found of physiological significance to this.

NEUROPSYCHIATRIC: Patient has been under psychiatric counseling a number of years and in order to resolve some of her personal problems
Continued---------------HISTORY AND PHYSICAL

PATIENT NAME

PATIENT NUMBER   165747751A

SIGNED

Figure 18-2
Continued.

6/9/81

HISTORY AND PHYSICAL

Page 2

She of course is suffering from low back pain secondary to her second laminectomy.

PHYSICAL EXAMINATION:

VITAL SIGNS: BP is 120/80, pulse is 80 and regular. She is afebvrile. The respirations are normal. The patient was found lying in bed, relatively comfortable at this time as long as she is in the prone position.

HEAD: Normocephalic with normal hair distribution. No disease of the scalp.

EYES: Pupils are equal and react to L&A well. fThe fundi are benign.

EARS: Normal and the TMs are clear.

NOSE: The septum is midline. There are no polyps.

MOUTH: She has her own teeth in good repair. There is no disease of the lips tongue or cheek. The tonsils are negative.

NECK: There is free range of motion. The thyroid is not enlarged. The trachea is midline. there are no bruits.

CHEST: The patient is heavy across the posterior chest wall with an obvious peniculus.

BREASTS: Breasts are 3 on a scale of 4. The patient has had a history of a breast abscess in the past which is now cleared and there are no masses present and no nodularity noted.

LUNGS: Clear to A&P. There are normal diaphragmatic excursions.

HEART: PMI is in the 4th intercostal space in the miclavicular line. There are no audible murmurs. The sounds are physiologic.

ABDOMEN: Patient is obese. There are 2 scars in the right lower quadrant, one being a Pfannenstiel incision from exploratory surgery when she had a hemorrhagic cyst, the other being an appendectomy scar and the third is a reopening of the pfannenstiel for her hysterectomy. LKS are normal. There is no area of
Continued----------------HISTORY AND PHYSICAL

PATIENT NAME

PATIENT NUMBER   165747751fA

SIGNED

Figure 18-2
Continued.

6/9/81

HISTORY AND PHYSICAL

Page 3

tenderness or rebound and there are no palpable masses.

PELVIC EXAMINATION: This is deferred at the present time.

RECTAL: Negative.

EXTREMITIES: The pulses are 2 plus at the brachial and radial. The femoral and the posterior popliteal and the dorsopedal. There is no weakness noted in the extremities that is obvious.

NEUROLOGIC: The cranial nerves II-XII are within normal limits. Reflexews remain 1 plus at the knee and 1 plus at the ankle but the patient does have to be distracted with the handgrips test in order to elicit good reflexes. Babinskis are negative bilaterally.

IMPRESSION:

1. Recurrent low back pain possibly secondary to adhesions because of pinpoint tenderness near the old incision line.

2. Radiation down the left leg shows muscle spasm, and som degree of neurological impairment may be present.

3. Post hysterectomy.

4. Chronic obesity now being tested and aleviated with a rather radical dietary regime in order to improve the patient's general physiology.

```
RN/io
D-6/9/81
T-6/15/81
CC:              M.D.
                 M.D.
    Medical Records
```

HISTORY & PHYSICAL

PATIENT NAME

PATIENT NUMBER   165747751A

SIGNED

diagnosis. Once the pathologist has examined the specimen, he or she dictates a report for the patient's records, a copy of which is often sent to the surgeon. The pathology report, which may be either typed in a block style or filled-in on a standard pathology report form, usually includes the name of the patient, his or her age and sex, the date on which the procedure or surgery was completed, the patient's hospital and/or surgical identification number, the names of the attending and operating physicians, the preoperative diagnosis and final diagnosis based on the specimen, the actual name of the specimen, and a gross description of the pathologic findings regarding the specimen (Figure 18-3).

Radiology Report

Reports describing x-rays and imaging findings are almost always generated by the radiology department and dictated by a radiologist. They may be typed either in a block style or may be filled-in on a standardized radiologic report form. Reports may be of traditional x-ray films, tomography, sonography, or magnetic resonance imaging, and generally include the patient's name, age, hospital identification number, date the x-ray was taken, a description of the procedure performed, and a brief statement of the radiologist's findings, impression, or conclusion (Figure 18-4).

Transcribing Consultation Reports

A consultation report is an account of medical and physical findings pertaining to a patient's specific health condition. It will be dictated by a physician to whom the patient's attending physican has referred him or her. These reports may either be typed in a block format or in narrative style, and generally include the date on which the patient was examined by the consulting doctor, the name of the referring physician, the reason for the consultation, a narrative discussion of the patient's present condition and why the consultation was sought, a review of the patient's past medical history, including any operations, a complete physical examination, and a brief discussion of the consulting physician's findings, impression, and potential treatment or recommendations (Figure 18-5).

Transcribing Discharge Reports

A discharge report is a summary of the cause, treatment, and probable prognosis of a patient's medical condition during his or her hospitalization. All physicians who hospitalize patients are required by law to complete a discharge report or summary upon writing the patient's discharge order. If more than one physician has provided care to the patient, the discharge summary is usually dictated by the attending physician. Included in the discharge summary, which may either be typed in block style or in narrative format, are the dates the patient was admitted and discharged, the name of the attending and, if any, the consulting physicians, a brief history of the patient's medical condition that prompted the hospitalization, the course of treatment the patient received during hospitalization, a list of any procedures and/or surgeries which may have been performed on the patient during his or her hospitalization, and the patient's diagnosis upon his or her discharge (Figure 18-6).

Transcribing Other Medical and Hospital Reports

Many reports and documents make up the patient's medical record. These include operative or surgical reports, progress records and physician's order sheets, laboratory and autopsy findings, and medico-legal and insurance reports. All hospitals, health care facilities, and physicians working under Joint Commission for the Accreditation of Hospitals (JCAH) guidelines are required to dictate these reports to provide the facility, physician, insurance company, attorney, or court with a legal document related to the patient's care, treatment, prognosis, and discharge. As a medical transcriptionist, your responsibility is to take these recorded and dictated reports and generate them into an accurate, error-free legal document. If you discover that you have made an error in your transcribing, you must never destroy the document. Rather, you should first draw one line through the error and then write or type the correction beside, above, or next to the original error. The error must then be initialed by the person who made it.

Summary

In this chapter, we discussed some of the more frequently transcribed reports and records generated by the hospital, health care facility, and physician. During our discussion, we talked about how some of these reports are transcribed, noting that the most frequently transcribed reports include patient history and physical reports, pathology and radiology reports, consultation reports, and discharge summaries.

Administrative Medical Services

133

DEPARTMENT OF PATHOLOGY

PATHOLOGISTS:

SURGICAL PATHOLOGY REPORT

PATIENT NAME: (Last) (First) (Middle)	DATE	SURGEON:	ROOM NUMBER
Heartburn, Audrey Jane	7/14/80	DR. D. Doolittle	247A

AGE	SEX	HOSPITAL NUMBER	SURGICAL NUMBER	ATTENDING PHYSICIAN
31	F	165747721A	S— 3745-80	DR. Longneck

DIAGNOSIS
Fibrohyaline cartilage and bone fragments- clinically, lumbar disc material.

PREOPERATIVE DIAGNOSIS:
Lumbar disc syndrome

SPECIMEN(S):
Bilateral L4

PATHOLOGIC FINDINGS:
GROSS DESCRIPTION:

Specimen submitted consists of multiple fragments of soft glistening greyish-white cartilaginous tissue along with spicules of bone. The largest tissue fragment is 1.8 cm in maximum dimension. Altogether weigh 5.8 grams. Multiple representative sections are submitted for decalcification.

RWW:ldw

SIGNED: _____ M.[ ]
PATHOLOGIST

70.06 R11/76

ATTENDING PHYSICIAN

Figure 18-3
Surgical pathology report. (Courtesy of Huntington Memorial Hospital, Pasadena, CA.)

RADIOLOGISTS:

DEPARTMENT OF RADIOLOGY

TELEPHONE
DIAG. RAD.
IN. PT.
OUT PT.
NUCL. MED.
THER. RAD.

AGE: 31   M.D.
           M.D.

7-11-80   247B   165747721A   05-43-88

LUMBAR MYELOGRAM

Pantopaque was injected into the lumbar subarachnoid canal by a needle inserted at L5/S1 by Dr. Northrop. The column is freely moveable. There is a waist-like bilateral narrowing of the opaque column at L4-5. Lateral views made in the erect and horizontal positions show an anterior extradural defect at that level and at L5/S1. The possibilities include a bilaterally herniated disc or an extradural collection of fluid or tissue, even neoplastic. This would be quite unlikely.

IMPRESSION

Bilateral extradural defect, L4-5, most probably due to disc herniation.

P. Phototaker   ., M.D., RADIOLOGIST
   D/T: 7-14-80/bc

PATIENT NAME: Audrey Heartburn
PATIENT NUMBER: 12345-78

SIGNED _____ RADIOLOGIST
FORM 7140.10 REV. 5/79

Figure 18-4
Radiology report. (Courtesy of Huntington Memorial Hospital, Pasadena, CA.)

Administrative Medical Services

RADIOLOGISTS:

DEPARTMENT OF RADIOLOGY

TELEPHONE
DIAG. RAD.
IN. PT.
OUT PT.
NUCL. MED.
THER. RAD.

```
                              AGE 30                    M.D.
9-26-79      348 A       130053012-2A      05-43-88
```

LUMBOSACRAL SPINE SERIES:

Examination is severely limited due to marked amount of barium within the colon, which obscures major portions of the lumbosacral spine. However, AP lateral views were attempted. No abnormality is identified of the osseous soft tissues, joint or disc spaces, but again, the studyis limited due to barium.

CONCLUSION:

Limited, negative study.

When barium is cleared from the colon, a full lumbosacral spine series is suggested.

```
                    M.D., RADIOLOGIST
D: 9-26-79      T: 9-27-79cs
```

PATIENT NAME

PATIENT NUMBER

SIGNED   RADIOLOGIST

FORM 7140.10 REV. 5/79

Figure 18-4
Continued.

RADIOLOGISTS: TELEPHONE

DEPARTMENT OF RADIOLOGY

DIAG. RAD.
NUCL. MED.
ULTRASOUND
CT
THER. RAD.

AGE: 31   M. D.
                                    M. D.
4/15/81   239   165747741A   05 43 88   M. D.
                          1232198

LUMBOSACRAL SPINE:

Residual of previous myelogram was again identified. Some slight narrowing of L5-S1 is seen. There is slight facet joint sclerosis at L4-5, L5-S1. No change appears to have occurred since 10/27/80.

CONCLUSION:

Mild degenerative spondylosis lumbosacral spine..

        , M. D., RADIOLOGIST

D: 4/15/81   T: 4/16/81   /amf

Figure 18-4
Continued.

DATE:

NEUROSURGICAL CONSULTATION NOTE

REFERRING PHYSICIAN:

REASON FOR CONSULTATION: This patient is a 32 y/o LVN, who was examined by me for the first time in the office today and admitted to the hospital.

The patient complains of severe pain in the low back with radiation into the left buttock and left leg. She relates this to an incident occurring while she was working in this hospital on 7-2-80.

She states that she was bending over to help a patient in getting off the toilet. The patient states that she was bent forward and that the patient she was assisting, put her arm around her neck and waist, and the patient attempted to lift her patient off of the comode. As the patient got to her feet, her gown dropped to the floor and the nurse bent over quickly to try to retrieve the gown. When she did this, she experienced a severe pain in the low back with radiation into the left buttock and this pain gradually spread down the left leg.

This patient consulted Dr.   and was sent home. Because of the approaching long holiday weekend, the patient was instructed to remain at home at bedrest as much as possible. She was given muscle relaxants and analgesics to take. In spite of this, the patient says that her pain has increased gradually.

The patient claims that this pain is now steady and she cannot maintain any one position very long and she finds that she sleeps a short period of time and finds that she has to get up and move around, and this is a similar situation when sitting for long periods of time. If she is up and having pain and lies down, the pain disappears for a couple of hours, but then returns.

She finds it difficult to clearly vocalize this pain. She has noticed numbness only in the great toe of the left. She does not think that there is any weakness in the leg, but really is not sure.

Coughing and sneezing aggravate the pain and so does any movement of her back.

The patient says that she has no symptoms in the right buttock or leg. She has no sphincter disturbances. A year or so ago, she helped move a patient, who was bedfast and developed some pain in the low back, but this did not persist and did not require
CONT'D.

PATIENT NAME

PATIENT NUMBER 165747721A

SIGNED

FORM 8702.02 R1/76

Figure 18-5
Consultation report. (Courtesy of Huntington Memorial Hospital, Pasadena, CA.)

DATE:

NEUROSURGICAL CONSULTATION NOTE Page 2.

any treatment.

PAST HISTORY: The patient says that she has no serious medical problems, except for obesity. She now weighs 175 lbs. At one time, she was considerably heavier.

OPERATIONS: Removal of a right ovarian cyst 2 years ago.
6 months ago, Because of her persistent abdominal pain, Dr. Lusk did a hysterectomy on her. The patient said that she had a partial small bowel obstruction and a chronic peritonitis. The abdominal pain has been relieved by this procedure. She denies any injuries, other than those already mentioned. She is single.

PHYSICAL EXAMINATION: The patient is a short and stocky 5'1" 175 lb 32 y/o Caucasian female.

She stands bearing most of her weight on the right leg. There is a tild of the lower back toward the right. She tends to keep her left hip, knee and ankle slightly flexed. When requested to do so however, she will stand on both legs but prefers to bear her weight on the right leg.

There is rather generalized left lower lumbar tenderness and some tenderness of the left medial puttock. With her leg straight, she flexes her backforcing it only a few degrees. She tolerates very little extension or lateral pending.

There is no atrophy of the legs or bone and joint abnormality noted.

The patient's reflexes in the arms are active and equal. Patellar and Achilles reflexes are also active and equal. Plantar responses are flexor. Straight leg raising is carried to about 45° on the left and 60° on the right, at which point, she gets pain in the left puttock.

There is sensory impairment, which is not terribly clearly defined, but appears to involve the 5th lumbar dermatome of the left leg. There are no similar changes to pain or touch on the right leg and none in the arms. Position sense and vibration are normal.

No clear-cut weakness was noted, although there might be slight difficulty with extension of the left toes and ankle. She stands on her toes fairly well and can stand on her heels, but seems to have some difficulty in elevating her left foot and toes off the floor. When the strength of the left extremity
CONT'D.

PATIENT NAME

PATIENT NUMBER 165747721A   247

SIGNED

FORM 8702.02 R1/76   M. D.

Figure 18-5
Continued.

DATE:

NEUROSURGICAL CONSULTATION NOTE Page 3.

is tested, the pain results and she does give way.

DISCUSSION: I think that this woman has the signs and symptoms of a ruptured and possibly extruded lumbar disc. She is admitted to this hospital for X-Rays of the lumbar spine and a brief course of physiotherapy. I think that if a few days of physiotherapy do not result in a real definite improvement, she should have a myelogram and further treatment be judged, at least in part, on the findings of that test.

```
PN:ke
CC:          M.D.
D
T-7-9-80
```

PATIENT NAME		
PATIENT NUMBER 165747721A 247	SIGNED	, M.D.
	FORM 8702.02 R1/76	

Figure 18-5
Continued.

CC:

DISCHARGE SUMMARY

DATE OF ADMISSION: 12-3-80
DATE OF DISCHARGE: 12-21-80

HISTORY: This 31 year old lady was readmitted to
Hospital because of recurrence of pain in the left lateral portion of
the leg which had become quite excruciating. Her last admission was
for a slipped lumbar disc at the L-4 level for which she had undergone
laminectomy. She had a relatively quiescent recovery period from that
problem and was well recuperated when she returned to work. Shortly
after return to work she began to complain of pain again in the left
leg. This began to crescendo and because of this Dr.   and Dr.
     felt that the patient should be admitted to
for repeat myelogram.

COURSE: Myelogram demonstrated the possibility of slipped lumbar disc
at L5 and the patient underwent a repeat laminectomy at the L5 area.
There was in fact a slipped portion of disc in that level and this was
corrected. The postoperative course was complicated only by severe
spasm in the left leg and a mild thrombophlebitis of the left calf.
This was corrected with Mini-Heparin and Coumadin and resolved itself
during the course of the hospitalization. The patient also had an
upper respiratory tract infection which was treated with antibiotics
and cleared over the course of her hospitalization.

She is discharged to the home setting on 12-21-80 and will be followed
by Dr.   and Dr.   as outpatient in the future. Condition
is improved. There was no infection present, no drug reaction, no
transfusions.

PROCEDURE: Laminectomy at L5.
                        Myelogram preoperatively.

RK:TT:ra
D: 1-5-81
T: 1-6-81
#946

DISCHARGE SUMMARY

PATIENT NUMBER 165747731-A

SIGNED   , MD

FORM 8702.02 R1/76

Figure 18-6
Discharge summary report. (Courtesy of Huntington Memorial Hospital, Pasadena, CA.)

Administrative Medical Services

```
                                        CC:   Business office
                                              MD
```

DISCHARGE SUMMARY

DATE OF ADMISSION: 4-11-81
DATE OF DISCHARGE: 4-16-81

HISTORY: This patient has had two lumbar disc operations, the first on 7-14-80, the second on 12-8-80. She was getting along reasonably well and working when she developed pain in the low back and left leg with complaints of weakness in the left leg. She said she fell on Saturday on 4-11-81 when her "leg gave out" and injured her back.

EXAMINATION: Examination reveals guarded, restricted movement of the low back with tenderness. Patellar tendon reflexes are active and equal while the Achilles reflexes were somewhat sluggish but equal. There was sensory impairment involving the entire left leg. The patient could stand on her toes and heels without apparent difficulty but testing the strength of leg caused complaints of pain and she gave way rather quickly.

X-rays of the lumbar spine show some narrowing at L4-5 disc spaces as would be expected.

Electromyogram was done by Dr.   and he reports normal right upper extremity. The left lower extremity showed evidence for acute denervative forms in the paralumbar spinous musculature, corresponding at 4 +2 and left lumbar 5 and sacral 1 and 2 to 3+ at left lumbar 3 and 4.

HOSPITAL COURSE: The patient was given physiotherapy and gradually improved. She still has some symptoms but is up and active and anxious to go home and continue her therapy on outpatient basis. She was discharged 4-16-81 and will take home Tylenol #3 to use for pain, Norflex to use 2 x day as muscle relaxant and Premarin 1.25 mgs which she takes daily. She will continue therapy daily next week and 3 x week for the following 2 weeks and will return to see Dr. Rainey in approximately 1 weeks.

PN:TT:ra
D: 4-20-81
T: 4-24-81
#524

DISCHARGE SUMMARY

PATIENT NAME

PATIENT NUMBER 165747741-A

SIGNED

FORM 8702.02 R1/76   , MD

Figure 18-6
Continued.

Review Questions

1. Give at least four examples of reference materials that might be used by the medical transcriptionist:

   a. _____
   b. _____
   c. _____
   d. _____

2. Briefly explain the purpose of a *patient history* and *physical report*.

3. Give at least five examples of what might be found on a history and physical report:

   a. _____
   b. _____
   c. _____
   d. _____
   e. _____

4. Briefly explain the purpose of a *pathology report*.

5. Briefly explain the purpose of a *radiology report*.

6. Why is it important for all physicians to complete a *discharge summary report?*

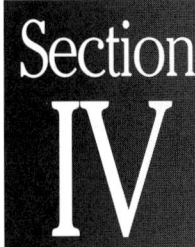

Section IV

Medical Records Clerk: Basic Concepts and Applications

19
Introduction to Medical Records

20
Processing Medical Records

21
Advanced Functions of the Medical Records Clerk

Introduction to Medical Records

Performance Objectives

Upon completion of this chapter, you will be able to:
1. Explain the function of medical records.
2. Discuss the role and identify various functions of the medical records clerk.
3. Discuss the training and career opportunities available to the medical records clerk.
4. Explain how medical records are filed and numbered.
5. Discuss the relationship of confidentiality and release of information to medical records.

Terms and Abbreviations

Abstracting the process of collecting information and data from a medical record that can later be used for planning and statistical purposes.

Analysis the process of reviewing a medical record to determine whether or not all required information, data, and documentation are present.

Assembly the process of arranging data found in the medical record in a specific order after a patient has been discharged.

Authorization generally a consent signed by the patient or his or her representative that gives a facility or a specific person the authorization to release confidential information.

Coding the assignment of numbers to specific diagnoses or procedures using a coding or classification system.

Condition of admission a signed document that specifies the legal obligations of the patient and the health care facility.

Custodian of Records a person who has been assigned responsibility for maintaining and retrieving records for a specific business.

Deficiency slip a document used to identify deficiencies or errors found in the medical record.

Discharge summary a medical report that has been dictated by the attending physician, at the end of a patient's hospitalization, which provides detailed information regarding the patient's diagnoses and treatment.

FOIA abbreviation for the *Freedom of Information Act*, which provides for the access to government records.

JCAHO abbreviation for *Joint Commission on the Accreditation of Healthcare Organizations,* which is responsible for establishing the standards of care for all health care institutions.

Medical record a document generally containing many forms and reports, which provides detailed information regarding the patient's diagnoses, tests and procedures, and treatment during his or her hospitalization.

Privacy Act of 1974 the federal act that provides citizens with regulations and information regarding confidentiality of records maintained on all individuals.

Quality assurance the process of evaluating the effectiveness of the treatments and services received while the patient is hospitalized.

Subpoena an order for a person to appear or for an item to be released, which is made by a court of law or an attorney.

Third party a person or agency other than the health care professional or the patient.

Tumor registry a listing or database of patients diagnosed with cancer.

Unit record a record containing information about and documentation of all encounters a patient has while hospitalized in a health care facility.

UR abbreviation for *Utilization Review,* which is a process used to evaluate or review the care and treatment of a patient during hospitalization.

One of the most important departments in the health care facility is the medical records department, for it is in this area that all records, reports, and documentation related to the care and treatment of hospitalized patients are kept.

Medical records are kept in all health care facilities that care for patients. These generally include large medical centers and acute care facilities, skilled nursing and extended care facilities, rehabilitation and specialized care hospitals, and ambulatory care facilities and home health agencies.

The Medical Records Team

Persons with several different levels of training and education make up the medical records staff. The individual generally in charge of this department is the Registered Record Administrator (RRA). This is a senior management person, generally with an advanced degree in business administration or public health as well as training in medical records, who is usually responsible for overseeing the entire daily workings of the department, as well as quality assurance and utilization review.

The Accredited Record Technician (ART) (Figure 19-1) usually works under the Registered Record Administrator, and is often asked to step into the supervisory position during the RRA's absence. Like the Registered Record Administrator, this person not only takes advanced college classes, but is also eligible to take a national certifying examination. In addition to supervising the medical records department, the ART may also work as an insurance coder, a tumor registrar, or an assistant medical records director.

A Medical Records Clerk may be an entry-level position or, in the case of someone with advanced training and experience, may be a position of Senior Records Clerk. As a beginning clerk, you may be responsible for tasks such as typing, filing, and basic indexing, abstracting, and coding of medical records, as well as for creating medical files and records (Figure 19-2). The senior clerk, who has additional training and has attended additional classes in medical records, usually performs some of the more advanced clerk jobs, such as assembly and analysis of medical records, interviewing patients, obtaining needed information and documentation, and securing needed data, such as physician's signatures on incomplete medical records and charts.

Figure 19-1
The accredited record technician.

Administrative Medical Services

Figure 19-2
The medical records clerk.

The Medical Records Department

The main purpose or function of the medical records department is to provide a location in which all information and documentation related to patients can be stored, kept, and, if necessary, retrieved. These records may be used for any number of reasons, including legal precedence; statistical information; substantiation of billing and collection of charges; abstracting, indexing, and coding of information; assembly and analysis of medical documentation; quality assurance and utilization review; teaching; and to provide a base-line for follow-up of patient care.

Keeping Medical Records

The primary purpose of keeping or maintaining medical records is to provide the facility or health care provider with a legal document to use as evidence of the evaluation and treatment received by a patient during his or her hospitalization or encounter with a health care facility or provider. These medical documents, or charts are a communication tool, that provides the reader with documentation regarding the patient's interaction with members of the health care team, including care, treatment, and evaluation of his or her medical condition.

Assemblying and Analyzing the Medical Record

One of the most important functions of the medical records department is the proper assembly and analysis of all medical records. Assembling the medical record means maintaining a specific order of all medical documents. Analysis of the medical record, which is often done at the same time as assembling of the record, deals with reviewing each page of the record to ensure that all entries have been properly signed by the physician and others who may have cared for the patient. During the analysis process, the clerk may note that something has been left out of a chart or that the physician has failed to sign the chart. If either is the case, a *deficiency slip* (Figure 19-3) must be filled out for each person who must complete the record.

According to the Joint Commission on Accreditation of Healthcare Organizations (JCAHO), all records must be completed within 30 days of a patient's discharge from a health care facility. If after careful analysis of the record you have noted any deficiencies, the medical chart is generally placed into the physician's incomplete area until such time as the incomplete items are corrected. Since many of the errors or deficiencies are associated with the physician, your responsibility often involves pulling out the charts and identifying the necessary corrections.

The Tumor Registry

Another important function of the medical records department is to maintain the hospital's tumor registry. The registry is a unit within the medical records department which is responsible for collecting information and documentation on all patients who have been diagnosed with and treated for cancer. All tumors are categorized according to their shape or morphology, cell type, location in the body, and reaction to treatment. If the hospital or health care facility runs a residency program for surgeons, maintenance of the registry is required by the American College of Surgeons.

According to law, all facilities are required to track at least 90 percent of their patients with cancer until death. Therefore, the registry is also responsible for obtaining follow-up information on an annual basis for these patients. To perform this task, patients and their family members are tracked according to care received and any recurrance of tumors until the patient dies.

Medical Record No. _____

MEDICAL RECORD DEFICIENCY SLIP

	Dictated	Signed	Date
Patient Registration Form	_____	_____	_____
History & Physical Form	_____	_____	_____
Progress Notes	_____	_____	_____
Physician's Orders	_____	_____	_____
Surgery Report	_____	_____	_____
Discharge Summary	_____	_____	_____

Other:

_____ _____ _____ _____
_____ _____ _____ _____

Patient's Name   Discharge Date

Attending Physician

Figure 19-3
Deficiency slip.

Administrative Medical Services

Utilization Review and Quality Assurance

Many health care facilities charge their medical records department with the responsibility of overseeing the process of utilization review and quality assurance. Utilization review has to do with determining whether or not a health care facility uses its resources in a manner deemed necessary and appropriate. This involves examining a patient's medical record while he or she is still confined in the hospital. The chart is reviewed to determine if prescribed treatment and care was appropriate for the patient's condition and would meet the standard of care that is usually provided to members of the community in which the patient was hospitalized. The utilization review process is necessary to protect the health care facility and the physician from any potential loss of revenue, and the patient from receiving any unnecessary or experimental care or treatment.

Providing a quality assurance program is a requirement of all hospitals accredited by JCAHO. The goal of any good quality assurance program is to make sure that the facility and its physicians maintain a certain standard of care and provide only those services that meet the standards and needs of the community in which the hospital, physician, and patient are located. While quality assurance records are often maintained by the medical records department, the actual process is usually undertaken by the medical staff, with the review and compilation of the actual medical information and data completed by hospital personnel.

Working in the Medical Records Department

Setting Up a Filing System

One of the major functions of the medical records clerk is to set up and complete the medical records files and patient's charts. This process is undertaken when the patient is admitted to a health care facility and completed when the patient is discharged from the facility. Before undertaking such a task, however, you must first have a basic understanding of how these files, or charts, are set up. This generally involves establishing a filing system.

Most medical records filing systems are set up in one of three ways: alphabetically, numerically, or by a specific color in a color coding system. The most popular method is the alphabetical method. In this system, all filing is done alphabetically by the patient's name.

A second method utilized by many health care facilities is numerical filing. In this system, all patient's medical records are assigned a specific number, usually beginning with "01" or "00."

There are several different types of numerical systems that can be implemented by a health care facility. The most basic of these is called *straight* numerical filing, according to which each record is assigned a number, usually beginning with the number "01" and continuing in numerical order. Records are filed sequentially depending upon the number they have been assigned.

Another type of numerical system is the *terminal digit* filing system, according to which, all files are divided into individual or "primary" sections, usually numbering 100, starting with 00 and ending with 99. To file the record, you must first go to the file section that corresponds to the primary digits in the patient's number. In that section, you would then look for a subsection which matches the secondary digits found in the number. The record is filed numerically according to where the number falls within the sequence of numbers, that is, according to its tertiary digits. For example, if a patient's medical file number is 07-16-49, 07 is defined at the tertiary number; 16 is defined as the secondary number; and 49 is defined as the primary number.

A third type of numerical filing employed by some facilities is called *middle digit filing*. In this method, the same six-digit numbers used in terminal digit filing are used. The major difference is that, in middle digit filing, the primary, secondary, and tertiary positions change. The middle pair of digits is the primary number, the last pair of digits is the tertiary number, and the first pair of digits is the secondary number. For example, if we consider the file number we used to demonstrate the terminal filing system example, the sequence changes to the following: 07 is the secondary number, 16 is the primary number; and 49 is the tertiary number. A major advantage of both the terminal digit and middle digit filing methods is that each provides the medical records department with an evenly distributed breakdown of records.

Two lesser used numerical filing systems are *unit numbering* filing and *serial number* filing. In unit number filing, all patients are assigned a medical record number on the very first encounter or visit to the facility. With any additional visits or hospitalizations, the patient is identified by the same medical record number. In serial numbering, which is not often used, the patient is assigned a completely different medical record number for every new visit or hospitalization. A major disadvantage of this system is that records for the same patient may eventually be stored in different locations throughout the facility.

One way in which many health care facilities are able to reduce the number of files that are misfiled and thus increase the accuracy and speed with which a file can be retrieved is to use a *color coding* system for all medical records. There are two standard color coding systems: the *Ames Color Files System* and the *Smead Corporation Color File System*. While many facilities choose to establish and use their own color coding systems, for those agencies that use the *Ames Color File* system, the numbers on the patient's medical record are color coded according to the following numerical and color breakdown:

- 0 *red*
- 1 *gray*
- 3 *orange*
- 4 *purple*
- 6 *yellow*
- 7 *brown*
- 9 *green*
- 2 *blue*
- 5 *black*
- 8 *pink*

If the facility in which you are employed uses the *Smead Corporation* method, the numbers will be broken down according to the following color:

- 0 *yellow*
- 4 *orange*
- 3 *purple*
- 7 *gray*
- 6 *green*
- 2 *pink*
- 9 *black*
- 5 *brown*
- 1 *blue*
- 8 *red*

Preparing the Records for Filing

Now that you have a fundamental understanding of the types of filing systems used by most medical offices and health care facilities, we will discuss some of the other "tools" used in the filing of medical records. When we use the word *tools*, we are referring to the supplies and equipment necessary to complete the task of filing. These generally include outguides, which are used by some departments to indicate that a specific medical record has been removed from the files, file folders, which are holders for placing entire medical records and other important documents; and fasteners and hole-punching equipment, for working with or filing loose documents.

Medical Records and the Law

In 1974, the federal government passed and enacted legislation setting forth regulations regarding the release of government records maintained on all individuals, including all medical records. This legislation, known as the *Privacy Act of 1974,* clearly states that, before any information, data, or documentation regarding a patient's health, care, treatment, progress, or discharge while under the care of a physician or confined in a health care facility may be released, the patient must give his or her written consent. The one exception is the dates of his or her treatment.

Like all other administrative and clinical professionals working in the health care industry, as a member of the medical records department, you must remember that you have a responsibility to maintain the confidentiality and privacy of all patients' records, charts, and documents encountered during your employment. This means never discussing or providing information or data found in the medical chart unless the patient has signed a *release of information* form (Figure 19-4) or the information has been subpoenaed by the courts, a law enforcement agency, an insurance company, or attorney's office. Medical records that have been requested by physicians, members of the health care team, and other health care facilities or hospitals may be released so long as proper documentation and release forms have been completed by the patient.

Allowing Patients Access to Medical Records

If you have ever been under a doctor's care or have been a patient in a hospital, you might remember asking a nurse or another health care worker about your care or about what was wrong with you. Well, if you asked before the mid-1980s, you might remember that these people were not allowed to tell you about your care or answer your questions, but rather referred you to your doctor. This was, and in some cases still is, because health care providers have always been told, or even trained, never to discuss the patient's care or treatment with anyone, other than the health care team, and especially not with the patient or his or her family and friends. The premise for this was that, if we were to tell the patient, for example, that his or her temperature or blood pressure was high, we might agitate or do further damage to the patient.

Back in the 1980s, not all health care professionals or government agencies believed that providing patients with information regarding their health care would be dangerous. The federal government, therefore, passed legislation allowing individual states, at their own discretion, to permit access to medical records by patients in federal hospitals. Under the guidelines of the *Freedom of Information Act (FOIA)*, the law also provided for the patient's right to amend certain information found in his or her medical record. So long as the patient's doctor agreed to such an amendment, the portion amended would be sent to all previous individuals or organizations that had requested the record. However, if the physician did not approve the amendment, according to the FOIA, the patient would be required to submit a statement

Administrative Medical Services 151

NAME OF FACILITY

Consent for Release of Information

DATE _____

1. I hereby authorize _____ to release the following information
   Name of Institution
   from the health record(s) of

   _____
   Patient Name

   _____
   Address
   covering the period(s) of hospitalization from:
   Date of Admission _____
   Date of Discharge _____
   Hospital # _____ Birthdate _____

2. Information to be released:
   ☐ Copy of (complete) health record(s)   ☐ Discharge Summary
   ☐ History and Physical   ☐ Operative Report
   ☐ Other _____

3. Information is to be released to _____

4. Purpose of disclosure _____

5. I understand this consent can be revoked at any time except to the extent that disclosure made in good faith has already occurred in reliance on this consent.

6. Specification of the date, event, or condition upon which this consent expires.
   _____
   _____

7. The facility, its employees and officers and attending physician are released from legal responsibility or liability for the release of the above information to the extent indicated and authorized herein.

   Signed _____
   (Patient or Representative)

   _____
   (Relationship to Patient)

   _____
   (Date of Signature)

Figure 19-4
Release of information form. (Reprinted with permission from Diehl, M. O. and Fordney, M. T. *Medical Typing and Techniques and Procedures,* 3rd ed. Philadelphia: W. B. Saunders, 1991, p. 16. Original figure © the American Medical Record Association [AMRA], Chicago, Illinois.)

regarding the information or data in question. Such a statement then would becomes part of the record and be sent with the rest of the released record.

Releasing Information by Telephone, by Mail, and Under Subpoena

So long as the patient has signed a release of information form, certain information may be released to patients, health care facilities, physicians, insurance companies, attorneys, and any other persons or businesses, by the telephone or mail and under subpoena. So long as state and federal guidelines are followed, the type of information that may be released is generally determined by the individual facility or practitioner. Among the more frequently released data are such items as the patient's name, address, date of birth, and sex; verification of dates of treatment or hospital confinement; and general medical or health condition.

In cases in which it has been requested that a patient's medical records be mailed to another facility, business, or individual, you should never, unless subpoenaed by law, send the original records. If at all possible, always send copies of the records, and, most important, never send the records unless the request is accompanied by a valid consent or authorization. Once you have determined the legitimacy of the request, and before you actually mail off the records, make sure you are sending the appropriate information, by following the procedure outlined below:

- Check for valid authorization.
- Place all documents in alphabetical order.
- Check the patient's medical record number.
- Decide if all records being requested actually exist.
- Make sure all records being sent have the correct patient medical record number.
- Pull out only those records that have been requested.
- Compare the patient's signature against his or her signature on the consent for release of information or authorization form.
- Copy only those records requested.
- Determine if the person or business requesting the records should be billed and do so if appropriate.
- Mail the requested copied records.

If cases in which you receive a request to release certain information, documentation, or medical records under a subpoena, you must proceed in a timely manner. However, before you accept the subpoena and release any information, you must make sure that the subpoena has not been made out to the custodian of records with a specific facility's name and that it has not been made out to a specific person, such as a doctor or hospital employee. You should also check to make sure that the date on which the records or custodian of records is scheduled to appear is clearly stated, and that the patient's name and dates of hospitalization or treatment have been specified.

Once you have checked to make sure all information has been clearly stated on the subpoena, you are ready to begin processing it. First, you should check to make sure that the records being requested do in fact exist. If they do, you must check the patient's medical record number against the subpoena. Once you have determined that the records exist and that the medical record number is correct, you may pull the records requested, copy them, complete the affidavit attached to the subpoena, and prepare an envelope for sending the records using certified mail. After you have mailed the records, attach the receipt you received for sending the copies by certified mail to the subpoena and file both in a predesignated location.

Summary

In this chapter, we discussed the function and purpose of the medical records department. We also discussed the various functions of each member of the medical records staff, as well as the training and career opportunities available to the medical records clerk. We also talked about the different methods and systems used in filing medical records. Finally, we discussed the confidentiality and release of information, and the relationship of each to medical records.

Review Questions

1. Briefly explain the function of the *medical records clerk*.
2. Define the following terms as each relates to medical records:

   a. analysis _____
   b. assembly _____
   c. abstracting _____

Administrative Medical Services

3. Define the following abbreviations:

   a. UR _____

   b. JCAHO _____

   c. FOIA _____

4. Briefly explain the concept of *quality assurance* and *utilization review*.

5. Briefly explain what a *tumor registry* is.

6. A(n) _____ ___ _____ is a person who has been assigned responsibility for maintaining and retrieving records for a specific business.

7. What is the name of a signed document that specifies the legal obligations of a patient and the health care facility?

8. Briefly explain the process by which medical records are filed and numbered.

9. Briefly explain the relationship between confidentiality and the release of information as each relates to medical records.

10. What is the name of the federal act that provides citizens with specific regulations and information regarding the confidentiality of records maintained on all individuals?

20

Processing Medical Records

Performance Objectives

Upon completion of this chapter, you will be able to:

1. Identify the three stages involved in processing medical records.
2. Identify and discuss the individual components of a medical record.
3. Discuss the role regulatory agencies play in medical records.
4. Identify the various reports found in a patient's medical record.
5. Define the function of a master patient index in medical records processing.
6. Explain the process involved in processing and discharging as each relates to processing medical records.
7. Discuss the purpose of analyzing a medical record after a patient has been discharged.

Terms and Abbreviations

Analysis   the process of reviewing a medical record to determine whether or not all required documentation is present.
Attending physician   the physician who is responsible for overseeing the total care of a patient during his or her hospitalization.
CPT   abbreviation for *Current Procedural Terminology*, which is a reference book used in coding procedures.
DRG   abbreviation for *Diagnosis Related Group(ings)*, which refers to a group of diagnoses requiring similar treatment.
ICD-9-CM   abbreviation for *International Classification of Diseases, Ninth Edition, Clinical Modification*, which is a reference book used to code diagnoses and procedures.

NCR   a form used to make duplicate copies without the use of carbon paper.
Physician privilages   those services to which a doctor is entitled by a health care facility.
Principal diagnosis   the diagnosis that proves to be the reason for a patient's admission into a hospital.
Record retrieval   the process of identifying the location of a medical record and then pulling that record from the files.
Suspension   a penalty imposed on a physician for having deliquent medical records, according to which his or her privileges have been temporarily restricted.

The processing of medical records by the medical records department involves three separate but related steps. These include the actual processing of the record, which involves working with a master patient index and processing hospital admissions and discharges; assembling the record, that is, putting it in a specific order as outlined by a specific health care facility, and according to JCAHO and other regulatory agencies; and finally, analyzing the medical record, which involves those steps and procedures used to review a record for completeness.

Processing the Medical Record

If you are working as a medical record clerk, there is a very good possibility that at some point you will be responsible for processing patients' medical records.

Administrative Medical Services

If you are required to do so, one of the things you will have to become familiar with is how to work with a *master patient index*, or *MPI*. The MPI is a type of record kept by all hospitals that provides a listing and documentation on all patients who are being treated in the facility. It also provides data on the number of visits and hospitalizations for individual patients.

Medical patient indexes can be kept manually or on a computer. Manual records are generally kept on 3 × 5 or 5 × 8 index cards, which are created at the time the patient is admitted into the hospital for the first time. Information kept on the manual MPI usually includes the patient's name, address, and telephone number; his or her date of birth; the name of the patient's primary or attending physician; and the patient's medical record number. Additional information kept on the MPI includes the date(s) of admission(s) to the facility, as well as the date(s) of discharge(s) (Figure 20-1).

If your facility uses some type of computer system, there is a good possibility that the MPI will also be computerized. The major advantages of using a computerized MPI are that it is capable of keeping much more information, it is easier and faster to use, and the medical records department will no longer have to maintain a manual indexing system on each and every patient who enters and is discharged from the facility.

Whether you are using a manual or a computerized MPI system, there may be instances when an error is discovered. Errors often occur in cases in which more than one patient has the same name or a similar date of birth or medical record number. To assure the accuracy of each card, make sure that you check each individual card very carefully. One way to avoid errors is to double-check the patient's date of birth against his or her medical record. If two patients have the same last name, check for a middle initial. Remember, the more patients a facility has, the greater chance there is for errors to occur.

Medical Record No. 4695-SJM-0
NAME: JONES, SAM ADDRESS: 123 Smithtown St.
CITY/STATE: Anywhere, PA 00000
TELEPHONE: (215) 555-5555 DATE OF BIRTH: 7-6-49
ATTENDING   DR: Dr. J.J. Smith

Date of Admission:	3-29-86	Discharge Date: 4-5-86
Date of Admission:	8-6-93	Discharge Date: 8-9-93
Date of Admission:	4-1-95	Discharge Date: 4-6-95

Figure 20-1
Manual master patient index card.

Processing Admissions and Discharges

Two of the most important procedures for which the medical records department is generally responsible are the completion and processing of medical records when the patient is admitted into the hospital and when he or she is discharged.

Processing Admissions

For most health care facilities, the procedure for generating and completing the admissions procedure by the medical records department is quite simple, and often takes little time to complete. It does, however, require accuracy and completeness, since any errors committed during this process can create inaccuracy and problems throughout the patient's hospitalization. If processing the medical record at the onset of the patient's admission is part of your responsibility, you may want to follow the steps outlined below:

- Check the admissions procedure manual for your facility to make sure you follow the correct steps.
- Before beginning the actual processing of the records, pick up copies of the patient's medical record, including the patient registration forms, from the admitting department.
- Check forms against the previous day's admissions and registrations to verify that you have received all of the registration forms.
- Verify the patient's name, address, telephone number, date of birth, attending physician, and medical record by checking the information against the MPI.
- Update the MPI by typing the new date of admission onto the card or computerized screen.
- Create a file folder for each patient being admitted.
- File all medical record folders according to the filing system your department uses.
- Depending on the system your department uses, file all separate or loose forms, such as registration data sheets and the MPI card, if manual index cards are used, in the appropriate manner.
- If required by your facility or the patient's attending physician, retrieve all previous patient records and deliver them to the nursing station on the ward that will be caring for the patient.

Processing Discharges

Processing the medical record at the time of the patient's discharge from your facility is just as important as processing his or her admission. It requires accuracy and completeness, and is often the responsibility of the medical records clerk. To complete the discharge process correctly, you should:

- Check the procedures manual your facility uses for processing the medical record once a patient has been discharged.
- Make sure you have the correct number of records and patients' names by verifying the number against the discharge list from the facility's admitting department or against registration data forms.
- Retrieve all the registration data forms, the MPI cards, and file folders from their location within the hospital, for example, nursing units.
- Stamp the discharge date on the registration data form and the MPI card.
- After you have placed each patient's medical record into an individual file folder, count the number of folders and registration data forms and verify them against all the records which have been picked up or retrieved.
- File or attach all loose or separate forms, such as the registration data card and the MPI card, in separate folders in the appropriate manner, according to the procedure your facility uses.
- File all records according to the filing procedure your facility uses.

Retrieving Medical Records

One of the tasks involved in processing medical records at admission and discharge is the retrieval of previous medical records and files. Because most health care facilities have specific regulations or policies regarding who may request previous information on a specific patient, it is extremely important that, before you begin retrieving medical records, you check with your supervisor or the department's employee manual regarding by whom and under what circumstances records may be retrieved.

Most health care facilities employ the use of a requisition to retrieve a patient's previous medical record (Figure 20-2). Such a form generally includes a space for the requester to write the patient's name,

---

REQUEST FOR MEDICAL RECORD
MEDICAL RECORD NO. _____
PATIENT: _____
DATE REQUESTED: _____ TIME: AM_____ PM_____
ASSIGNED TO: _____
DEP'T./EXTENSION: _____

Figure 20-2
Requisition form for retrieving a medical record.

his or her medical record number, the date and time the records are being requested, the person or department that is requesting the record, and where the record is to be sent. Once you have identified all this information, you may pull the records.

Assembling the Medical Record

As we have already discussed, the second step in processing the patient's medical record deals with the actual assembling of all the forms contained in the record. Usually, when the patient has been discharged from the hospital, all of the forms making up the medical record must be removed from the binder or holder in which they have been kept and brought to the medical records department. Once the medical records department receives the records, which are often still in the order used by the nursing unit, they must be rearranged in the order the facility uses for storing records. While the order may differ according to each individual hospital, the forms to be assembled are often the same for all facilities.

The forms contained within a patient's medical record generally include a *face sheet* or *patient registration data form;* a *conditions of admission* form; an *admission* or *nursing assessment admission summary;* a *discharge summary sheet;* a *patient transfer* form, if the patient was transferred to another facility; an *emergency room record,* if the patient was admitted into the hospital from the emergency room; a *history and physical* form; a *consultation report(s),* if the patient was seen by physicians consulted by his or her primary doctor; *laboratory reports;* and *radiology reports; respiratory therapy* and *physical therapy reports,* if the patient received any treatments from these two departments (Figure 20-3A–J).

Additional forms generally found in the patient's medical record include *nurses notes; graphics sheets;* a *preoperative nursing assessment; consent for surgery; anesthesiology record; pathology report; surgical report,* if the patient underwent any surgery during his or her hospitalization; *physician's progress notes; physician's orders* form, which may or may not be preprinted or standardized orders; *medication records,* and a *record of death* and/or an *autopsy report,* if the patient died during his or her hospitalization (Figures 20-4A–H).

Part of the process involved in assembling the medical record is making sure that each individual form or report has been carefully checked for accuracy, completion, and, when necessary, authorization or signature from the appropriate person before the record is disassembled and stored. Steps often taken at the time the record is assembled include:

Administrative Medical Services

1. Ensuring that all medical records have been properly "signed off" by the patient's admitting or attending physician.
2. Verifying that the record is complete and documents the patient's history and physical report.
3. Making sure that all consents for treatment, specialized procedures, and surgeries have been properly signed and dated by the appropriate physician(s) and technicians.
4. Making sure that all medical records were checked for timeliness and frequency of notation by the patient's doctor, as noted in the physician's progress notes and nurse's notes, both of which must have been written out in long hand.
5. Ensuring that the medical record has been properly assembled by making sure all forms and reports are placed in chronological order before the completed medical record is stored. This step is always performed last.

Analysis of the Medical Record

The final step involved in processing the medical record, as we previously stated, is checking the record to make sure it has been properly analyzed for documentation, to ensure it meets the criteria and legal requirements set forth by the Joint Commission on the Accreditation of Healthcare Organizations (JCAHO) and other regulatory bodies, such as the hospital medical staff. According to JCAHO, all hospitals that have been accredited under their regulations are granted up to 30 days from the time a patient has been discharged to complete the medical record. Most hospitals agree, however, that if a record has not been completed by the fourteenth day after the patient's discharge, the record is delinquent, which puts the attending physician's hospital privileges in jeopardy of being suspended or even terminated.

To avoid the possibility of the physician losing his or her privileges to admit and treat patients in a hospital, most medical records departments set up a standard policy that provides for notification of a physician that his or her chart has been assigned delinquency status. At this time, the physician is also notified regarding how long and on what date the chart must be cleared of any descrepincies or delinquencies to ensure he or she may maintain hospital privileges.

Performing the Actual Analysis

Analyzing a medical record means that each and every report, form, and part of the patient's medical record or chart must be individually analyzed and checked for any missing documentation or signatures. The process also involves checking each page to make sure that it contains the patient's full name, medical record number, and the date on which a specific documentation was made.

While the person performing the analyzis of the record is responsible for checking each and every report found in the chart, there are a few reports which are often checked more closely and express for completeness within a specific time frame. These may include the history and physical forms, consultation reports, operative and anesthesia reports, physician's orders and progress notes, nurse's notes, consultation reports, and test results reports. For example, JCAHO requires that the admitting physician write or dictate the patient's history and physical (H&P) report within 24 hours after the patient has been admitted to the hospital. The best, and often the easiest, way to check that this has been done is to look at the last page of the H&P and make sure that the date entered or dictated has been recorded. If more than one day has elapsed from the date of admission, it will be apparent that the doctor did not complete the history and physical.

Another report that must be dictated or written by the physician within 24 hours from the time it occurred, is the operative or procedure report. As with the history and physical form, you must first check the date of the report against the actual date of the surgery or procedure. All operative reports must be checked to make sure they contain both a preoperative and postoperative diagnosis, the reason why the surgery or procedure was performed or indicated, a description of the surgery or procedure performed, and the findings at the time of the procedure.

Analyzing the Physician's Orders, Progress Notes, and Nurse's Notes

According to JCAHO regulations, all physician's orders must be signed by the physician giving the order. This is extremely important when the orders were given to the nurse verbally, either by telephone or in person. Most hospitals require that, whenever a nurse takes a doctor's order over the telephone, the physician must sign the orders within 24 hours after the order has been issued.

When you are checking the physician's orders, you must also check to make sure that the physician's progress notes, including those made by the admitting physician at the time of the patient's admission to the hospital, have been properly signed, even if the patient only stayed for a few hours.

The Joint Commission on the Accreditation of Healthcare Organizations also requires that any

(text continues on page 180)

1. **NURSING CARE:** This hospital provides only general nursing care unless the patient's physician orders more intensive nursing care. If the patient's condition is such as to need the service of a special duty nurse, it is agreed that such must be arranged by the patient or his/her legal representative. The hospital shall in no way be responsible for failure to provide the same and is hereby released from any and all liability arising from the fact that said patient is not provided with such additional care.

2. **MEDICAL AND SURGICAL CONSENT:** The patient is under the care and supervision of his/her attending physician, and it is the responsibility of the hospital and its nursing staff to carry out the instructions of such physician. The undersigned recognizes that all physicians and surgeons furnishing service to the patient, including the radiologist, pathologist, anesthesiologist, emergency room physician, and the like, are independent contractors and are not employees or agents of the hospital. The undersigned hereby consents to X-ray examination, laboratory procedures, anesthesia, emergency treatment, medical or surgical treatment, or hospital services rendered the patient under the general and special instructions of the physician.

   INITIALS ____ DATE ____

3. **RELEASE OF INFORMATION:** To the extent necessary to determine liability for payment and to obtain reimbursement, *the hospital or attending physicians may disclose portions of the patient's record,* including his/her medical records, to any person or corporation which is or may be liable, for all or any portion of the hospital's charge, including but not limited to, insurance companies, health care service plans, or workers' compensation carriers.

   Special permission is needed to release this information where the patient is being treated for alcohol or drug abuse.

4. **PERSONAL VALUABLES:** It is understood and agreed that the hospital maintains a safe to protect the patient's personal property, money and valuables. The hospital shall not be liable for any loss or damage to the patient's personal property, money or valuables unless those items have been deposited within the hospital safe.

   INITIALS ____ DATE ____

5. **SAFE ENVIRONMENT FOR PATIENT CARE:** Weapons or other dangerous objects, illegal drugs, and drugs not prescribed by the patient's physician are not permitted in the patient's room. The Medical Center's obligation to provide a safe environment for patient care must override the patient's right to privacy. The Medical Center reserves the right to search the patient and room and to confiscate such objects upon reasonable probable cause.

6. **FINANCIAL AGREEMENT:** The undersigned agrees, whether he/she signs as agent or patient, that in consideration of the services to be rendered to the patient, he/she hereby individually obligates himself/herself to pay the account of the hospital in accordance with the regular rates and terms of the hospital and/or as set forth by the terms of managed care contracts entered into by _____ and/or applicable Workers' Compensation regulations. Should the account be referred to an attorney for collection, the undersigned shall pay actual attorney's fees and collection expense. All delinquent accounts shall bear interest at the legal rate. Emergency room patients will be billed separately by the emergency room physician and all terms and conditions of this paragraph apply to emergency room bills.

7. **ASSIGNMENT OF INSURANCE BENEFITS:** The undersigned authorizes, whether he/she signs as agent or patient, direct payment to the hospital of any insurance benefits otherwise payable to the undersigned for this hospitalization at a rate not to exceed the hospital's regular charges. It is agreed that payment to the hospital, pursuant to this authorization, by an insurance company shall discharge said insurance company of any and all obligations under a policy to the extent of such payment. It is understood by the undersigned that he/she is financially responsible for charges not covered by this assignment. The terms and conditions above also apply to emergency room treatment which does not require hospital admission.

8. **MEDICARE INSURANCE BENEFITS AND EXCLUSIONS:** I certify that the information given by me in applying for payment under Title XVIII of the Social Security Act is correct. I authorize any holder of medical or other information about me to release to the Social Security Administration or its intermediaries or carriers any information needed for this or a related Medicare claim. I request that payment of authorized benefits be made on my behalf. Some services may not be covered by Medicare, such as the following: 1) Worker's Compensation, 2) Dental, 3) Cosmetic Surgery, 4) Custodial Care, 5) Personal Comfort items, and any service determined to be unnecessary or unreasonable by Medicare.

9. **HEALTH CARE SERVICE PLANS:** This hospital maintains a list of health care service plans with which it has contracted. A list of such plans is available upon request from the financial office. The hospital and/or emergency department has no contract, express or implied, with any plan that does not appear on the list. The undersigned agrees that he/she is individually obligated to pay the full cost of all services rendered to him/her by the hospital if he/she belongs to a plan which does not appear on the above mentioned list.

10. **CONSENT TO PHOTOGRAPH:** The taking of pictures of medical or surgical progress and the use of same for scientific, education, or research purposes is approved.

The undersigned certifies that he/she has read the foregoing, receiving a copy thereof, and is the patient, or is duly authorized by the patient as patient's general agent to execute the above and accept its terms.

CONDITIONS OF TREATMENT AND/OR ADMISSION

DISTRIBUTION: ORIGINAL – Medical Records
CANARY – Business Office
PINK – Patient

60-0208-R0594/8560

Figure 20-3
Examples of forms used in medical records.

Administrative Medical Services

Huntington Memorial Hospital
ADMISSION INFORMATION SHEET

DATE: _____

TIME: _____

(ADDRESSOGRAPH)

INFORMATION OBTAINED FROM: _____

ADMITTED FROM: ☐ Home ☐ ECF ☐ ER ☐ Doctor's Office ☐ Dispensary
☐ Other _____

VIA: ☐ Wheelchair ☐ Guerney ☐ Ambulance ☐ Ambulatory ☐ Carried

ACCOMPANIED BY: _____ Identification Band Checked: ☐

REASON FOR HOSPITALIZATION: _____

_____

_____

_____

_____

_____

_____

VALUABLES

☐ None

Dentures: ☐ Upper ☐ Partials ☐ Cane ☐ Glasses ☐ Electrical Appliance: _____
☐ Lower ☐ Other ☐ Crutches ☐ Contacts _____
☐ Denture cup at bedside ☐ Walker ☐ Hearing Aid
☐ W/C ☐ Prosthesis: _____

Check here if no other valuables: ☐

Other Valuables: Description | Disposition

_____ | _____
_____ | _____
_____ | _____
_____ | _____

ORIENTATION TO: ☐ Call Lights ☐ Bed ☐ Side Rails ☐ Bathroom ☐ Room Lights
☐ Phone ☐ Meals ☐ Visiting Hours ☐ TV & Educational Channels

☐ Unable to orient patient. Explain: _____

VITAL STATISTICS	Height	Weight	T	P	R	BP	
	☐ By Report	☐ Bed scale				R	L

ADMITTED BY: _____
Signature and Title

Figure 20-3
Continued. (Figure courtesy of Huntington Memorial Hospital, Pasadena, CA.)

Huntington Memorial Hospital
Nursing Admission Assessment
Adult Data Base

Date: _____ Time: _____

(Addressograph)

Vital Statistics ☐ From Admitting Information Form	Height ☐ By report	Weight ☐ Bed scale	T	P	R	BP
						R L

Admitting Diagnosis:

Information given by: ☐ Family member ☐ Friend ☐ Patient ☐ Other: _____
☐ Unable to obtain interview/subjective data. Explain: _____

Pertinent Medical/Surgical History:

History of blood transfusion: ☐ No ☐ Yes - Reaction: ☐ No ☐ Yes: _____

Health Maintenance/Risk Factors:
Use of tobacco: How much?_____ How long?_____ No (never)____ No (Quit)____ When: _____
Use of alcohol: How much?_____ How long?_____ No ____
At risk for falls: No ____ Yes ____ Explain: _____
At risk for skin breakdown: No ____ Yes ____ Explain: _____

Signs/Symptoms, Events Leading to Present Hospitalization:

Prescription/ Non-Prescription Medications

Name	Dosage & Frequency Prescribed	Time of Last Dose	Comments

Disposition of medication: ☐ Sent home ☐ To pharmacy with form ☐ Copy of form in chart ☐ No meds brought in

Sensitivities/ Allergies   ☐ No known allergies/sensitivities

Substance	Effect

Health Perception and Health Management

Patient lives with: _____ ☐ Lives alone
In emergency, contact person (name, phone): _____
Occupation: _____ Diversional activities: _____
Would you like chaplain to visit? ☐ No ☐ Yes (if yes, notify patient services) Religious affiliation: _____
Concerns about how your illness affects you? ☐ No ☐ Yes, explain: _____

Psychosocial

Vision: ☐ No difficulty ☐ Impaired _____   ☐ Glasses ☐ Contacts
Hearing: ☐ No difficulty ☐ Impaired _____   ☐ Uses hearing aid
Communication: ☐ Speaks English ☐ Other language _____
Speech: ☐ Clear ☐ Impaired _____   ☐ No verbal response
Orientation: ☐ Person ☐ Place ☐ Time   LOC: ☐ Awake ☐ Restless ☐ Drowsy ☐ Responds to pain ☐ No response
Discomfort/Pain: ☐ No ☐ Yes, where? _____   Severity |—|—|—|—|—|—|—|—|—|—|
                                                                         0   5   10
                                                                       None   Severe
How do you manage your pain? _____

Cognitive/Perceptual

N - 42, 2/91   **This form must be completed by an RN.**   mc

Figure 20-3
Continued. (Figure courtesy of Huntington Memorial Hospital, Pasadena, CA.)

Administrative Medical Services

Sleep/Rest
Sleep Routine _____
Sleeping aids: ❏ No ❏ Yes: _____
Sleep problems: ❏ None ❏ Trouble falling asleep ❏ Staying asleep ❏ Early a.m. waking ❏ Other: _____

Activity/Exercise
Activities of daily living I = Independent A = Assist D = Dependent
___ Feeding ___ Bathing ___ Dressing ___ Toileting ___ Grooming ___ Ambulation
Assistive devices: ❏ Cane ❏ Walker ❏ Wheelchair ❏ Other: _____
Balance and gait: ❏ Steady ❏ Unsteady ❏ Non-ambulatory ❏ Other: _____
ROM: ❏ Full ❏ Other: _____
Extremity strength: RUE: _____ RLE: _____
                   LUE: _____ LLE: _____

Ventilation
Short of breath: ❏ No ❏ Yes, when?: _____
Uses accessory muscles: ❏ No ❏ Yes: _____
Cough: ❏ No ❏ Yes → Productive: ❏ No ❏ Yes → Sputum appearance: _____
Lung sounds: _____

Circulation
Apical pulse: _____ Pacemaker: ❏ No ❏ Yes: _____
Edema: ❏ No ❏ Yes, describe: _____
Peripheral pulses: _____

Nutrition/Metabolic
Special diet: ❏ No ❏ Yes: _____ ❏ Nausea ❏ Vomiting ❏ Difficulty swallowing ❏ Recent weight changes
Abdominal assessment: _____
Skin → Color/temp.: _____
Skin integrity: ❏ Intact ❏ Other: _____
❏ Skin Care Record initiated ❏ Admission picture taken

Elimination
Bladder elimination: ❏ No problem ❏ Incontinent ❏ Frequency ❏ Nocturia ❏ Dysuria ❏ Other: _____
❏ Urinary catheter: _____
Bowel elimination: ❏ No problem ❏ Incontinence ❏ Diarrhea ❏ Constipation ❏ Other: _____
Last B.M.: _____ Usual B.M. frequency: _____ Bowel aids: ❏ No ❏ Yes: _____
❏ Ostomy: _____

Reproductive
❏ Not applicable (male patient) Last menstrual period: Date _____ ❏ Post-menopause
Obstetrical data: ❏ Not applicable (non-obstetrical patients) Gravida _____ Para _____ Abortion _____
Estimated due date: _____ Weeks of gestation: _____ Previous C-section: ❏ No ❏ Yes
Uterine contractions: ❏ None ❏ Yes, description: _____
Membranes: ❏ Intact ❏ Ruptured, when: _____ Description of fluid: _____
Fetal activity: ❏ None ❏ Present Fetal heart rate: _____ Post-delivery: Delivery date: _____ ❏ Vaginal ❏ C-section

Learning Needs/Discharge Planning
Anticipated LOS: [_____] Days Discharge destination: ❏ Home ❏ Rehabilitation ❏ Convalescent care ❏ Other: _____
Self-care knowledge deficits: ❏ No ❏ Yes, describe: _____
Priority learning needs: _____
  None identified. Explain: _____
Family/significant other to include in teaching: ❏ None ❏ Yes, who: _____
Anticipated equipment needs for home: ❏ No ❏ Yes, explain: _____
Anticipated referrals: ❏ No ❏ Yes → Patient Services ❏ No ❏ Yes ❏ Other: _____
Anticipated follow-up after D/C: ❏ No ❏ Yes → ❏ VNA ❏ Dispensary ❏ Other: _____

R.N. Signature _____ Later subjective/interview data base additions: Date added data & initial.

Initials	Signature	Initials	Signature	Initials	Signature

Figure 20-3
Continued.

Huntington Memorial Hospital
Skin Integrity Risk Factor Assessment Tool

All patients that are immobile or who require assistance with repositioning, will be assessed *every 72 hours*. Indicate the appropriate number for the six categories in the right hand column. If the total score is 16 or less, implement the prevention protocol.

Write date in box above column. Indicate appropriate numbers in boxes below.

Category	1	2	3	4		
SENSORY PERCEPTION Ability to respond meaningfully to pressure-related discomfort.	**1. Completely limited:** Unresponsive (does not moan, flinch or grasp) to painful stimuli, due to diminished level of consciousness or sedation. Or limited ability to feel pain over most of body surface.	**2. Very limited:** Responds only to painful stimuli. Cannot communicate discomfort except by moaning or restlessness. Or has a sensory impairment which limits the ability to feel pain or discomfort over 1/2 of body.	**3. Slightly limited:** Responds to verbal commands but cannot always communicate discomfort or need to be turned. Or has some sensory impairment which limits ability to feel pain or discomfort in 1 or 2 extremities.	**4. No impairment:** Responds to verbal commands. Has no sensory deficit which would limit ability to feel or voice pain or discomfort.		
MOISTURE Degree to which skin is exposed to moisture.	**1. Constantly moist:** Skin is kept moist almost constantly by perspiration, urine, etc. Dampness is detected every time pt. is moved or turned.	**2. Very moist:** Skin is often but not always moist. Linen must be changed at least once a shift.	**3. Occasionally moist:** Skin is occasionally moist, requiring an extra linen change approximately once a day.	**4. Rarely moist:** Skin usually dry, linen only requires changing at routine intervals.		
ACTIVITY Degree of physical activity.	**1. Bedfast:** Confined to bed.	**2. Chairfast:** Ability to walk severely limited or non-existent. Cannot bear own weight and/or must be assisted into chair or wheelchair.	**3. Walks occasionally:** Walks occasionally during day, but for very short distances, with or without assistance. Spends majority of each shift in bed or chair.	**4. Walks frequently:** Walks outside the room at least twice a day and inside room at least once every 2 hours during waking hours.		
MOBILITY Ability to change and control body position.	**1. Completely immobile:** Does not even make slight changes in body or extremity position without assistance.	**2. Very limited:** Makes occasional slight changes in body or extremity position but unable to make frequent or significant changes independently.	**3. Slightly limited:** Makes frequent though slight changes in body or extremity position independently.	**4. No limitations:** Makes major and frequent changes in position without assistance.		
NUTRITION Usual food intake pattern.	**1. Very poor:** Never eats a complete meal. Rarely eats more than 1/3 of any food offered. Or is NPO and/or maintained on clear liquids or IV's for more than 5 days.	**2. Probably inadequate:** Rarely eats a complete meal and generally eats 1/2 of food offered. Or receives less than optimum amount of liquid diet or tube feeding.	**3. Adequate:** Eats over 1/2 of most meals. Or is on a tube feeding or TPN regimen which meets most of nutritional needs.	**4. Excellent:** Eats most of every meal. Never refuses a meal. Does not require supplements.		
FRICTION AND SHEAR	**1. Problem:** Requires moderate to maximum assistance in moving. Complete lifting without sliding agnist sheets is impossible. Frequently slides down in bed or chair, requiring frequent repositioning with maximum assistance. Spasticity, contractures or agitation lead to almost constant friction.	**2. Potential problem:** Moves feebly or requires minimum assistance. During a move skin probably slides against sheets, chair restraints, or other devices. Occasionally slides down in bed or chair.	**3. No apparent problem:** Moves in bed and in chair independently and has sufficient muscle strength to lift up completely during move. Maintains good position in bed or chair at all times			
Date:	Signature:				**TOTAL SCORE:**	
Date:	Signature:					

Figure 20-3
Continued. (Figure courtesy of Huntington Memorial Hospital, Pasadena, CA.)

Skin Care Record

KEY - Show location of each ulcer by capital letter.

Stage I: Redness of the skin not resolving within 30 min. of pressure relief. Skin is intact.

Stage II: Superficial circulation & tissue damage which involves excoriation, vesiculation (blister), or skin break.

Stage III: Full thickness loss of skin which may or may not include the subcutaneous tissue level & which produces serosanguinous drainage.

Stage IV: Full thickness loss of skin with invasion of deeper tissues and/or structures such as fascia, connective tissue, muscle or bone.

Stage III/IV: Any area with necrosis, before debridement as staging cannot be confirmed without visualization of wound base.

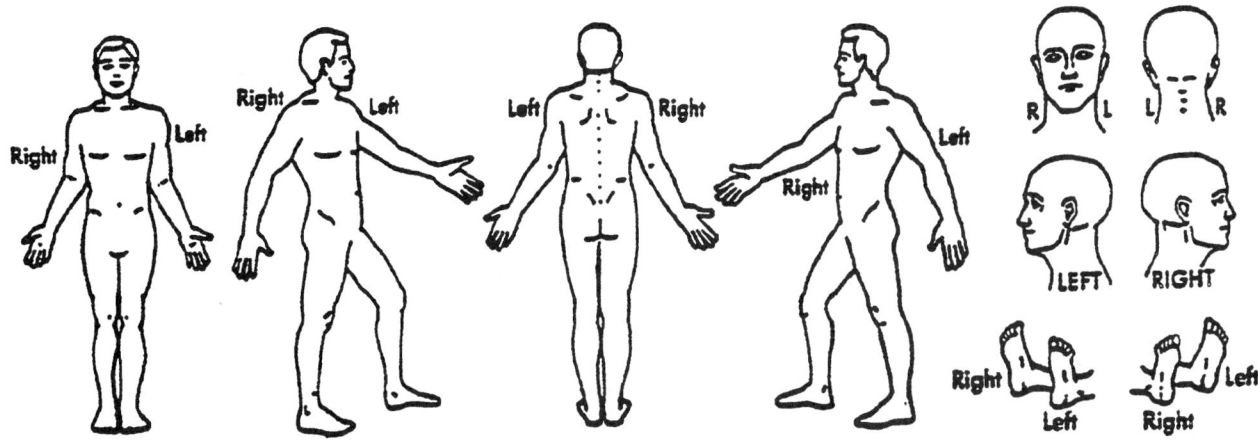

Assessments						
Date						
Type						
Location (letter)						
Stage						
Length x width (cm)						
Depth (cm)						
Necrotic tissue %						
Drainage (color & amt)						
Tunneling (✓)						
Incontinence (✓)						
Intervention Plan						
Pressure relief						
Diet supplement						
Cleansing routine						
Treatment						
Comments						
Signature						

Figure 20-3
Continued. (Figure courtesy of Huntington Memorial Hospital, Pasadena, CA.)

Huntington Memorial Hospital
FLUID BALANCE SHEET

	INTAKE	OUTPUT	BALANCE
PREVIOUS 24 HOUR TOTALS			
24 HOUR RENAL OUTPUT			

DATE: /   /

TIME	INTAKE				OUTPUT				BALANCE
	Intravenous Solutions	ORAL	Tube Feeding	Miscellaneous (Label)	RENAL	STOOL	Miscellaneous (Label)		
23-24									
24-01									
01-02									
02-03									
03-04									
04-05									
05-06									
06-07									
8 HR. TOTAL									8 HR. BALANCE
8 HR. COMBINED INTAKE					8 HR. RENAL				
					8 HR. COMBINED OUTPUT				
07-08									
08-09									
09-10									
10-11									
11-12									
12-13									
13-14									
14-15									
TOTAL									8 HR. BALANCE
8 HR. COMBINED INTAKE					8 HR. RENAL				
					8 HR. COMBINED OUTPUT				
15-16									
16-17									
17-18									
18-19									
19-20									
20-21									
21-22									
22-23									
TOTAL									8 HR. BALANCE
8 HR. COMBINED INTAKE					8 HR. RENAL				
					8 HR. COMBINED OUTPUT				
*24 HOUR INTAKE					*24 HOUR RENAL OUTPUT				
					*24 HOUR COMBINED OUTPUT				
					24 HOUR BALANCE				

N39 REV.5/88

*RECORD 24 HR. TOTALS ON THE NEXT FLUID BALANCE SHEET

Figure 20-3

Continued. (Figure courtesy of Huntington Memorial Hospital, Pasadena, CA.)

Administrative Medical Services 165

DATE: 24 HOUR INTRAVENOUS SOLUTION VOLUME   *See MAR

	TIME	AMOUNT	SOLUTION	WITH ADDI-TIVE*	RATE	SIGNATURE	AMOUNT ABSORBED
2300 — 0700							
							TOTAL
0700 — 1500							
							TOTAL
1500 — 2300							
							TOTAL

CHART: Catheter Insertions (Start/Restart/Size/Location) Conversion (IV/Lock)
D/C (Document status of catheter & Site), Central Line Care, Dressing Change

Time	Comments	Signature

Tubing Change_____ Condition of Site_____

Figure 20-3
Continued. (Figure courtesy of Huntington Memorial Hospital, Pasadena, CA.)

Huntington Memorial Hospital
Nursing Service
Falls Prevention Protocol

Key: Yes = ✓ (applies to patient)

* = Focus Note

(Assess Patient Q 8 hrs + PRN)

Date

Time

A. Moderate Risk Assessment Factors
  I. Medications: sedatives, narcotics, antihypertensives, diuretics, laxatives, psychtrophics, anesthetics
  II. Sensory limitations
  III. Impaired cardiovascular status
  IV. Age 70 or over
  V. Use of ambulatory devices (crutches, walkers, etc.)

B. Interventions for Moderate Risk: MUST institute interventions 1-6 if 1 or more assessment factors are present.
  1. Call light in reach
  2. Bed in low position with brakes locked
  3. Side rails up 2 or up 4
  4. Night light
  5. Toileting
  6. Options:

C. High Risk Assessment Factors
  I. History of recent falls
  II. Confusion, transient confusion states and/or temp >103°F
  III. Impaired neurological status
  IV. Unsteady gait with assistive devices
  V. Attitude (resistent, belligerent, combative, fearful, anxious)

D. Interventions for High Risk: If 1 or more assessment factors are present, nurse must institute interventions 1-8 & one of 9-12. If interventions 9-12 not selected, nurse must document why interventions not appropriate for this patient.
  7. Orange armbands and Falls labels
  8. Reorient to person/place/time every 2 hours
  9. Bedside commode at bedside
  10. Safety devices
  11. Sitter if interventions still not adequate
  12.

☐ Family/significant other requests to be notified of any changes in the patient's risk assessment factors and interventions.

Initials	Signatures	Initials	Signatures	Initials	Signatures

Figure 20-3

Continued. (Figure courtesy of Huntington Memorial Hospital, Pasadena, CA.)

Administrative Medical Services

Huntington Hospital
Discharge Instructions

Discharge Date: _____ Time: _____

(ADDRESSOGRAPH)

Accompanied by: _____ Discharge Questionnaire Given: ❏ Yes ❏ No ❏ N.A.

Discharge to: ❏ Home ❏ Other _____ ❏ A.M.A.

Per: ❏ Car ❏ Ambulance ❏ Other: _____ **Via:** ❏ Wheelchair ❏ Ambulatory ❏ Other: _____

Valuables: ❏ None
Dentures: ❏ Upper ❏ Partials ❏ Cane ❏ Glasses ❏ Electrical appliance: _____
❏ Lower ❏ Other ❏ Crutches ❏ Contacts ❏ Clothing
❏ Pre-admit drugs ❏ Walker ❏ Hearing aid ❏ Pressure mattress
❏ Valuables in admitting ❏ W/C ❏ Prosthesis: _____
❏ Other belongings: _____

Follow-up: M.D. office: _____
Dispensary: _____ Date: _____ Time: _____
Referrals: _____ Phone: _____

Special Instructions (activity, diet, bath, wound care, dressing, cast, equipment): _____

Instructed by:

Medication Instructions: _____

Pharmacist:

I have received and understood these instructions.   ❏ Received valuables

_____   _____   _____
Patient/Other   Relationship   Signature of HMH Staff

Figure 20-3
Continued. (Figure courtesy of Huntington Memorial Hospital, Pasadena, CA.)

Huntington Memorial Hospital
Discharge Summary

Admission Date: _____

Discharge Date: _____

Reason for Hospitalization: _____
_____
_____

Pertinent Physical, Laboratory & X-ray Findings: _____
_____
_____
_____

Surgical and/or Medical Treatment: _____
_____
_____
_____

Course of Hospitalization (including complications): _____
_____
_____
_____
_____
_____

Condition on Discharge: _____
_____

Principal Diagnosis: _____
_____

Secondary Diagnosis/Complications: _____
_____
_____

DISCHARGE INSTRUCTIONS

Diet: _____

Physical Activities and/or Limitations: _____
_____
_____

Medications: _____
_____
_____
_____

Follow-up: _____

Other: _____
_____

Date: _____ Attending Physician: _____ MD

N-125 115827   12/94 mc

Figure 20-3
Continued. (Figure courtesy of Huntington Memorial Hospital, Pasadena, CA.)

Administrative Medical Services

Huntington Memorial Hospital
Hematology/Oncology Discharge Instruction Sheet

On _____ you received your chemotherapy treatment consisting of:
_____
_____
_____

Your next appointment is _____ at _____

Doctor: _____ Phone #: _____

Instructions

1. Drink eight to 10 glasses of liquid (not caffeinated) each day for the next three to four days.
2. Allow time during your day for rest periods.
3. Avoid crowds, people and children with colds and infections.
4. Avoid commercial mouthwashes, since they contain alcohol that can dry out sensitive tissue.
5. Rinse mouth out six times a day with 1/2-cup of a solution made of one quart water, one teaspoon salt and one teaspoon baking soda. Replace this solution daily.
6. Use sun block and protective clothing when outdoors.
7. Call your doctor for:
    a) Mouth sores that prevent you from eating.
    b) Nausea and/or vomiting in which you can't keep food or liquids down during a 24-hour period.
    c) Diarrhea or loose, watery stools that last for more than 24 hours.
    d) Temperature greater than 100° Fahrenheit or chills.
    e) Any unusual bruising or bleeding.
    f) Shortness of breath or difficulty breathing.
    g) Weight gain or loss of five pounds or more in a two-week period.
    h) Any skin changes where your IV's were removed or around Hickman/PICC catheter skin site or around your port.
8. Other information: Refer to "Chemo and You" booklet, "Tips" and "Eating Hints" for other helpful suggestions. (Circle those provided.)
9. Community support referrals: The Wellness Community, "I Count Too," VNA, "Reach for Recovery."
10. Prescription instructions: _____

I hereby accept, understand and can verbalize these instructions:

Patient/S.O. _____ Relationship _____

Nurse _____ Date _____ Time _____

White Copy to Chart   Yellow Copy to Patient   5/94

Figure 20-3
Continued. (Figure courtesy of Huntington Memorial Hospital, Pasadena, CA.)

Huntington Memorial Hospital
CASE HISTORY

(Addressograph Plate)

CHIEF COMPLAINTS

AGE

PRESENT ILLNESS

PERSONAL
- MEDS.
- KNOWN ALLERGIES
- HABITS

SYSTEMIC REVIEW
- SKIN
- HEENT
- CARDIAC
- RESP.
- G.I.
- G.U.
- GYN.
- MUSCULOSKELETAL
- NEUROPSYCHIATRIC

PAST HISTORY
- SURGICAL
- MEDICAL

FAMILY HISTORY

Figure 20-3 Continued. (Figure courtesy of Huntington Memorial Hospital, Pasadena, CA.)

Administrative Medical Services

171

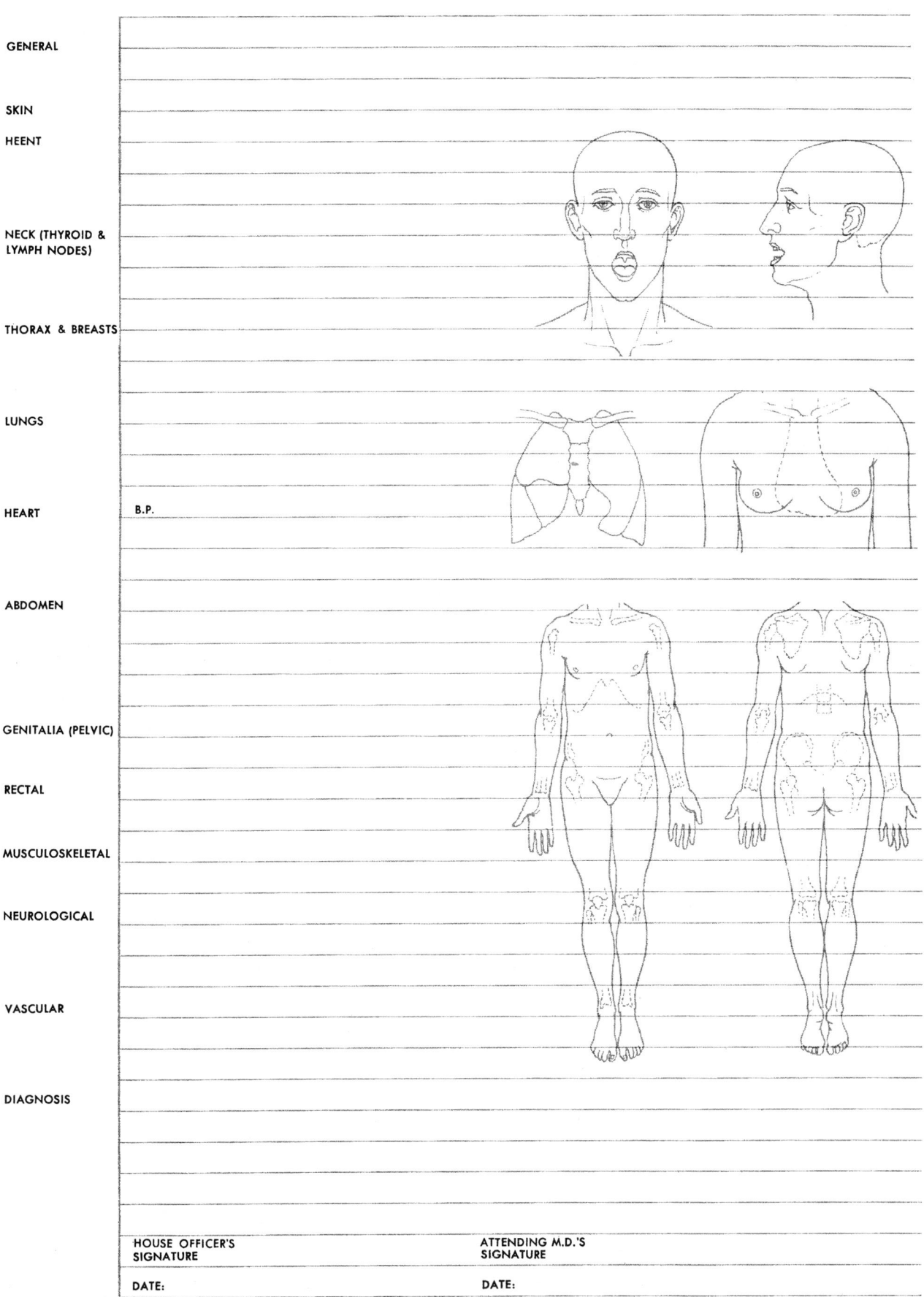

Figure 20-3
Continued. (Figure courtesy of Huntington Memorial Hospital, Pasadena, CA.)

HOLD

HUNTINGTON MEMORIAL HOSPITAL

PHYSICIAN'S ORDERS

872-002 REV. 6/92 VHA +PLUS TM

HOLD

R.N. SIG. M.D. SIG. **1.**

HOLD

R.N. SIG. M.D. SIG. **2.**

HOLD

R.N. SIG. M.D. SIG. **3.** **19**

Figure 20-4
Examples of forms used in medical records. (Courtesy of Huntington Memorial Hospital, Pasadena, CA.)

PRINTED BY STANDARD REGISTER U.S.A. FOR VHA +PLUS TM

HUNTINGTON MEMORIAL HOSPITAL

DISCHARGE PRESCRIPTION ORDERS

PRESS HARD —
YOU ARE MAKING
3 COPIES
PLEASE USE
BALLPOINT PEN

DISTRIBUTION:
ORIGINAL: PATIENT-PHARMACY
WHITE: MEDICAL RECORDS
PINK: PHYSICIAN

DISCHARGE PRESCRIPTION/S (818) 397-5176
HUNTINGTON MEMORIAL HOSPITAL • 100 W. CALIFORNIA BLVD. • PASADENA, CA 91109-7013

ADDRESSOGRAPH

PATIENT NAME ______ DATE ______

ADDRESS ______ AGE ______

℞

PHYSICIAN ______ M.D. CALIF. LIC. NO. ______ DEA NO. ______

SIGNATURE ______ TELEPHONE ______

ADDRESS ______

DISCHARGE PRESCRIPTION/S (818) 397-5176
HUNTINGTON MEMORIAL HOSPITAL • 100 W. CALIFORNIA BLVD. • PASADENA, CA 91109-7013

ADDRESSOGRAPH

PATIENT NAME ______ DATE ______

ADDRESS ______ AGE ______

℞

N-32-#104455

PHYSICIAN ______ M.D. CALIF. LIC. NO. ______ DEA NO. ______

SIGNATURE ______ TELEPHONE ______

ADDRESS ______

Figure 20-4
Continued. (Figure courtesy of Huntington Memorial Hospital, Pasadena, CA.)

Huntington Memorial Hospital

Physician's Order - Code Status
Do Not Thin From Chart

Date: _____   ❑ Family Consult Date: _____

Time: _____   ❑ Advanced Directive in effect

Code Status:
❑ Do Not Resuscitate
❑ Do Not Resuscitate (partial): Indicate Procedures for This Patient

    Intubation:
        ❑ Yes
        ❑ No
    ❑ Chemical Code Only **(applies to unmonitored patients only)**:

Other Specific Instructions: _____

_____

_____

                    MD Signature: _____

Arrhythmia Protocol (monitored patients only):
    ❑ Yes
    ❑ No

Other Specific Instructions: _____

_____

_____

                    MD Signature: _____

Telephone orders from the physician require two (2) nursing signatures.

T/O: _____ / _____
                                                      MD Signature

RN Signature: _____

2nd Licensed Nurse Signature, Witness: _____

Noted by RN: _____

DISCONTINUATION OF DNR

Date: _____   Time: _____

❑ Discontinue Above DNR Orders: _____
                                                         MD Signature

Noted by RN: _____

REV. 4/93 N - 120

Figure 20-4
Continued. (Figure courtesy of Huntington Memorial Hospital, Pasadena, CA.)

Administrative Medical Services

Huntington Memorial Hospital
Progress Record

Date	Time	N.B. Frequent Periodical Notations should be made concerning progress of case, complications, etc.

Figure 20-4
Continued. (Figure courtesy of Huntington Memorial Hospital, Pasadena, CA.)

Huntington Memorial Hospital

GRAPHIC RECORD

Key for Activity: 0 = No Stool $\frac{+}{1}$ = One Stool

TEMPERATURE: • ... Oral o ... Rectal Ax ... Axillary

DATE																														
TIME →	24	04	08	12	16	20	24	04	08	12	16	20	24	04	08	12	16	20	24	04	08	12	16	20	24	04	08	12	16	20
F C																														
105 - 40.6																														
104 - 40.0																														
103 - 39.4																														
102 - 38.9																														
101 - 38.3																														
100 - 37.8																														
99 - 37.2																														
NORMAL - N — 37.0																														
98 — 36.7																														
97 — 36.1																														
96 — 35.5																														
Pulse																														
Respirations																														
VITAL SIGNS: Blood Pressure	24		04		08		24		04		08		24		04		08		24		04		08		24		04		08	
	12		16		20		12		16		20		12		16		20		12		16		20		12		16		20	
Weight																														
Defecation																														

TEMPERATURE: • ... Oral o ... Rectal Ax ... Axillary

DATE																														
TIME →	24	04	08	12	16	20	24	04	08	12	16	20	24	04	08	12	16	20	24	04	08	12	16	20	24	04	08	12	16	20
F C																														
105 - 40.6																														
104 - 40.0																														
103 - 39.4																														
102 - 38.9																														
101 - 38.3																														
100 - 37.8																														
99 - 37.2																														
NORMAL - N — 37.0																														
98 — 36.7																														
97 — 36.1																														
96 — 35.5																														
Pulse																														
Respirations																														
VITAL SIGNS: Blood Pressure	24		04		08		24		04		08		24		04		08		24		04		08		24		04		08	
	12		16		20		12		16		20		12		16		20		12		16		20		12		16		20	
Weight																														
Defecation																														

N3 Rev. 5-88

Figure 20-4

Continued. (Figure courtesy of Huntington Memorial Hospital, Pasadena, CA.)

Administrative Medical Services

Huntington Memorial Hospital

CHEMOTHERAPY ADMINISTRATION CHECKLIST

Date: _____

Diagnosis: _____

1. PHYSICIAN'S ORDERS	2. PROTOCOL	3. DOSE GIVEN	4. PATIENT DATA
_____	_____	_____ INITIAL, DATE, TIME	Ht. ____ Wt. ____
_____	_____	_____	Ideal Wt. ____
_____	_____	_____	Wt. Change ____
_____	_____	_____	BSA ____
_____	_____	_____	Dose Verified ☐

5. LABORATORY VALUES

Total WBC ____	Absolute WBC ____	RBC ____	Calcium ____	Lab Checked ☐
Bands ____	Platelets ____	Hgb ____	Uric Acid ____	Abnormal Lab Verified With MD ☐
Polys ____	Cr (Serum) ____	HCT ____	____	
Bilirubin ____	BUN ____	K^+ ____		
Magnesium ____	Urine Cr Cl ____	Na^+ ____	Chemotherapy Held ____ Initial Time	

6. ADRIAMYCIN ADMINISTRATION: EKG _____ HR _____ Ejection Fraction _____

Cumulative (Total Life) Dose _____ mg/m^2 _____ N/A
(Include present mg/m^2)

7. BLEOMYCIN ADMINISTRATION: RR _____ CXR _____ PFT _____ Test Dose _____
(Circle IM/IV/SQ)
Cumulative Dose _____ Units _____ N/A

8. NURSING ASSESSMENT/PATIENT PREPARATION/PATIENT RESPONSE: Consent Signed at HMH _____ MD Office _____
Status of Veins (Circle one or more) Soft, Pliable, Fragile, Thin-walled, Thready, Hard, Knotty, Rolling, Valves
Unable to Start _____ Number of times IV start attempted _____ Blood return present _____ Size of catheter _____ ga.
Ⓡ Atrial Catheter Used _____ IV RN Initial _____ Other Routes: _____ Comments: _____

9. CIRCLE APPROPRIATE EXTREMITY AND ASPECT _____ N/A
(Example: (Right) Left Medial Upper Forearm

10. MARK WITH • LOCATION OF IV(s):
Right/Left Lateral Upper Forearm   Right/Left Lateral Ankle   Right Ventral
Right/Left Medial Upper Forearm   Right/Left Medial Ankle   Left Dorsal
Right/Left Lateral Lower Forearm   Right/Left Foot   _____ Finger/Thumb (L/R
Right/Left Medial Lower Forearm
Right/Left Dorsal Metacarpal   I.V. Restarted _____
S.Q. Access Port   SITE   DATE   TIME   INITIAL

11. EXTRAVASATION PROTOCOL:
Initiated _____ N/A _____

Signatures	Witness

ED. MATERIALS GIVEN:	Initials:
Chemo & You	☐
"Tips"	☐
Other	☐

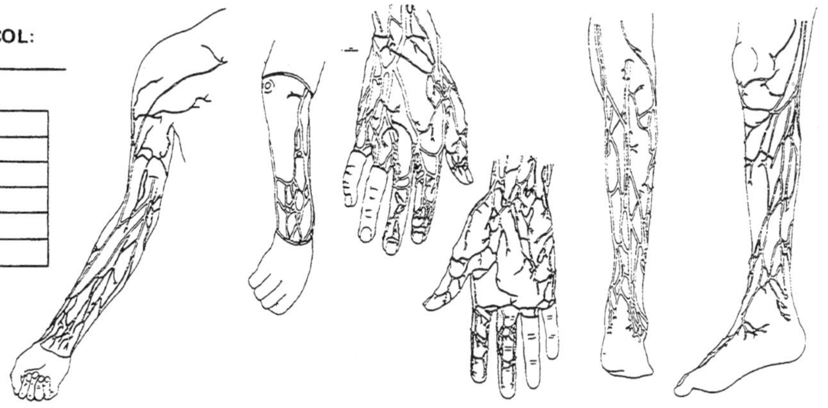

NS 257.001

WHITE — CHART COPY   CANARY — NURSING COPY

Figure 20-4
Continued. (Figure courtesy of Huntington Memorial Hospital, Pasadena, CA.)

HUNTINGTON MEMORIAL HOSPITAL
MEDICATION RECORD

PAGE: _____ OF _____

Injection Site Codes:
- ① Rt. Deltoid
- ② Lt. Deltoid
- ③ Rt. Gluteus Medius
- ④ Lt. Gluteus Medius
- ⑤ Rt. Ventrogluteal
- ⑥ Lt. Ventrogluteal
- ⑦ Rt. Anterolateral Thigh
- ⑧ Lt. Anterolateral Thigh
- ⑨ Rt. Abdomen
- ⑩ Lt. Abdomen

For Pediatric Use Only
- ⑪ Rt. Anterior Thigh
- ⑫ Lt. Anterior Thigh

TRANSFER TIME: _____

ALLERGIES

MEDICATION DATE Rx#	GENERIC NAME DESCRIPTION COMMENTS (TRADE NAME) ROUTE FREQ.	2300 - 0700 (11-7)	0700 - 1500 (7-3)	1500 - 2300 (3-11)

SIGNATURES (11-7)	SIGNATURES (7-3)	SIGNATURES (3-11)

854,002
REV. 1/87

MEDICATION ADMINISTRATION KEY
- ⑨ - MEDICATION NOT GIVEN (EXPLANATION ON REVERSE SIDE)
- 𝟎𝟓̸ - MEDICATION GIVEN

Figure 20-4
Continued. (Figure courtesy of Huntington Memorial Hospital, Pasadena, CA.)

I, ______________________________, authorize and direct Drs. ______________________________

my surgeon, and/or physicians, surgeons, or assistants of his choice, to perform an operation or special diagnostic or therapeutic procedure upon me and to any other diagnostic and therapeutic procedure that their judgment may dictate to be advisable for my well being.

My doctor has fully explained to me the nature and purpose of the following mentioned surgical operation or special diagnostic or therapeutic procedures and that such surgical operation or special diagnostic or therapeutic procedures may involve calculated risks of complications, injury or even life threatening consequences, from both known and unknown causes, and no warranty or guarantee has been made as to result or cure. I recognize that I have a right to be informed of the nature and purpose of the operation or procedures, the risks of complications, and the alternative methods of treatment, if applicable. I further recognize that I have the right to consent to or refuse any proposed operation or special procedure based upon the description or explanation received. Further I recognize that this form is not intended to be a substitute for the explanations of the nature and purpose of the operation or procedures, the risks of complications, and the alternative methods of treatment, if applicable, which have been provided by the physician mentioned above. I understand and consent to the performance of the following stated operation or special diagnostic or therapeutic procedure:

LUMBAR EPIDURAL STEROID INJECTION

I hereby authorize and direct the physicians and surgeons of the medical staff of **Saint Joseph Medical Center** to provide such additional services for me as they may deem reasonable and necessary, including but not limited to, the administration and maintenance of the anesthesia and the performance of services involving pathology and radiology, and I consent thereto.

I understand that the person or persons in attendance at such operation or procedure, as indicated above, for the purpose of administering anesthesia, and the person or persons performing other specialized professional services, such as radiology, pathology and the like, are not the agents, servants or employees of the above named medical center or of the above named physician and surgeon, but are independent contractors performing specialized services on my behalf and, as such, are the agents, servants or employees of myself.

I hereby authorize the presence of observers and supportive personnel authorized by my physician or surgeon

I hereby authorize the hospital pathologist to use his discretion in the disposal of any tissue, member, or organ removed during the procedure set forth above.

TRANSLATION (If necessary) – I have accurately and completely read the foregoing document to the signator identified below in the patient's/patient representative's primary language. He/she understood all terms and conditions and acknowledged his/her agreement by signing this document in my presence.	PATIENT/PATIENT REPRESENTATIVE'S PRIMARY LANGUAGE: SIGNATURE OF TRANSLATOR

SIGNATURE OF PATIENT / PATIENT'S REPRESENTATIVE	DATE	TIME	REASON PATIENT DID NOT SIGN
X	2/22/95	2:30	
RELATIONSHIP TO PATIENT ☐ PATIENT ☐ SPOUSE ☐ CHILD ☐ SIBLING ☐ GUARDIAN ☐ OTHER (specify)	WITNESS		TITLE/DEPARTMENT

AUTHORIZATION FOR AND CONSENT TO SURGERY OR SPECIAL DIAGNOSTIC OR THERAPEUTIC PROCEDURES

Patient Name

M.D. Name ADM 02/22/95

ACCT 25173238

Medical Record No:

DOT

Admit Date:

60-0712-R1193/8560 DISTRIBUTION: ORIGINAL - Medical Record

Figure 20-4
Continued.

(text continued from page 157)
department, including nursing and other ancillary services, that provides care or treatment to a patient provide written documentation as to what type of treatment was provided and what type of outcome was achieved. Many facilities complete the analysis of these departments through their utilization review process. However, in those cases when the medical records department is responsible for analyzing such documents, great care must be taken to check for proper documentation as to the reason and outcome of the treatment. Such analysis usually includes checking for deficiencies, such as failure to sign off on an entry or provide the date on or time at which a procedure or entry was made.

Analyzing the Discharge Summary

Obviously, the best time for a physician to write or dictate his or her patient discharge report would be at the time of discharge. However, because most physicians generally admit, and consequently see, more than one patient in the hospital at any given time, this is usually not done. Because physicians are often occupied with caring for patients who are still hospitalized, making sure that a summary has been written or dictated regarding a patient who has been discharged is sometimes forgotten or assigned to a "back burner": Hence the reason for including the discharge summary as part of the analysis procedure.

Most hospitals require that the patient's attending physician complete the discharge summary. When completing your analysis, you must make sure that the physician has signed the discharge summary, noted the admitting diagnosis, identified any surgery or procedures that may have been performed during the hospitalization, and described the patient's progress during hospitalization, as well as the final outcome that was achieved during the patient's stay in the hospital. The dates of admission and discharge must also be noted.

Summary

During this chapter, we discussed the three stages, or steps, involved in processing medical records. We noted that these steps, which include processing, assembling, and analyzing, are all regulated, not just by the medical records department, but by other agencies and regulatory bodies, such as JCAHO and members of the medical staff. We also talked about how to use a master patient index to verify various reports found in the patient's medical record. We discussed how to assemble medical records properly for storage. And, finally, we talked about how to complete the actual process of analyzing the medical record for proper documentation and completeness.

Review Questions

1. List the three stages involved in the processing of medical records:
   a. _____
   b. _____
   c. _____

2. Briefly explain the role regulatory agencies play in the processing of medical records.

3. Briefly explain the function of a *master patient index*.

4. Define the following abbreviations:
   a. CPT _____
   b. DRG _____
   c. ICD-9-CM _____

5. Briefly explain the following:
   a. physician privileges _____
   b. record retrieval _____
   c. suspension _____

6. Explain the purpose of analyzing a medical record after a patient has been discharged.

7. A(n) _____ _____ is the person responsible for overseeing the total care of a patient during his or her hospitalization.

8. Briefly explain the steps involved in admissions and discharges as each relates to the processing of medical records.

9. List at least four forms that might be found in a medical record:
   a. _____
   b. _____
   c. _____
   d. _____

21

Advanced Functions of the Medical Records Clerk

Performance Objectives

Upon completion of this chapter, you will be able to:

1. Discuss the role and identify the functions of the medical records clerk as they relate to working with incomplete and delinquent medical records.
2. Discuss the role and identify the functions of the medical records clerk as they relate to working with specific advanced functions, including the following: processing birth certificates, performing quality assurance and utilization tasks, registering data and contacting patients through a facility's tumor registry, and performing tasks involved in diagnostic-related groups.

Terms and Abbreviations

Accreditation a process by which approval is granted to an institution that has met criteria set forth by the agency providing the accreditation. For example, *JCAHO Accreditation* provides such approval for three years to hospitals meeting their standards for providing healthcare.

Attestation a statement signed by a physician, which is mandatory for all hospitalized patients receiving benefits under the Medicare program, in which the physician attests to the truthfulness of all information recorded regarding the patient's diagnosis and treatments.

Data slip a strip of thermal paper containing a physician's identification number and the time of and date on which the physician dictated information into a cassette.

Delinquent report a medical record or report that remains incomplete for a period longer than that allowed by a facility's medical staff bylaws, rules and regulations, and policies.

DRG abbreviation for Diagnosis Related Group(s), a group of diagnoses with similar treatment required.

Incomplete record any medical record that contains deficiencies in its documentation.

Quality assurance the process by which a health care facility evaluates the effectiveness of treatment and services provided to patients, and recommends any actions needed to improve quality.

Suspension list a list of physicians who have temporarily lost their privileges to practice in a given hospital, due to failure to complete their delinquent medical records.

Utilization review a process by which a health care facility reviews the length of stay and the use of hospital resources by physicians and other staff members as they relate to each patient hospitalized in the facility.

Many of the more advanced responsibilities required of the medical records clerk involve additional training and understanding of the role of the medical records department. Once you have adjusted to your position and performed some of the more basic skills necessary in your job, you may find your supervisor asking you to get involved in some of the more advanced tasks required of your department. These may include more complex functions involving issues dealing with the facility's accreditation, physician suspension, maintenance of statistical information, obtaining and researching data required for legal documents, and working with the hospital's other ancillary departments, such as utilization review and quality assurance.

Meeting Accreditation Standards

As we have already discussed, in order to provide high standards in the quality of care, the majority of hospitals and health care facilities operating throughout the United States must meet specific criteria such as those set forth by the Joint Commission on the Accreditation of Healthcare Organizations (JCAHO). In addition to meeting the requirements established by JCAHO, in many instances health care facilities are also obligated to follow set guidelines and regulations that have been instituted by individual states and in-house medical staff committees. Part of your role as a member of the medical records staff is to make sure these rules, policies, and regulations are adhered to and followed.

Advanced Functions of the Medical Records Clerk

As a member of the medical records team, you may be asked to take part in some of the more advanced functions overseen by your department. These functions generally involve tasks related to ensuring the timely completion of delinquent medical records by physicians and the facilitation and maintenance of accurate legal records, such as birth certificates and other important documents.

Notifying Physicians of Incomplete Medical Records

One of the areas in which you will probably become involved is notifying physicians with regard to incomplete or delinquent records. Most facilities have a set policy according to which the medical records department contacts doctors who have incomplete records, with most usually waiting a set number of days after patient discharge before beginning the notification process. After this period has elapsed, the department sends a "friendly reminder" to the physician, notifying him or her that, within a specified number of days, the medical record will become delinquent. Since physicians realize that this first reminder is sent as a courtesy, most follow up on their obligation and complete the record before it becomes delinquent. There may be some instances, however, in which the physician does not meet his or her responsibility and allows the medical record to remain incomplete for a longer period. If this occurs, the medical records department will usually send a "stronger" letter of delinquency giving the physician a set time to complete the record. If the physician does not comply with the requirements necessary to complete his or her records within the period set forth by the delinquency letter, the final step often involves the physician being notified both by phone and in writing that he or she is pending suspension if the record is not completed.

Suspension letters are generally sent out when a medical record remains incomplete 21 days after the patient has been discharged. The letter will usually state that, unless the physician meets his or her obligation to complete the medical record, all hospital privileges will be temporarily suspended until such time as the obligation has been met. Once a doctor has been formally suspended from hospital privileges, his or her name is placed on a suspension list, which is then routed to all departments within the hospital notifying all members of the staff that the physician is no longer allowed to admit patients, perform consultations, or perform any procedures or surgeries on patients in the hospital. A physician will remain suspended until he or she completes all outstanding medical records and, in some cases, has been reinstated by the medical staff committee.

One task which you may be required to perform to ensure that physicians complete their medical records, is to pull the individual record or report forms needing completion and file these forms while they await completion by the physicians. If this job becomes one of your responsibilities, there are two methods or procedures that may follow which are acceptable to most health care facilities.

The first system used for filing incomplete reports is called the *doctor's box* method. It involves filing all records or reports for an individual physician by that physician's name, and then allowing them to remain there until the doctor completes them. A major disadvantage in using this method is that it requires a great deal of cross-checking, which can lead to errors

A more efficient method for filing outstanding medical records and reports, which is used by most health care facilities, deals with filing records according to a specific terminal digit order. A major advantage of this system, over using a doctor's box, is that it allows you much more ease in pulling any record at any time and for any reason.

If you are responsible for pulling incomplete medical records, the department in which you work will probably have a set procedure or policy for doing so. Following your department's system is extremely important, since many physicians tend simply to show up to complete their records without prior notification. Because most physicians are very busy, often many of them drop into the medical records department when they have a few moments between seeing patients or between their lunch and late afternoon office hours. Because most medical records departments must accommodate the scheduling needs of busy doctors, often health care facili-

Administrative Medical Services

ties follow the guidelines outlined below for pulling incomplete medical records:

- Pull all deficiency slips from the physician's file and place them in terminal digit order.
- Pull all incomplete medical records, starting with the most readily found ones.
- Search records pulled for other physicians.
- After you have searched all records, document any that cannot be located.
- If you are unable to locate all of the physician's records, ask the physician to return at a later time.

Because of the doctor's time constraints, your department will be much more successful in ensuring the completion of medical records if a method is established that requires the physician to be called in advance of coming to the medical records unit. Once the doctor has notified the department that he or she is coming in to complete patients' medical records, it becomes the responsibility of your department to pull all records and reports requiring completion. All records should be placed in the physician's incomplete section of the department, with a note or memorandum indicating the name of the physician and the date and time the records were pulled affixed to the top of them.

Increasing Physician's Productivity in Completing Medical Records

While completing delinquent medical records may seem to be of great importance to you and the members of your department, to the physician it may seem like nothing more than a bothersome "inconvenience." You must remember that, sadly, the only reason many doctors take the time to complete their delinquent charts and records is to avoid suspension, which ultimately leads to a loss of revenue. Therefore, one of the most important roles you may find yourself playing as a member of the medical records team is to understand why the physician ignores or forgets to complete his or her records. Many doctors just don't see the completion of the medical record as an important task. Others are simply too busy or stressed out and don't have the time, however short it may seem to you, to come into your department to complete their charts and records.

A well-trained, professional member of the medical records department generally knows how to deal with physicians when it comes to getting them to complete their records. Often it involves providing the doctors with incentives or just plain "stroking." It always means that you must be familiar with the records that need correction, as well as any of the "quirks" of the physicians whose records need to be corrected. Being skilled and knowledgeable in your job also means being able to communicate well and speak clearly and distinctly with the physicians. Above all, it means always remaining calm, courteous and respectful toward all physicians. Remember, the bottom line is that you must ensure completion of the delinquent records. And short of committing fraudulent or deceitful acts to meet this important objective, you must do whatever is necessary.

Updating Completed Medical Records

Once a medical record has been completed, most facilities require that it be checked a second time to make absolutely sure that no inconsistencies or errors itemized on the delinquency slip exist. If you are required to do this, an easy way of double-checking yourself is to pull the indicator slip from the section that was corrected immediately after you have checked the completed item. If, after you have checked the completed record, any indicator slips are still remaining in the record, you will know immediately what corrections are still needed. You may then place a new delinquincy slip on the front of the record, with only those items needing corrections checked. When the doctor comes in to complete the record a second time, he or she will know immedately what items must be completed, based on the new delinquency slip.

Completing Legal Documents

Processing Birth Certificates

Even though the patient's entire medical record is considered a legal document, the document that requires the highest degree of accuracy because of its usefulness throughout one's life is a certificate of live birth. All states require births to be registered, and completion of the birth certificate is often the responsibility of the medical records department.

The key to completing a birth certificate is accuracy. While states differ as to the appearance of the document, the information required is generally the same. All data documented on the certificate are important, and include the parents' place of residence and circumstances surrounding the birth.

There are generally four steps in completing a birth certificate: interviewing the parent(s), processing the certificate, mailing the certificate, and maintaining or logging the certificate. If you are given the responsibility of completing a birth certificate, begin by interviewing the patient who has delivered the infant. Most facilities use a preprinted worksheet to gather the necessary information. This method is the most widely used because of its simplicity and because the data are obtained directly from the parent of the newborn.

Once you have obtained all of the information from the parent(s) and completed the document by processing the data and transferred it to the birth certificate, the next step is to mail the original certificate to the local health department, which is responsible for registering all live births in a given state. Since most hospitals have several certificates to mail at any given time, the easiest way to ensure all certificates are picked up by the postal service is to batch all the documents together. However, since the post office is not infallible, there is always the possibility that the certificates will get lost during the mailing or delivery process. To solve this potential problem, before mailing the certificates, type an itemized list of all the documents, enclose the original list with the certificate(s) in the mailing envelope, and keep a copy for the department. Once the health department has received the certificate(s), it will return the original list, generally with each individual name checked off and the date on which it was received noted.

The final step in completing the birth certificate is to enter the birth into the birth certificate log, which is often maintained by the medical records department. The log consists of a list of all births, by the newborn's name which was provided by the parents, and the date the birth took place. If the mother's last name is different from the father's, it is also listed.

Working with Other Departments

As you become more proficient in your work as a medical records clerk, you may be asked to take on additional responsibilities involving work that is integrated with other ancillary departments in the hospital. This often involves working with members of the utilization review and quality assurance teams by providing pertinent information related to diagnosis related groupings (DRGs), tracking patients from the time of admission to the hospital until discharge, attesting to and obtaining signatures from physicians, and working in the tumor registry. As with all your responsibilities, when working with any of these departments, the most important element is to make sure all the information and data obtained are accurate and meet the requirements and standards of care set forth by your individual hospital and the accrediting agency under which your facility has been authorized to function.

Summary

In this chapter, we discussed some of the more advanced functions of the medical records clerk. We identified some of these functions, including working with incomplete and delinquent medical records, processing birth certificates, and working with other departments in the hospital environment.

Review Questions

1. Briefly discuss the function of the medical records clerk as it relates to working with and processing incomplete and delinquent medical records.

2. Briefly explain the concept of *accreditation* as it relates to a health care facility.

3. Define the following terms:

   a. attestation _____

   b. quality assurance _____

3. Briefly define what is meant by an *incomplete record*.

4. Briefly explain the concept of DRGs as they relate to medical records.

5. What is the name of the report used to indicate that a medical record remains incomplete for a time longer than is allowed?

6. A(n) _____ _____ refers to a list of all physicians who have temporarily lost their privileges to practice within a given hospital because of failure to complete their delinquent medical records.

7. What is the name of the process by which a health care facility reviews the length of stay and the use of hospital resources by physicians and other staff members?

8. What does the abbreviation JCAHO stand for?

9. List at least three advanced functions of the medical records clerk:

   a. _____

   b. _____

   c. _____

10. Briefly define the four steps involved in processing a birth certificate:

    a. _____

    b. _____

    c. _____

    d. _____

Section V

Medical Coding and Claims Worker: Basic Concepts and Applications

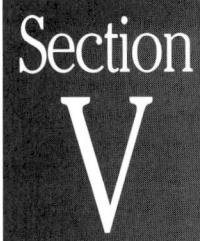

22
Introduction to Medical Coding

23
Coding Diseases

24
Medical Coding and Claims Processing

Introduction to Medical Coding

Performance Objectives

Upon completion of this chapter, you will be able to:

1. Discuss the purpose and function of data collection and information coding in the health care industry.
2. Identify the main differences between nomenclature and classification systems as each relates to billing and coding.
3. Identify and discuss terminology and reference materials used by the medical information coder.
4. Discuss the purpose and function of CPT coding.
5. Discuss the purpose and function of ICD-9-CM coding.
6. Explain the basis and function of V-codes.

Terms and Abbreviations

AHIMA   abbreviation for the *American Health Information Management Association*, which is a national organization of professional medical record administrators and technicians.

ART   abbreviation for *Accredited Record Technician*, which is a person who has completed a program in medical record technology and passed a national certification examination.

DRG   abbreviation for *diagnosis-related groups*, which is used to subdivide patients admitted into a hospital according to their medically related diagnoses, treatments, and lengths of stay.

HCFA   abbreviation for *Health Care Financing Administration*, a federal agency located within the Department of Health and Human Services, which is responsible for implementing laws governing Medicare and Medicaid health care programs.

ICD-9-CM   abbreviation for *Clinical Modification of the World Health Organization's International Classification of Diseases*, which is published for use in health care facilities throughout the United States.

MDC   abbreviation for *major diagnostic categories*, which is a system used to cluster patients into their medically related groups.

While many of us who work in the health care industry don't really want to admit it, one of the most important responsibilities we have in fulfilling our jobs is to make sure that the physician or institution in which we are employed gets paid for the services rendered to the patient. It's unfortunate, but many consumers are misinformed regarding payment for health care services they have received. Doctors and health care institutions must be paid so that they can go on caring for and treating other patients. That is why the role of the medical coding specialist (see Figure 22-1) is vitally important. For it is this person who has the very important responsibility of making sure that all diagnoses, services, treatments, and procedures have been properly documented to make sure that the people and instututions that provide the patient with his or her care are paid.

Figure 22-1
The medical coding specialist.

Purpose and Function of Coding

For the health care provider to be properly paid for administering care and treatment to the patient, the facility or person requiring the payment must first make sure that the diagnosis, care, or treatment is legally acceptable according to the National Center for Health Statistics, which is part of the United States Department of Public Health Services. We refer to this process as *coding*. It usually involves first obtaining specific information about the patient and his or her diagnosis, and then translating that information into a number, which is part of an acceptable numerical coding system.

Information which has been coded can be used for any number of things. It may be used to retrieve specific data which might be necessary to determine the patterns of care or the success of treatments provided by a facility. It may also be used to predict staffing and other specific needs of a facility. It may be used to determine the incidence and clusters of specific diseases. And it may also be used to provide the federal government and private insurance carriers with specific information for the purpose of reimbursement.

Because of coding systems, most health care providers and institutions generally require basic information regarding the patient. This usually includes the patient's hospital or medical record number, date of birth, sex, race or ethnicity, zip code or code for a foreign country, and an expected pay source; the facility's identification number; the date on which the patient was admitted and discharged from the facility; the number of the attending physician and surgeon, if the patient had surgery; the disposition of the patient when he or she was discharged; the principal diagnosis and any others which may have been made during the patient's hospitalization; and the name of any significant procedures which may have been performed on behalf of the patient.

Once personal information has been properly obtained, the provider or facility is responsible for taking this documentation and transferring it into coded data, which can be expanded into both a disease index and a procedures index. The diseases and procedures numbers are listed numerically according to the ICD-9-CM coding system, which identifies every disease or condition and procedure performed during a particular time period. These codes are accompanied by other basic data items, such as the patient's identification number, date of birth, and sex. This information allows the provider to obtain medical records which might be necessary for audits, committee meetings, or other specific research.

Nomenclature and Classification Systems

The purpose of any nomenclature or classification system is to provide health care facilities, institutions, physicians, and government agencies with standardized statistical information. A system which provides the unique name for each disease process is called a *nomenclature* system, whereas a system which groups like or related disease and procedures is called a *classification* system.

International Classification of Diseases

Many years ago, the World Health Organization decided it was necessary to devise an international system providing governments and health care facilities with a listing of all diseases which could be used throughout the world. Originally referred to as the ICD-9 (International Classification of Diseases) system, the primary purpose was to serve as a listing of potential causes of death. However, over the years the ICD-9 became the tool of choice in providing information regarding all diseases, conditions, and procedures that patients may have undergone.

The ICD-9 is revised every 10 years and published by the United States National Center for Health Statistics. By adding fourth and fifth digits, categories were made more specific for use throughout the United States. The latest, that is, the ninth, revision is known as the ICD-9-CM, with the "CM" standing for "clinical modification." It is the recommended and in some cases required standard for most U.S. inpatient coding purposes. When the fourth and fifth digits are dropped, the coding system is compatible with the World Health Organization's system.

Components of the ICD-9-CM Coding System

To use the ICD-9-CM coding system properly, you must be able to meet three very important standards. All of your entries must be accurate, be complete, and follow proper sequencing. Once you have met these requirements, you must make sure that the diagnosis, disease, treatment, condition, or procedure falls under one of four specific categories. These include *principal diagnosis*, which is the condition that has been determined after diagnostic tests and workups and is the chief purpose for the patient's admission to a health care facility; *other diagnoses*, which are generally conditions that coexist at the time the patient has been admitted or which may have developed during the patient's hospitalization; *principal procedure*, that is, the one that is performed for treatment rather than for diagnostic purposes; and *other significant procedures*, that is, procedures which may have carried an operative or anesthesia risk or which may have required highly trained personnel to perform.

Guidelines for Working with the ICD-9-CM Coding System

If you are responsible for working with the ICD-9-CM coding system, there are some general guidelines to follow which will ultimately make your work a great deal easier, as well as help you to make sure everything you code is accurate. Many of these guidelines were first established by the Health Care Financing Administration, which is responsible for overseeing administration of the Medicare program throughout the United States. These basic guidelines include:

- Reading through the introduction section of the ICD-9-CM to make sure that you thoroughly understand any color coding and any use of symbols, abbreviations, or terms.
- Using both Volume I and Volume II while you are coding.
- Making sure you use the principal diagnosis.
- Coding only diagnoses which have been determined by a physician.
- Coding the diagnosis to the highest number of digits possible.
- Making sure you always code the main diagnosis, condition, or reason for the encounter first, followed by all others.
- Using the appropriate V-Code when a patient is seen for any ancillary diagnostic services.
- Coding diagnosis codes for chronic diseases or conditions as often as necessary.
- Using the correct code for billing surgical procedures.
- Looking up all adjectives and synonyms and cross-referencing them accordingly.

Working with the ICD-9-CM Coding System

There are three volumes that make up the ICD-9-CM. *Volume I* is a tabular or numerical listing that includes two books. *Volume II*, is an alphabetical listing that also contains two books. *Volume III* is a numerical listing that includes all surgical procedure codes.

When you are working with Volume I, it's important to understand that it is structured numerically according to individual body systems, and it should only be used when an ICD-9-CM code is provided and no linguistic description of the diagnosis exists. Table 22-1 shows you how Volume I is categorized:

The most commonly used book in ICD-9-CM coding is Volume II, which is an alphabetical listing of diagnoses. It is divided into four separate sections that include an alphabetical index of diseases and injuries, a table of drugs and chemicals, an alphabetical index of external causes of injuries and poisonings, and a listing of factors affecting the health status of an individual. The index that comprises the external causes of injuries and poisonings includes all accidents, and entries are referred to as *E-Codes*. The listing, which provides the factors affecting the health status of individuals, is referred to as *V-Codes*.

Volume III of the ICD-9-CM is primarily used for coding all diagnoses and procedures performed in

Table 22-1
ICD-9-CM Volume I Categories

Number	Body System/Classification
00–13	Infectious and parasitic diseases
14–23	Neoplasms
24–27	Endocrine, nutritional, metabolic diseases
28	Diseases of the blood and blood-forming organs
29–31	Nervous and mental health disorders
32–38	Diseases of the nervous system and sensory organs
39–45	Diseases of the circulatory system
46–51	Diseases of the respiratory system
52–57	Diseases of the digestive system
58–62	Diseases of the genitourinary system
63–67	Complications of pregnancy, childbirth, and in the puerperium
68–70	Diseases of the integumentary system and subcutaneous tissue
71–73	Diseases of connective tissue and the musculoskeletal system
74–75	Congenital anomalies
76–77	Specific causes of perinatal morbidity and mortality
78–79	Symptoms, senility and ill-defined conditions
80–86	Sprains, dislocations, fractures, and internal injuries
87–90	Lacerations
91–99	Other accidents, poisoning and violent injuries
VO–Y24	Miscellaneous codings; no particular diagnosis is noted

the hospital setting. It contains both a tabular listing, which itemizes all procedures according to body systems, and an index, which lists procedures in alphabetical order, thus making it easier to look up a procedure.

Body systems found in Volume III are arranged in the following manner:

- Operations on the nervous system
- Operations on the endocrine system
- Operations on the eye
- Operations on the ear
- Operations on the nose, mouth, and the pharynx
- Operations on the circulatory (cardiovascular) system
- Operations on the respiratory system
- Operations on the hemic and lymphatic system
- Operations on the urinary system
- Operations on the digestive system
- Operations on the female genital organs
- Obstetrical procedures
- Operations on the male genital organs
- Operations on the musculoskeletal system
- Operations on the integumentary system
- Miscellaneous diagnostic and therapeutic procedures

CPT Coding

CPT coding is the most commonly used system implemented by medical billing and claims processing personnel in the health care industry. As we have already stated, the purpose of the CPT is to provide a listing of descriptive terms and specific codes which are used for reporting medical services and procedures performed by physicians. Simply stated, CPT coding uses a five-digit code to identify procedures and services, which not only simplifies reporting but also provides for accurate compilation of data by a computer.

The RVS, or Relative Value Study, is often used with the CPT because it provides the coder with a listing of procedures and their appropriate codes, along with a unit value that has been assigned to the procedure. Often the CPT and the RVS are used interchangeably. However, the CPT, not the RVS, should be used when coding procedures.

Both the CPT and the RVS are divided into six major categories. These include *evaluation and management* (99201–99499); *surgery* (10000–69999); *anesthesia* (0010–0199 CVT; 99100–99140 RVS); *radiology and nuclear medicine* (70000–79999); *pathology and laboratory tests* (80000–89999); and *medicine* (90701–99199).

Evaluation and Management Codes

The purpose of evaluation and management codes is to designate procedures which are used to evaluate a patient's condition and to assist the patient in managing his or her own care. The most important components in determining the proper code include history, examination, medical decision making, counseling, coordination of the patient's care, the nature of the presenting problem, and time. Codes listed under evaluation and management include the following:

- *99201–99215*   Office visits
- *99221–99238*   Acute care
- *99241–99263*   Consultations
- *99271–99275*   Second surgical opinion consultations (SSO)
- *99281–99288*   Emergency department services
- *99301–99313*   Skilled nursing care (snf)
- *99321–99333*   Custodial care

- *99341–99353* Home services
- *99381–99429* Preventive medicine
- *99401–99492* Nervous and mental health care and treatment

Surgery Codes

Codes listed under surgery are categorized in the CPT book according to body systems. Within each of the body systems, the surgeries are arranged as to their individual anatomical positions from the head down. The arrangement of the surgical codes is listed in the following breakdown:

- *10000–17999* Integumentary system (skin, hair, and nails)
- *19000–19499* Breast
- *20000–23929* Musculoskeletal system (trunk, head, and body)
- *23930–26989* Musculoskeletal system (upper limbs)
- *26990–28899* Musculoskeletal system (lower limbs)
- *29000–29799* Musculoskeletal system (casting and strapping, all areas)
- *29815–29883* Musculoskeletal system (arthropscopy, all areas)
- *30000–32999* Respiratory system
- *33010–37799* Cardiovascular system
- *38100–38999* Hemic and lymphatic system
- *39000–39599* Diaphragm
- *40490–49999* Digestive system
- *50010–53899* Urinary system
- *54000–55899* Male reproductive system
- *56000–58999* Female reproductive system
- *59000–59899* Obstetrics, maternity care, and abortions
- *60000–60699* Endocrine system
- *61000–64999* Nervous system
- *65091–68899* Sensory system: Eye
- *69000–69979* Sensory system: Ear (divided into the outer, middle, and inner ear)
- *93501–93553* Cardiac catheterization

Block and Asterisk Procedures

Procedures requiring multiple surgical interventions during the same operative session, and which are performed in the same operative area, are called *block procedures*. These procedures are coded with a primary code and then subsequent codes. When a surgical procedure is accompanied by an asterisk, it generally means that special handling is required during the coding process. If the asterisk follows a CPT code, you should follow the guidelines listed below:

- Always make sure that the listed value is for the surgical procedure only.
- Make sure that all preoperative and postoperative care is added on a fee-for-service basis only.
- Use the procedure number 99025 whenever the asterisk procedure was carried out at the time of a new patient's initial visit.
- Never code a service visit separately when the asterisk procedure was carried out at the time of an established patient's follow-up visit.
- If an asterisk procedure required hospitalization, under the CPT, the appropriate hospitalization visit should be coded in addition to the asterisk procedure and its follow-up care.

Anesthesia Codes

Coding for anesthesia ranges from 00100 to 01999, and coding anesthesia claims can be completed using either the RVS or the CPT codes. In the RVS book, anesthesia unit values are listed for all procedures requiring anesthesia administration by a licensed anesthesiologist. You should also know that local anesthesia is never allowed separately. Thus, all benefits for anesthesia are those that have been allowed for procedures requiring more than a local anesthetic. These units are only used when the anesthesia was personally administered by a licensed physician, and when the physician remained in constant attendance during the entire procedure for the sole purpose of rendering the anesthesia.

Basic and Base Units, Time Units, Modifiers, and Qualifying Circumstances

Whenever you are responsible for coding surgical procedures and anesthesia, you must also have an understanding of how specific codes are determined. This has to do with basic and base units, time units, modifiers, and qualifying circumstances.

The *basic* or *base* anesthesia units are those that have been designed to allow for the usual preoperative and postoperative care, the administration of the anesthetic, and the possible administration of other fluids, such as blood, which may be necessary during the surgery. Surgical unit values include surgery, local infiltration, digital block, and topical anesthesia.

The length of time during which an individual remains under anesthesia determines the amount of money that will be allowable for that particular procedure. Coding for these services is referred to as *time units*. Anesthesia time begins in the operating room when the anesthesiologist first starts to prepare the patient for the induction of the anesthesia and ends when the anesthesiologist is no longer in constant attendance, that is, usually when the patient is ready for postoperative care and supervision. This time is always indicated on the claim form when billing for anesthesia services.

All anesthesia services must be billed according to a five-digit procedure code plus the addition of a physical status *modifier*. These modifiers are represented by the initial "P," followed by a single digit from one to six. For example:

- *P1*   A normal healthy patient
- *P2*   A patient with a mild systemic disease
- *P3*   A patient with a severe systemic disease
- *P4*   A patient with a severe systemic disease that is a constant threat to life
- *P5*   A patient who is not expected to survive without the surgical intervention
- *P6*   A patient who has been declared "brain dead," and whose organs are being removed for organ donation

In some cases, certain optional modifiers may be used to denote special circumstances. Some of the more frequently used include the following:

- *−22*   An unusual service was performed
- *−23*   An unusual anesthesia was used
- *−32*   A specific service was mandated
- *−47*   The anesthesia was administered by a surgeon.
- *−51*   Multiple procedures were performed during the surgery

In some cases, anesthesia may have to be administered under very difficult or unusual circumstances. In these cases, we refer to the determination of specific codes for services rendered as *qualifying circumstances*. In essence, this means that procedures which fall under this category would not be reported alone, but would be reported as additional procedure numbers qualifying an anesthesia procedure or service. Some examples of codes listed under qualifying circumstances include the following:

- *99116*   Anesthesia which was complicated by the use of total body hypothermia
- *99100*   Anesthesia for a patient of extreme age, generally one who is either younger than one year or older than 70 years of age
- *99135*   Anesthesia which has been complicated by the use of controlled hypotension
- *99140*   Anesthesia which has been complicated by an emergency condition

Radiology, Pathology, and Lab Codes

Radiology codes are categorized according to the anatomical position, that is, by individual body part, starting at the patient's head and moving downward toward the patient's feet. All radiology codes are classified into four individual categories: *diagnostic radiology* (70010–76499), *diagnostic ultrasound* (76506–76999), *therapeutic radiology* (77299–77799), and *nuclear medicine* (78000–79999).

Examination of chemical or tissue type make-up which have to do with analyzing the substances that make them up is a function of the pathology department. CPT codes which range from 80002 to 80092 refer to various types of laboratory panel tests. A panel test is a test composed of multiple examinations which are combined and then run from one individual specimen. In most health care facilities, it is not uncommon for the provider to "unbundle" the billing. Unbundling has to do with billing separately for each individual test performed even though all the tests may have come from the same specimen and been done at the same time. If you do receive a bill that was unbundled, you will need to rebundle it.

Many of the more routinely performed laboratory tests are listed in the CPT book and fall under codes from 80002 to 80092. The most frequently encountered of these include:

- *81000*   Urinalysis
- *82947*   Glucose
- *84702*   Pregnancy testing
- *88155*   Pap smear
- *87181–87192*   Culture and sensitivity (culture only)
- *87040–87163*   Culture and sensitivity (both culture and sensitivity)
- *93000–93350*   Electrocardiogram and other cardiac procedures
- *99000–99002*   Collection and handling charges

To bill the remaining laboratory tests, you must first determine whether or not these tests fall under the subsections of the original laboratory test. These subsections include the following:

- *81000–81099*   Urinalysis
- *82000–84999*   Chemistry and toxicology
- *85000–85999*   Hematology
- *86000–86999*   Immunology
- *87001–87999*   Microbiology

- *88000–88299*   Anatomical pathology
- *88300–88399*   Surgical pathology
- *89050–89399*   Miscellaneous

In terms of coding for the pupose of billing, whenever the physician has ordered a laboratory test or an x-ray for a patient, there are two distinct services that are actually being performed. The first service, called the *technical component*, is the actual taking of the specimen or x-ray. This charge is always included in the expense for the personnel actually performing the test and the cost of the necessary equipment, and is denoted by adding the modifier –27 to the CPT code. The second service, referred to as the *professional component*, has to do with the cost of reading the test's results. This charge is denoted by adding the modifier –26 to the CPT code.

Medicine Codes

Services billed for nonsurgical or medical care services fall under the medicine codes category. Nonsurgical services generally include optometry care, chiropractic care, acupuncture, physical therapy, and hospital care. The corresponding codes in the CPT range from 90701 to 99199.

Codes that fall under medical care services are classified into 10 specific sections. These include the following:

- *90701–90749*   Immunizations
- *90782–90799*   Therapeutic injections (subcutaneous, intradermal, intramuscular, intraarterial, and intravenous)
- *90900–90915*   Biofeedback (pain management)
- *90935–90999*   Dialysis
- *92002–92499*   Ophthalmology care and treatment
- *92502–92599*   Otorhinolaryngologic services (ear, nose, and throat)
- *92950–93979*   Cardiovascular services (EKG—93000; EKG interpretation and report only—93010; cardiovascular stress testing—93015; 24-hour EKG monitoring—93224)
- *97010–97799*   Physical therapy
- *97010–97145*   Physical medicine
- *97260; 97261*   Physical medicine

Miscellaneous services:

- *91000–91299*   Gastroenterology
- *94010–94799*   Pulmonary
- *95000–95199*   Allergy and immunology
- *95805–95999*   Neurology
- *96400–96549*   Chemotherapy
- *96900–96999*   Special dermatological procedures
- *99361–99373*   Case management services
- *99000–99199*   Special services and reports

Using Modifiers in CPT Coding

Modifiers are two-digit codes that may be added to CPT codes to denote an unusual circumstance. They are also used to more fully describe a specific procedure which may have been completed. Modifiers may also be used to alter the valuation of the procedure by increasing or decreasing the allowed amount.

Most modifiers are used with surgical codes. Modifiers –22, –23, –32, –47, and –51 may also be used with anesthesia coding. The most frequently used modifiers include the following:

–20   Microsurgery
–22   Unusual services
–23   Unusual anesthesia
–26   Professional component
–32   Mandated services
–47   Anesthesia by surgeon
–50   Bilateral procedure
–51   Multiple procedures but not bilateral
–52   Reduced services
–54   Surgical care only
–55   Postoperative management only
–56   Preoperative care only
–90   Reference laboratory
–64   Co-surgeons
–80   Assistant surgeon
–66   Surgical team
–58   When an asterisk(*) procedure is carried out at the time of the initial office visit

Using Unlisted Codes

In some cases, you may have to use unlisted codes. These are for procedures that are thought to be unusual or new and are, therefore, not designated with a descriptive code. They are located at the end of each CPT section and subsection, are often referred to as "junk codes," and always end with the numbers "99."

Terminology and Reference Materials

As you begin working with medical codes, you will soon realize that it is next to impossible for to do your work without having an understanding of the basic terms, abbreviations and symbols, and reference materials used in medical coding and claims processing.

Insurance Terms and Provider Abbreviations

The majority of the terms which you will encounter and must become familiar with have to do with medical insurance information. The most common of these terms include the following:

- *Accumulation period:* That period of time, usually from January 1 through December 31, which is used to satisfy a specific deductible amount and accumulate COB credit reserves.
- *Adjustment:* A specific correction of an application for benefits or for coding a claim.
- *ASO* Abbreviation for *Administrative Services Only*, which is the time in which services such as actuarial benefit plan, claim processing, and analysis must be performed.
- *Administrator:* The "third party" responsible for handling all functions related to the operation of a group or self-funded insurance plan.
- *Assignment of Benefits:* Used in the processing of health insurance coverage, it is a method according to which a claimant requests that his or her benefits be paid to a designated person or institution.
- *Audit claim:* The process by which the accuracy of a claim is checked according to the provisions of the contract of the plan.
- *Basic benefits:* The portion of a plan of benefits which is not usually subject to a deductible.
- *Beneficiary:* The person who has been designated by a policy to receive the proceeds or benefits from a policy upon the death of the insured.
- *COB:* Abbreviation for *Coordination of Benefits*, that is, a process by which claim payors are able to determine who are the primary, secondary, and tertiary payors.
- *COBRA:* Abbreviation for *Consolidated Omnibus Budget Reconciliation Act of 1985*, which was established to provide for the continuation of coverage for all employees who are no longer considered eligible by their employer's insurance plan.
- *Coinsurance:* The arrangement between a member and a plan to share a specific ratio of covered losses under a policy.
- *CoPay:* The amount which a member or insured person is responsible for paying out of his or her pocket; this is normally 20% of the charge for services.
- *Covered expenses:* Those expenses that are allowable under a covered plan.
- *Deductible:* The amount of covered expenses that must be paid by the insured or member before any benefits can be paid by the plan.
- *DEFRA:* Abbreviation for *Deficit Reduction Act of 1984*, which requires that all employers with 20 or more employees to offer spouses, age 65 or older, of active employees the same coverage offered to younger spouses.
- *Eligibility period:* That period, generally 31 days, following the eligibility date, during which an individual member of a specific group will remain eligible to apply for insurance coverage.
- *EOMB:* Abbreviation for *Explanation of Medicare Benefits*, which are issued by the Medicare claims payor to explain the application of benefits on all claims submitted on behalf of Medicare participants.
- *Evidence of insurability:* A health questionnaire used to determine whether a person's physical and mental condition affects the acceptance of the applicant's application for health or life insurance coverage.
- *Exclusion:* A term used to specify conditions or types of services for which a policy does not provide benefits.
- *Extended benefits:* The continued entitlement of a member, under certain circumstances, to receive benefits after coverage has terminated.
- *HMO:* Abbreviation for *Health Maintenance Organization*, that is an alternative form of health care delivery that refers all patients to a specific location and provides a full range of health care services.
- *Last quarter carry-over:* Amounts that have been applied to a deductible during the last three months of a calendar year which are carried over to the next year.
- *Major medical:* That portion of a health care plan that pays benefits at a set percentage, and which is generally subject to a deductible.
- *Maximum(s):* The maximum amount payable by a plan.
- *OOP:* Abbreviation for *Out-of-Pocket Expenses*, which are the amounts applied to expenses for which the insured is held responsible and which he or she must pay.
- *Peer review:* The process by which an independent review by a group of physicians is performed to determine if the services submitted by another physician are usual, customary, and reasonable for the condition being treated.

- *Policy:* That document issued by an insurance company to the policy holder that outlines the terms and conditions of the insurance coverage.
- *Preexisting exclusion:* A condition that has been specified by a plan prior to the effective date of coverage, and which is not covered by that plan.
- *PPO:* Abbreviation for *Preferred Provider Organization*, which is a medical group made up of physicians that negotiates discounted fees, and does utilization review and quality assurance in exchange for a referral of a population of patients or monthly per capita payments.
- *Subrogation:* A term used when contract provision is provided by a plan that gives the plan authority to recover monies paid on claims that are covered by a third party.
- *Third party liability provision:* A provision by a plan that, when an injury is covered by a third party, reimbursement will be provided to the plan should recovery be made through that coverage.
- *TIN:* Abbreviation for *Taxpayer Identification Number*, which is a number required by the Internal Revenue Service for a provider or business for tax purposes.
- *UCR:* Abbreviation for *Usual, Customary and Reasonable*, which refers to a specific amount determined to be the maximum allowable by a plan, based on the contract provisions, procedure, date of service, and geographical location of the provider.

Since many of the types of insurance benefits paid to institutions and individual providers are based upon the provider's type of licensing and the type of service being provided, it is also very important for you to learn and understand how to use the many provider abbreviations used in coding and insurance claim processing. The most common of these include the following:

- *Clinical psychologist* A person who is licensed to perform psychological testing and therapy.
- *C.R.N.A.* Abbreviation for a *Certified Registered Nurse Anesthetist*, or a registered nurse certified to administer anesthesia under the direction and supervision of a medical doctor.
- *D.C.* Abbreviation for *Doctor of Chiropractic*, a person responsible for performing manual manipulations of the spine and other musculoskeletal areas.
- *D.D.S.* Abbreviation for *Doctor of Dental Surgery*, a person who is licensed to perform all dental care including dental surgeries and surgeries of the face and jaw.
- *D.O.* Abbreviation for *Doctor of Osteopathy*, a person licensed to perform any service that a medical doctor may.
- *D.P.M.* Abbreviation for *Doctor of Podiatry*, a person licensed to provide care, treatment, and surgery of the feet.
- *E.M.T.* Abbreviation for *Emergency Medical Technician*, a person licensed to administer emergency care and procedures such as CPR.
- *L.C.S.W.* Abbreviation for a *Licensed Clinical Social Worker*, a person licensed to provide psychological counseling and to whom referral must be made by a medical doctor.
- *L.V.N.* Abbreviation for *Licensed Vocational Nurse*, a licensed nurse, trained to perform patient-care treatments, and administer certain medications, except those requiring intravenous push and the infusion of blood and blood by-products, who generally works under the supervision of a registered nurse; *L.P.N.*, which is the abbreviation for *Licensed Practical Nurse*, is the equivalent of an L.V.N.
- *M.D.* Abbreviation for *Medical Doctor*, a person licensed to perform any and all medical care, treatments, and procedures.
- *M.F.C.C.* Abbreviation for *Marriage, Family, and Child Counselor*, a person licensed to provide psychological counseling and marriage and family counseling, referral to whom must be made by a medical doctor.
- *Midwife* A registered nurse who has received additional training and is certified as a nurse midwife. He or she must be associated with a medical doctor or doctor of osteopathy; this person is usually responsible for handling only routine maternity cases.
- *M.S.W.* Abbreviation for *Medical Social Worker*, a person licensed to provide family and psychological counseling, referral to whom must be made by a physician.
- *M.T.* Abbreviation for *Medical Technologist*, a licensed person who generally works in a medical lab and who is often responsible for drawing blood and performing clinical laboratory tests.
- *N.A.* Abbreviation for *Nurse Assistant (Aide)*; upon certification, this person is generally responsible for providing basic care in skilled nursing facilities, in acute care hospitals, and through home health agencies; may also be known as a *C.N.A. (Certified Nurse Assistant)*.
- *N.P.* Abbreviation for *Nurse Practitioner*, a licensed registered nurse who has received additional training to become certified as a nurse practitioner and who works under the direction of a physician or D.O.
- *O.D.* Abbreviation for *Doctor of Optometry*, also

known as an Optometrist. This person usually performs eye refractions, dispenses glasses and contact lenses, and is not licensed to perform any type of eye surgery.
- *O.T.* Abbreviation for *Occupational Therapist*, a person licensed to perform occupational therapy generally involved with retraining muscles and nerves to perform basic routine daily activities and movements.
- *P.A.* Abbreviation for *Physician's Assistant*, a person who has attended a specialized program that allows him or her to assist with routine care and services under the direction of physicians.
- *Paramedic* A person who has received extensive training as an E.M.T., is licensed to administer emergency care and treatments, and normally works on an ambulance or for a local fire department.
- *R.N.* Abbreviation for *Registered Nurse*, a person who has undergone the highest level of nurse training and who is licensed to practice nursing under the direction of a physician.
- *R.P.T.* Abbreviation for *Registered Physical Therapist*, a licensed person who has received training necessary to perform physical therapy when services are prescribed by a physician or D.O.

Using Reference and Supplementary Coding Books

In addition to having a basic understanding of some of the fundamental terms and abbreviations used in medical coding and claims processing, it is also important for you to be able to use some of the more frequently used reference books (see Figure 22-2). All of these books should be used as sources of information when you are required to perform coding or billing tasks. The most common books used for billing purposes include the *International Classification of Diseases–9th Revision Clinical Modification (ICD-9)*, the *Physician's Current Procedure Terminology* (CPT), the *Relative Value Study* (RVS), the *Physicians' Desk Reference* (PDR), a *Medical Dictionary*, and the *Merck Manual*.

The ICD-9, which consists of three volumes, is used to index conditions and is generally used by the medical biller and claims examiner to convert descriptions of illnesses, injuries, or other conditions

Figure 22-2
Reference books used in coding.

Administrative Medical Services

into numerical codes. It also provides for the classification of diseases for statistical purposes and identifies symptoms, diseases, injuries, and routine services according to a three-, four-, or five-digit code that can be either entirely numerical or an alpha-numeric combination.

The *Physician's Current Procedure Terminology* (CPT) and the *Relative Value Study* (RVS) are both books used for coding procedures or services performed by physicians. The purpose of both the CPT and the RVS is to provide the coder with a uniform method for accurately describing medical, surgical, and diagnostic services. They may be used interchangeably even though the rules guiding their use are somewhat different.

There are six major sections that make up both the CPT and the RVS. They include the following:

- *99201–99499*   Evaluation and Management
- *90001–99199*   Medicine
- *10000–69999*   Surgery
- *0010–0199 (CPT)*   Anesthesia
- *99100–99140 (RVS)*   Anesthesia
- *70000–79999*   Radiology and Nuclear Medicine
- *80000–89999*   Pathology and Laboratory Tests

A PDR, or *Physicians' Desk Reference*, enables the person processing the claim to determine if a pharmaceutical is a prescription or nonprescription drug. When billing for payment, this is an extremely important distinction, since most health care insurance plans do not cover nonprescription drugs.

The PDR, which is often used by lay people wishing to look up specific drug interactions and determine whether these drugs are covered by their insurance plan, is divided into six major categories:

- Manufacturers' Index.
- Product Identification Index
- Product Name Index.
- Product Information Section
- Product Category Index.
- Diagnostic Product Information

Another important reference book used by the medical coding worker is a medical dictionary. While there are many different medical dictionaries currently on the market, the purpose of all of them is to provide the worker with a listing of medical terms and their definitions, synonyms, illustrations, and supplemental information, all of which can be used to assist the biller or claims examiner in identifying specific diagnoses and their symptoms, prognosis, and common treatment protocols. For information providing the greatest detail with regard to symptoms and treatment protocols, the *Merck Manual* is considered the most definitive.

In addition to providing you with basic definitions, most medical dictionaries include other data or information regarding the word or term. These usually include the etymology, or original language or meaning, of the word, pronounciation of the word, biographical information on specific diseases, symtoms, conditions, procedures, or cures that have been given a specific name; synonyms; abbreviations; etiology of diseases; common medical treatments for specific conditions or diagnoses; cross-referencing data; prognosis for specific diseases; nursing implications and nursing diagnosis, cross-referencing; subentries that contain more specific information regarding a term or condition; illustration cross-referencing; and cautions or warnings, such as side effects or adverse reactions that may occur from use of a specific drug or treatment.

If you find yourself having to look up information regarding, for example, specific symptomology, prognosis, treatment, protocols, or etiology regarding a specific diagnosis, you will probably have to rely upon the *Merck Manual*. This reference book, which is divided into two sections, a listing of diseases and an index, is generally used to assist the medical coder in identification of treatment that may or may not be appropriate for a reported diagnosis or symptomology. To look something up, the coder simply turns to the index to find the page number with information on the disease. This information includes the diagnosis, symptoms, prognosis, and treatment.

Working with V-Codes and E-Codes: Supplementary Reference Books

As you continue working with coding and claims processing, you may come into contact with supplementary listings of specific data that may ultimately affect how a patient's medical condition is coded for billing and payment purposes. Two of these supplementary listings include V-Codes, which provide the coder with additional information or factors affecting the health status of patients, and E-Codes, which are a supplementary classification of external causes of injury and poisoning.

Using the V-Code Supplementary Listing

There are three instances relating to a specific disease or condition, but not considered to be a disease, in which V-Codes are used. The first is when a non-ill person visits a health care provider for a specific purpose for which there is no specific disease.

This can be a visit for a vaccination or a check-up or to discuss a problem that is not considered a disease or a condition. The second instance in which V-Codes are used is when a patient with a chronic or recurring condition visits a physician or a health care provider for a service that may be associated with treatment of that condition. This may include times when a patient has come in to have his or her pacemaker monitored or checked, have a cast changed, or have dialysis for renal disease.

When a situation occurs that may influence a person's health, but the condition is not considered a disease, you may also use V-Codes. These types of conditions generally include animal bites that require vaccination, suspected infectious diseases, allergies, or exposure to potential health hazards.

V-Codes are arranged into eight categories:

- V01–V09   Hazards related to communicable diseases
- V10–V19   Hazards related to personal and family history
- V20–V29   Circumstances related to reproduction and development
- V30–V39   Liveborn infants categorized according to type of birth
- V40–V49   Persons with a medical condition which may influence their health status
- V50–V59   Persons who use health services for specific procedures and aftercare
- V60–V68   Persons who use health services in other circumstances
- V70–V82   Persons who have no reported diagnosis who are seen during an examination and investigation of persons and populations

Using the E-Code Supplementary Listing

As we have already stated, E-Codes are used when information is needed with regard to external causes of injuries and poisoning. They may also be used to indicate a cause of the injury but not the injury itself. This is why an E-Code is always considered a supplementary code and not a primary diagnosis code.

When using E-Codes, you must always consult the index located in Section 3 of Volume 2, behind the table of drugs and chemicals. This index is used in the same manner in which that for diseases and conditions are used. You must first locate the cause in the index and then turn to the tabular listing to confirm your choice and insure the correct fourth and fifth digit subclassifications.

Whenever you are coding transport accidents, you should use the following E-Code categories. Always remember that, if more than one type of vehicle has been involved, you follow the order according to how it is listed below:

- E840–E845   Aircraft and spacecraft
- E830–E838   Watercraft
- E810–E825   Motor vehicles
- E800–E807   Railway
- E826–E829   Other road vehicles

For any accidents involving machinery, that is, nontransport vehicles, you use category E919.x. This allows you a much broader description of the type of machinery and activity that caused the injury. If you would rather provide a more detailed description of the injury, you may also use the *Classification of Industrial Accidents According to Agency* reference book, which was developed by the Internal Labor Office. However, this classification should be used only as a supplement to the appropriate E-Code and not in its place.

Summary

In this chapter, we discussed the function of coding, including the main differences between nomenclature and classification systems as each relates to billing and coding procedures. We also talked about coding systems using the International Classification of Diseases, that is, the ICD-9-CM coding system, CPT coding, and V-Codes and E-Codes. We also identified some of the most frequently used terms, abbreviations, and reference manuals used in basic coding procedures.

Review Questions

1. Define the following abbreviations:
    a. MDC _____
    b. ART _____
    c. DRG _____
    d. AHIMA _____
    e. HCFA _____

Administrative Medical Services

2. Briefly explain the purpose and function of *medical coding*.

3. What organization was responsible for establishing the International Classification of Diseases?

4. Briefly define the following as they relate to ICD-9-CM coding:

   a. principal diagnosis _____

   b. other diagnosis _____

   c. principal procedure _____

   d. significant procedure _____

5. The ICD-9-CM includes three volumes. Describe what is contained in each of the volumes:

   a. Volume I _____

   b. Volume II _____

   c. Volume III _____

6. For the codes listed below, describe the services identified for each series of codes:

   a. 99201–99215: _____

   b. 99321–99333: _____

   c. 99301–99313: _____

7. All radiology codes are classified into four individual categories. Name them.

   a. _____

   b. _____

   c. _____

   d. _____

8. Briefly explain what an *unlisted code* is.

9. Define the following abbreviations:

   a. COBRA _____

   b. DEFRA _____

   c. EOMB _____

   d. COB _____

10. Briefly explain the following two concepts:

    a. coinsurance _____

    b. evidence of insurability _____

Coding Diseases

Performance Objectives

Upon completion of this chapter, you will be able to:

1. Define the purpose and function of the integumentary system and briefly explain how diseases of this system are properly coded for billing.
2. Define the purpose and function of the musculoskeletal system and connective tissue and briefly explain how diseases of each are properly coded for billing.
3. Define the purpose and function of the digestive, urinary, and reproductive systems and briefly explain how diseases of each are properly coded for billing.
4. Define the purpose and function of the cardiovascular system and blood-forming organs and briefly explain how diseases of each are properly coded for billing.
5. Define the purpose and function of the nervous and endocrine systems and briefly explain how diseases of each are properly coded for billing.
6. Discuss various types of infections and neoplasms and briefly explain how each are properly coded for billing.
7. Identify various types of unusual situations and briefly explain how each is properly coded for billing.

Terms and Abbreviations

Alcoholic psychosis an organic psychotic state due to the excessive consumption of alcohol that can include delirium, amnesia, hallucinosis, withdrawal, and paranoia.
Benign something, usually a tumor or neoplasm, not returning or tending to progress; not malignant.
CBC abbreviation for *complete blood count*, which is a routine test used to assess the cellular components of blood.
Contusion an injury that does not break the skin, is usually caused by a blow to the area, and is characterized by swelling, discoloration, and pain.
CVA abbreviation for *cerebrovascular accident*, which is an abnormal condition involving the blood vessels of the brain and which may also include thrombosis, embolism, and hemorrhage.
Diabetes Insipidus a metabolic disorder caused by insufficient production of the antidiuretic hormone by the pituitary gland.
Diabetes Mellitus a disorder of the digestive system in which the ability to use carbohydrates is lost as a result of disturbances in insulin production by the pancreas.
Glaucoma a disorder of the eye with which there is an elevated pressure within the eye due to an obstruction of the outflow of the aqueous humor.
Hallucination a sensory perception not resulting from any external stimulus that is generally caused as a result of a drug overdose or mental health disorder.
Hydrocephalus a pathological condition of the newborn with which there is an abnormal accumulation of cerebrospinal fluid in the ventricles of the brain.
Hypertension a disorder of the cardiovascular system that is characterized by an elevation in the blood pressure.
Hypothyroidism a condition of the endocrine system that is characterized by decreased activity of the thyroid gland.
Ischemic heart disease a pathological disorder of the myocardium of the heart that is caused by a lack of oxygen reaching the tissues.
Kyphosis an abnormal condition of the musculosketal system that is characterized by a convexity of the thoracic spine: also known as "*hunchback.*"
Metastasis the tranfser of a disease from one organ to another,

usually at distant sites, through the lymphatic system, the bloodstream, or body cavities.

M.I. abbreviation for *myocardial infarction*, which is an occlusion of a coronary artery caused by atherosclerosis or an embolism.

PKU abbreviation for *Phenylketonuria*, which is a metabolic disorder of the newborn caused by the absence or deficiency of phenylalanine hydroxylase.

Psychosis a major emotional disorder that involves the loss of contact with reality.

Spina bifida a congenital defect in the posterior vertebral arch that may also involve a herniation of the spinal cord or meninges, thus resulting in neurological deficits.

Ulcer a crater-like lesion of the mucous membrane that results from an inflammatory or infectious process.

Urinalysis a physical, microscopic, or chemical examination of urine that is used to identify various structures found in the urine, including sugar, protein, ketones, and blood.

UTI abbreviation for urinary tract infection.

If you are the person responsible for coding, billing, or processing the patient's medical insurance claim form, it is extremely important that you have an understanding of the didactic theory and hands-on skills involved in medical coding procedures, as well as an awareness of the most common diseases and disorders of the individual body systems. Once you are knowledgable in both areas, you will find yourself in the unique position of having many choices open to you as to where and for whom you work. One of the greatest advantages to working in medical coding and insurance claims processing is that health care workers employed in these areas seem to be higher paid and in greater demand. Much of this is due to the role this person plays in maintaining the facility's or physician's fiscal well-being, that is, making sure that the institution or provider gets paid for services rendered.

As we discussed in Chapter 22, all diseases, disorders, infections, neoplasms, and emergency medical conditions are identified, according to individual body systems, by a specific code in the ICD-9-CM coding book. Diseases itemized in this book are listed according to the following systems: integumentary, musculoskeletal, digestive, urinary, male and female reproductive and obstetrics, cardiovascular and blood-forming organs, nervous and sensory, and endocrine. Infections, neoplasms, and emergency and unusual medical situations are also identified. In this chapter, we will discuss how to code some of the more common diseases.

Diseases of the Integumentary and Musculoskeletal Systems

The Integumentary System

The integumentary system, which is composed of the skin, subcutaneous tissue, hair, and nails, is responsible for protecting our bodies against injury and the invasion of bacteria. It also helps in the regulation of body temperature, thus aiding in the prevention of dehydration.

While there are many disorders of the integumentary system, those you will encounter most frequently generally include diseases causing some type of skin hypersensitivity or irritation, often referred to as *dermatitis*; those causing a type of scarring of the skin, called *keloids* and *cicatrixes*; those leading to an excision from the subcutaneous tissue, called *lesions*; and those causing a general breakdown of the layers of the skin, commonly referred to as *ulcers*.

When coding diseases of the integumentary system, especially those listed under dermatitis, you must first determine what type of substance the patient has come into contact with. For example, if the patient has come into contact with a substance taken internally, the code will be listed under category 693. If the patient came into physical constact with a substance which caused the hypersensitive reaction, the code will be found under category 692. Also, skin disorders identified as ulcers, that is, those that cause a breakdown of the layers of the skin, can be found under category 700. For example, an ulcer diagnosed as gangrenous will be coded as 785.4.

Keloids and cicatrixes are much easier to code. Each can be found in the CPT. Keloids are found under the 700 codes, and cicatrixes are located under the 900 codes. All excisions of tumors or lesions can best be located by first looking under the site in which the lesion is located, and then noting that specific code. For example, a lesion that has been excised from a woman's breast would be coded as 85.21.

The Musculoskeletal System

When we talk about the musculoskeletal system, we are actually talking about those individual structures, such as bones, joints, muscles, ligaments, and tendons, that are necessary for our bodies to be supported and protected. These structures are also important and necessary for us to move and produce and store certain mineral deposits, such as calcium and phosphorus.

Diseases of the musculoskeletal system and those associated with connective tissue disorders are classified from 710 through 739 of Volume I in your CPT book. Those which you will enounter most frequently including the following:

- *710–719*   Arthropathies and related disorders
- *721*   Spondylosis and allied disorders
- *736*   Acquired deformities
- *737.2*   Acquired lordosis
- *737.3*   Scoliosis and kyphoscoliosis
- *737.4*   Curvatures of the spine associated with other conditions
- *739*   Nonallopathic lesions, somatic dysfunctions, and segmental dysfunctions

Diseases of the Digestive, Urinary, and Reproductive Systems

The Digestive System

The digestive system aids us in the breakdown and metabolism of foods and rids our bodies of by-products that are not necessary to the body's metabolism. Like the integumentary and musculoskeletal systems, there are many disorders of the digestive system. Most of those which you will be responsible for coding can be found under category 500. The most common of these include the following:

- *520*   Congenital anomalies of the teeth
- *534*   Gastrojejunal ulcers
- *560*   Diseases involving the paralytic ileus (intestinal peristalsis) and paralysis of the intestine or colon
- *550*   Inguinal hernias and their repair
- *562*   Diseases of the diverticulum

The Urinary System and Reproductive Systems

The major function of the urinary system is to produce and eliminate urine. Diseases of this system that you will most often encounter generally involve the structures of the urinary system, including the kidneys, ureters, urethra, and urinary bladder, as well as the reproductive organs of both the male and female. Because both the urinary and the reproductive systems are involved, when discussing disorders coded under this section, we often refer to these systems as the genitourinary system. The most common disorders under this category include those of the ureter and urethra, endometriosis, fistulas, renal failure and renal dialysis, and urinary tract infections.

Endometriosis, which is when endometrial tissue from the female uterus moves from its original or normal place of origin to other parts of the body, is coded according to where the endometriosis is located. While the actual category of this disease is listed under 617, the individual subcodes are used to identify where the endometrial tissue has moved. For example, if the uterus is involved, it is coded as 617.0; if the ovary is involved, it is coded 617.1; and if the fallopian tube is involved, it is coded as 617.2.

A common disorder of the female genitalia, which affects both the urinary and the reproductive systems, is the fistula, which involves the build-up or development of an abnormal passageway, either between two organs or between one organ and the surface of the body. All fistulas involving the organs of the female are listed under category 619.

Disorders that have to do with renal failure, that is, the inability of the kidneys to excrete waste, thus resulting in an accumulation of concentrated urine, are all coded under category 585. However, if the failure of the kidneys has to do with the patient suffering from hypertension, or high blood pressure, and the condition becomes chronic, the disease is coded under category 403.

One of the most common disorders of the urinary system is a urinary tract infection or UTI. The most important aspect for you to remember when coding UTIs is that you must make sure that, when you code this disease, you are specific as to whether the UTI is a result of an infection or is secondary to another disorder. For example, a UTI that results from a trichomonas infection would be found under category 131, while a UTI that results from a urethral fistula would be located under category 599.1.

Diseases of Pregnancy and Childbirth

Whenever we discuss diseases of the reproductive system, we must also address those disorders classified under pregnancy and childbirth. The majority of these conditions are classified according to ectopic and molar pregnancies, abortion, complications related to pregnancy, malpresentations, fetal abnormalities, Rh factor incompatability, death of the fetus in utero, elderly primigravida, trauma to the perineum and vulva during childbirth, postpartum hemorrhage, cesarean delivery, sterilization, disorders of the breast and lactation, and high-risk pregnancies. Following is a list of the codes involving conditions during pregnancy and childbirth which you will encounter most frequently:

- *630*   Hydatidform mole
- *633*   Ectopic pregancy
- *639*   Complications of abortion and ectopic and molar pregnancies
- *640*   Complications related to pregnancy
- *650*   Normal delivery
- *651*   Multiple pregnancies (gestation)
- *652*   Malpositions and malpresentations of the fetus
- *654*   Abnormalities of the organs and soft tissues of the pelvis
- *656*   Rh factor incompatabilities and other fetal and placental complications
- *658*   Premature rupture of membranes
- *659*   Elderly primigravida
- *663*   Complications of the umbilical cord
- *664*   Perineal and vulva trauma during delivery (lacerations)
- *666*   Postpartum hemorrhage
- *669*   Cesarean delivery
- *676*   Disorders of the breast and lactation associated with childbirth
- *V22*   Normal pregnancy
- *V23*   High-risk pregnancy
- *V24*   Postpartum care and examination
- *V30–V39*   Liveborn infants according to type of birth
- *763*   Forceps delivery
- *765*   Conditions related to short term gestation and low birth weight
- *767*   Birth trauma
- *768*   Fetal distress
- *770*   Other respiratory disorders of the fetus and newborn
- *771*   Congential conditions in the newborn
- *773*   Hemolytic conditions in the newborn due to isoimmunization
- *741*   Spina bifida
- *749*   Cleft palate and cleft lip in the newborn
- *331.4*   Hydrocephalus in the newborn

Diseases of the Cardiovascular System and Blood Forming Organs

When we talk about diseases of the cardiovascular system and blood forming organs, we are actually discussing three separate types of structures. The first group of structures is comprised of those organs found in the circulatory system, including the heart and the blood vessels. The second group of structures is comprised of those organs associated with the manufacturing of blood. And finally, the third group of organs found in the cardiovascular system is comprised of those structures associated with the process of respiration. They include the nose, larynx, pharynx, bronchus, and lungs.

While there are many different diseases and conditions associated with the cardiovasuclar system and its associated blood forming organs, those which you will encounter most frequently include the following:

Diseases of the circulatory system:

- 796.2   Hypertension (high blood pressure), as a symptom and not a disease
- 401   Hypertension
- 402   Hypertensive heart disease
- 403   Hypertensive renal disease
- 404   Hypertensive heart and renal disease
- 410–414   Ischemic heart disease (acute myocardial infarction (MI): 410; old MI: 412)
- 413   Angina
- 414   Coronary artery arteriosclerosis (ASHD)
- 429.2   Ateriosclerotic cardiovascular disease (ASCVD)
- 426   Disorders of the heart's conduction system
- 440   Arteriosclerosis with no other associated condition
- 786   Chest pain (as a symptom: 786.50; as precordial pain: 786.51)
- 428   Congestive heart failure (not associated with hypertensive heart disease)
- 436   Cerebrovascular disease (acute but ill-defined)

Procedures used in diagnosing diseases of the heart and blood vessels:

- 36.1 — Bypass anastomosis for heart revascularization
- 37.21–37.23 — Cardiac (heart) catheterization
- 37.8 — Insertion, revision, replacement, and removal of pacemaker devices (leads: 37.7)
- 88.5 — Angiocardiography
- 39.61 — Cardiopulmonary bypass
- 996 — Complications of coronary arterial bypass

Diseases of the blood and blood-forming organs:

- 280–282 — Anemia
- 648 — Blood disorders associated with pregnancy and childbirth

Diseases of the respiratory system:

- 461 — Sinusitis (acute)
- 474 — Tonsillitis (chronic); tonsillitis with adenoids involved: 474.1
- 482 — Pneumonia (bacterial)
- 493 — Asthma
- 496 — Chronic obstructive pulmonary disease (COPD)
- 500–508 — Lung diseases due to external agents (pneumoconiosis and asbestosis)

Diseases of the Nervous System and Sensory Organs

When we discuss diseases of the nervous system and the sensory organs, we are actually discussing two distinct categories. The first, called the central nervous system (CNS), is made up of the brain and the spinal cord. The second part of the nervous system is called the peripheral nervous system. It includes all of the cranial nerves, as well as the nerves of the autonomic nervous system (ANS), and generally includes all of the disorders and conditions affecting the eyes and the ears.

Like all of the other body systems, there are many diseases and disorders which affect each and every individual structure within the two parts of the nervous system. Therefore, we will list only those which you are most likely to encounter during your experience as a medical coding worker. These include the following:

Diseases of the nervous system:

- 320–320.3 — Meningitis (bacterial: 320; hemophilus: 320.0; pneumococcal: 320.1) streptococcal 320.2; staphylococcal: 320.3
- 323.5 — Encephalitis (following immunization)
- 344 — Paralytic syndromes (quadriplegia: 344.0; paraplegia: 344.1)
- 345 — Epilepsy involving seizures (grand mal: 345.1; grand mal status: 345.3)
- 780.3 — Epilepsy as a symptom, no seizures
- 784 — Headache (sick headache: 346.1)

Diseases of the eyes and ears:

- 369 — Profound impairment, both eyes (blindness)
- 378 — Esotropia (convergent concomitant strabismus)

Mental Health Disorders

Diseases classified as mental health disorders are found in Appendix B of Volume I in the ICD-9-CM book. Many physicians and psychiatrists also use the DSM or Diagnostic and Statistical Manual of Mental Disorders to code their diagnoses. Most of the codes found in the DSM are identical to those listed in the ICD-9-CM.

Mental health disorders and conditions which are most frequently diagnosed by licensed psychiatrists include the following:

- 290–294 — Organic psychotic diseases (presenile dementia: 290.10)
- 295 — Schizophrenic disorders
- 296 — Manic-depressive disorders
- 301 — Personality disorders
- 303 — Alcohol dependence
- 304 — Drug dependence
- 305 — Alcoholic psychosis
- 306 — Psychogenic condition caused by physical symptoms and/or conditions
- 308 — Acute reaction to stress (anxiety attack)
- 310 — Organic brain syndrome
- 311 — Depression
- 317–319 — Mental retardation
- 331 — Alzheimer's disease

Disorders of the Endocrine System, Infectious Diseases, and Neoplasms

Disorders identified and classified under diseases of the endocrine system are diseases and conditions affecting the various glands of the body and the nutritional and metabolic processing of food as it makes its way throughout the body's systems. Conditions that affect our immunity to specific diseases are also listed under endocrine diseases.

Most disorders and medical conditions classified as endocrine, nutritional and metabolic, and immunity disorders are represented by codes 240 through 279, with the exceptions being those disorders and disturbances specifically related to the fetus and newborn. They are coded as 775.0 through 775.9. The disorders you will most frequently encounter during your coding experience include the following:

- 242   Goiters
- 243   Hypothyroidism (congenital)
- 244   Hypothyroidism (acquired)
- 245   Thyroiditis
- 250   Diabetes mellitus (gestational diabetes mellitus: 648.8)
- 251   Hypoglycemic coma (insulin coma)

Infectious and Parasitic Diseases

Most of the infections and parasitic diseases which you will encounter are listed in Volume I of the ICD-9-CM and are coded from 001 through 139. These include the following:

- 003–003.8   Intestinal infections
- 005   Bacterial food poisoning
- 009   Intestinal infections that cannot be defined (gastroenteritis, colitis, enteritis)
- 010–018   Tuberculosis
- 027   Other zoonotic bacterial conditions
- 041; 079   Used to identify bacterias and viruses not otherwise identified
- 042–043   Human immunodeficiency virus (AIDS with specified condition: 042; AIDS-related complex [ARC]; pre-AIDS: 043)

Neoplasms

Neoplasms, which are growths or tumors, are found in Volume II of the ICD-9-CM. The site in which the neoplasm is located determines which specific code number should be used. The site may either be primary, that is, located in the place of origin of a malignancy, or it may be secondary, meaning those which have the potential for becoming invasive and, eventually, lethal.

Most of the neoplasms you will be responsible for coding will fall under the following categories: nonsolid tumors, benign neoplasms, and malignant neoplasms. Tumors which are nonsolid use codes 200 through 208. The most common of these is leukemia, which is coded as 204.

Benign neoplasms include those which are noninvasive and can be surgically removed without further threat to the patient's life or limb. Those codes range from 210 through 229.

Malignant tumors or neoplasms are potentially most lethal. These are tumors which may begin in a primary location, but always have the ability to move into a secondary site through a process known as metastasis. Codes for malignant neoplasms range from 140 through 239. Those most frequently seen include the following:

- 153.9   Metastic carcinoma of the colon
- 164   Malignant neoplasms of the heart, thymus, and mediastinum
- 171.9   Metastatic sarcoma
- 197   Secondary site, lung
- 200–208   Malignant neoplasms of lymphatic and hematopoietic tissue

Coding Unusual Situations and Nondefined Conditions

As you continue working in the field of medical coding, you may be given additional responsibilities and tasks that require you to code disorders or conditions which have no real defined sign or symptom. We generally refer to these instances as unusual situations: unusual, because not easily found or identified in the ICD-9-CM.

Before we can identify which codes are used for which situations, we must first define the difference between a sign and a symptom. A *sign* refers to a specific objective or observable incident or evidence of a disease or dysfunction. Such observations can be seen or measured by a physician or another member of the health care team, such as a nurse or a therapist. *Symptoms*, on the other hand, refer to changes which eventually cause a patient to seek medical care.

Coding Frequently Seen in Unusual Situations

Unusual situations which you may encounter include the following:

- 780.1   Hallucinations as a result of drug abuse
- 798   Sudden infant death syndrome (SIDS)
- 920–924   Injuries of the skin, such as contusions in which the skin may or may not be broken
- 800–829   Fractures
- 836   Dislocation of the knee
- 840–848   Sprains and strains
- 845   Sprains and strains of the ankle and the foot
- 942; 948   Burns
- 958   Complications following a trauma (hemorrhage, wound infections, shock, etc.)

Unusual situations caused by external causes (traffic accident):

- E810   Accident caused as a result of a collision with another object (train)
- E812   Accident involving collision with a motor vehicle
- E814   Accident involving collision with a pedestrian
- E815   Accident involving other collision on a highway (safety island, tree, etc.)
- E816   Accident involving loss of control without a collision
- E818   Other noncollision accidents (excludes explosion, fire, object thrown in vehicle)

Codes used to identify the place or location of an unusual situation (accident):

- E849   Identifies the place of occurrence: used with E850–E869; E880–E928 (cannot be used with adverse effects, homicides, acts of war, suicides)

Unusual situations caused as a result of misadventure during a patient's medical or surgical care:

- E870   Accidental cuts and punctures during surgery or a medical or surgical procedure
- E871   Foreign object left in a patient's body during a procedure
- E872   Failure to provide for sterile precautions
- E873   Failure to administer the right dosage (medications, fluids, heat and cold, etc.)
- E874   Mechanical failure of an instrument or machine used during care or treatment
- E875   Use of contaminated or infected blood, fluids, drugs, or biological substances
- E876   Other unspecified misadventures, including administering unmatched blood, suture failure, and performing an inappropriate surgical procedure

Summary

In this chapter, we discussed the purpose and function of all the individual body systems and identified the various medical codes which would be used to bill specific diseases, disorders, medical conditions, and care and treatment related to each of these systems. We also talked about the various types of infections, neoplasms, and unusual situations that the medical coder might encounter, including the identification of codes used for billing these types of situations.

Review Questions

1. Define the following terms:
   a. benign _____
   b. malignant _____
   c. metastasis _____

2. What is another term used to define an abnormal increase or elevation in the blood pressure?

3. Briefly explain the difference between the following two diseases:

a. diabetes mellitus

   b. diabetes insipidus

4. Define the following abbreviations:

   a. MI _____

   b. PKU _____

   c. CBC _____

   d. UTI _____

   e. CVA _____

5. What is the medical term used to describe a hunchback condition?

6. Identify the musculoskeletal condition for each of the following CPT codes:

   a. 739: _____

   b. 721: _____

   c. 737.4: _____

   d. 737.2: _____

7. What do the CPT codes V30 through V39 represent?

8. What does the CPT code 331.4 represent?

9. What do the CPT codes 410 through 414 represent?

10. What mental health disorders are represented by codes 317 through 319?

24

Medical Coding and Claims Processing

Performance Objectives

Upon completion of this chapter, you will be able to:

1. Identify the four forms which must be completed by a patient before his or her insurance claim can be processed.
2. Identify the various types of billing forms used in claims processing.
3. Discuss the role of the law as it relates to billing, collections, and claims processing.
4. Explain the process involved in processing Medicare claims.
5. Explain the process involved in processing Workers' Compensation benefits and claims.
6. Discuss the process involved in processing the HCFA 1500 and UB-92 claim forms.

Terms and Abbreviations

Assignment of benefits form a form used to request that payments from payers to be directed to the provider holding the assignment.
COB abbreviation for *Coordination of Benefits*.
DEFRA abbreviation for the *Deficit Reduction Act of 1984*.
ESRD abbreviation for *End Stage Renal Disease*.
HCFA abbreviation for *Health Care Financing Administration*.
Ledger card a card used to indicate a chronological record of all services rendered to a patient.

Release of information form a form used to request additional information from other providers of service.
Skip a person who has received services without payment and who has moved without leaving a forwarding address.
TEFRA abbreviation for the *Tax Equity and Fiscal Responsibility Act of 1982*.
UB abbreviation for *Uniform Billing*.

As we have previously discussed, there are three steps involved in medical coding. The first, as we have already stated, deals with gaining an understanding of the purpose, function, and use of medical coding, as well as learning how to use various types of coding and reference books. The second step involves learning about the various diseases, medical conditions, care and treatments, and medical procedures, and each one's identifying codes, so that insurance claims and billing procedures can be properly completed. And the third step involves the actual processing of the claim forms, so that your employer can eventually be paid for services rendered to the patient.

Processing Insurance Claims and the Law

Whenever you are required to bill patients and process insurance claims, there are important legal implications. These issues have to do with privacy, rules and regulations regarding collections procedures, and fraud.

Understanding the Concept of Privacy

If you have ever consulted a doctor or been a patient in a hospital, you should understand the importance of insuring the patient's privacy: not just physi-

Administrative Medical Services

cal privacy but, just as important, privacy of personal information. In health care, one question which we must always remember to ask ourselves is: "Should the needs of the organization be more important than the individual's right to privacy?"

All medical information is considered to be privileged and confidential in the context of the physician–patient relationship, and any unauthorized disclosure of such information could represent a violation of that confidentiality. To maintain the patient's privacy, particularly as it relates to his or her medical record, observe the following guidelines:

- When working with medical records or billing, gather only that information which is necessary and relevant to the billing or processing of the claim.
- Always make every reasonable effort to assure that the information on which an action is based, is relevant, timely, complete, and, most important, accurate.
- Only use legal and ethical means to collect the information required to perform your task.
- Upon request, the insured or claimant should be given the opportunity to correct or clarify information provided to him or her, and the file should then be amended to reflect that correction.
- Release information to a third party only after the patient has signed an authorization to release such information.
- Never make a disclosure of diagnosis or any other information to a member of a patient's family or ex-spouse unless expressly required to do so by the patient, and then only after a release of information form has been signed by the patient.

Collection Procedures and the Law

If you are required to perform collection procedures for your employer, you should understand that the law is quite specific as to what you can and cannot do. For example:

- When collecting past due payments, never make frightening or abusive calls to someone.
- Always make your collection calls during business hours, usually between 8:00 a.m. and 5:00 p.m.
- Never request collection of payments through a third-party.
- Never harass a person you are trying to collect from.

In some cases, particularly those in which a patient may owe a great deal of money, a provider may allow a patient to set up a payment plan. According to *Regulation Z* of the *Truth and Lending Act*, if the plan requires that more than four installment payments be made to pay the bill, you must provide the patient with a written disclosure. This disclosure must include the amount of the debt, the amount which will be used as a down payment, the estimated date on which the patient will make the final payment, the amount of each of the payments, and the date each payment is due.

Fraud

Fraud is most simply defined as any intentional misrepresentation of a fact with the intent to deprive a person of his or her property or legal rights. Unfortunately, health care is one industry which has seen more than its share of fraud. The most common instance of fraud in our industry has to do with physicians and other providers of service billing for goods or services that they did not provide. This is a serious offense and constitutes fraud. If convicted, the penalties are extremely stiff, both for the physician and the person processing the fraudulent claim.

Because of the frequent instances of fraudulant behavior, everything that is billed for must be documented in the medical record in order to prove that the services were actually provided to the patient. Those of us who work in this industry know all too well about medical billing fraud and, therefore, are keenly aware of the unwritten "law of documentation." This "law" clearly states that, "if it wasn't documented, it just didn't happen!"

Obtaining Patient Information

If you are working for a private physician, you will soon find out that, before your employer can even think of being paid for services he or she provided, you will have to obtain specific documentation directly from the patient (Figure 24-1). This documentation, which will eventually be transferred into the patient's medical chart or record, generally includes a *patient information sheet*, a signed *release of information form* and *assignment of benefits form*, and a completed *patient medical history form*.

The Patient Information Sheet

The patient information sheet has general information regarding the patient. Information usually found on this form includes the patient's name, address, and telephone number; the patient's date of birth and social security number; a business address, telephone number, and occupation, if the person is

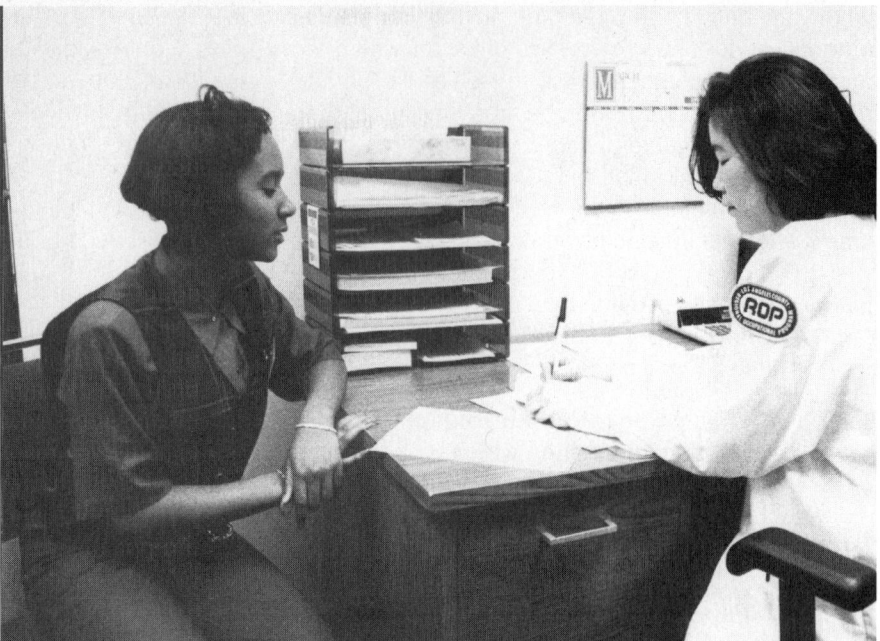

Figure 24-1
Obtaining information from the patient.

employed; the name of the patient's spouse, if he or she is married; the name of the person who is responsible for the account or, if the patient is insured, the name of the insured person and any information pertaining to insurance billing information; a driver's license number; the name of a close relative or friend who could be contacted in case of an emergency; and, if applicable, the name of the person or physician who referred the patient to your facility.

The Release of Information Form

The release of information form requests information from other providers that may be necessary in order to treat the patient. Whenever a physician or a health care facility submits insurance claims for a patient, the patient is required to sign the release of information form before any information can be given out to an insurance company, an attorney, or any other third party.

The Assignment of Benefits Form

The assignment of benefits form is used to formally request that all payments from payers be directed to the provider who is holding the assignment. Most health care facilities and individual providers consider this extremely important, especially in those cases in which a patient has medical insurance. The main purpose of the assignment of benefits form is to make sure that all payments go directly to the provider of services.

Patient History Form

The patient history form is of greatest importance to the physician or caregiver, since it helps to identify any previous incidents of illness that may be important in treating the patient's present condition. The form is generally quite detailed, asking basic health questions usually requiring only yes or no answers.

Billing for Services Rendered

Assuming your patient is covered by health insurance or a government plan such as Medicare, before you can begin to process the claim form, you must first take the proper and necessary steps involved in billing the patient for services which he or she received.

The first step in billing is usually to obtain a charge slip, which is a form the provider generally fills out after he or she has seen the patient. The form, which usually lists the services provided by the physician along with the corresponding billing codes, is given to the patient after he or she has completed the visit. The patient then returns the completed slip to the receptionist or office manager prior to leaving the facility. If the patient does not have insurance, the receptionist is usually required to collect all fees at the time the services were provided. For those patients who are covered by insurance, the office may choose to collect a small payment before the patient leaves the office. A receipt is then issued to the patient and the form of payment is attached to the charge slip.

Administrative Medical Services

The majority of health care facilities who do their own billing generally use individual ledger cards which provide a chronological record of all services provided to a specific patient. This card can also be used at a later time to record any individual payments and adjustments to the patient's account.

Processing Insurance Claims

There are two ways in which a claim can be processed in order to make sure that your employer is paid for services. The first involves the processing of the claim manually, which requires you to use an individual claim form. The second, and most accurate method, is to process the claim electronically, with data sent through a computer system.

Manual Insurance Claims Processing

If your employer requires insurance claims to be processed manually, you will use one of three claim forms: the *HCFA-1500* (Figure 24-2), which is a basic form prescribed by the Health Care Financing Administration for billing services from physicians and suppliers; the Uniform Bill *UB-92* (Figure 24-3), which is used for billing for hospital services; and the *superbill*, which is a type of billing form, usually carbonless, that can be used by many providers of services and suppliers. The only insurance program that will not accept superbills for payment is Medicare.

In addition to using the three standard billing forms listed above, many health care facilities and providers may also ask the patient to complete a patient claim form, which is usually provided by the insurance carrier for the purpose of updating the patient's or insured's records.

Processing the HCFA-1500 Billing Form

As we have already noted, the HCFA-1500 claim form is a basic, standardized form which has been approved for use by the American Medical Association, due to its availability as a "universal" claim form for billing for professional medical services. It is also the only form which is acceptable for use by physicians and other individual health care providers to bill Medicare for services.

Processing the UB-92 Hospital Billing Form

The Uniform Bill UB-92 billing claim form is a standardized billing form which can be used by all hospitals and other inpatient health care facilities. Its format has been designed to provide the health care facility with the basic data and information needed by most payers to adjudicate the greater majority of insurance claims.

Processing Electronic Claims

Many health care facilities have started to use electronic claims processing, primarily because doing so ensures a much more accurate and, hence, quicker return of payments. Submitting claims through an electronic process involves sending information directly through a computer system (Figure 24-4) to the insurance company. Since many health care facilities generate large numbers of claims on a daily basis, submitting claims electronically is much easier than generating numerous pieces of paper and signing and mailing them.

Most claims processed through the computer are completed on a weekly basis, usually in the early morning hours or late in the day, after all the patients have been seen. At this time, the medical biller calls the insurance carriers on the computer and downloads all the necessary information. If any of the claims are rejected or if there is a request for additional information, the insurance carrier can then contact the facility by telephone, by fax, or directly through the computer's modem.

Processing Medicare and Medicaid Claims

If you find yourself responsible for processing Medicare and Medicaid claims, it is extremely important that you first have a good understanding of these two government programs.

The Medicare Program

Medicare is a health insurance program that was established by the Federal Health Insurance Benefits Plan for the Aged and Disabled under Title XVIII of Public Law 89-97 of the Social Security Act, and was intended for all people aged 65 years or older and for certain individuals who are totally disabled.

In order for an individual to be eligible to receive benefits under Medicare, that person must meet certain requirements enacted by the Social Security Administration. If a person has applied for benefits based on his or her age, coverage commences on the first day of the month in which he or she reaches age 65. Hence, persons born on the first day of the month are eligible on the first day of the month preceding their birth date.

If a person applies for benefits based on total disability, coverage begins on the first of the 25th month from the date approved for Social Security Disability or Railroad Retirements benefits. People who are eli-

PLEASE DO NOT STAPLE IN THIS AREA

PICA

HEALTH INSURANCE CLAIM FORM

PICA

CARRIER

PATIENT AND INSURED INFORMATION

1. MEDICARE (Medicare #) MEDICAID (Medicaid #) CHAMPUS (Sponsor's SSN) CHAMPVA (VA File #) GROUP HEALTH PLAN (SSN or ID) FECA BLK LUNG (SSN) OTHER (ID)

1a. INSURED'S I.D. NUMBER (FOR PROGRAM IN ITEM 1)

2. PATIENT'S NAME (Last Name, First Name, Middle Initial)

3. PATIENT'S BIRTH DATE MM DD YY SEX M F

4. INSURED'S NAME (Last Name, First Name, Middle Initial)

5. PATIENT'S ADDRESS (No., Street)

6. PATIENT RELATIONSHIP TO INSURED Self Spouse Child Other

7. INSURED'S ADDRESS (No., Street)

CITY STATE

8. PATIENT STATUS Single Married Other

CITY STATE

ZIP CODE TELEPHONE (Include Area Code) ( )

Employed Full-Time Student Part-Time Student

ZIP CODE TELEPHONE (INCLUDE AREA CODE) ( )

9. OTHER INSURED'S NAME (Last Name, First Name, Middle Initial)

10. IS PATIENT'S CONDITION RELATED TO:

11. INSURED'S POLICY GROUP OR FECA NUMBER

a. OTHER INSURED'S POLICY OR GROUP NUMBER

a. EMPLOYMENT? (CURRENT OR PREVIOUS) YES NO

a. INSURED'S DATE OF BIRTH MM DD YY SEX M F

b. OTHER INSURED'S DATE OF BIRTH MM DD YY SEX M F

b. AUTO ACCIDENT? YES NO PLACE (State)

b. EMPLOYER'S NAME OR SCHOOL NAME

c. EMPLOYER'S NAME OR SCHOOL NAME

c. OTHER ACCIDENT? YES NO

c. INSURANCE PLAN NAME OR PROGRAM NAME

d. INSURANCE PLAN NAME OR PROGRAM NAME

10d. RESERVED FOR LOCAL USE

d. IS THERE ANOTHER HEALTH BENEFIT PLAN? YES NO *If yes*, return to and complete item 9 a-d.

READ BACK OF FORM BEFORE COMPLETING & SIGNING THIS FORM.

12. PATIENT'S OR AUTHORIZED PERSON'S SIGNATURE I authorize the release of any medical or other information necessary to process this claim. I also request payment of government benefits either to myself or to the party who accepts assignment below.

SIGNED ____ DATE ____

13. INSURED'S OR AUTHORIZED PERSON'S SIGNATURE I authorize payment of medical benefits to the undersigned physician or supplier for services described below.

SIGNED ____

FOLD HERE / USE ENVELOPE NO. 1500E

PHYSICIAN OR SUPPLIER INFORMATION

14. DATE OF CURRENT: MM DD YY ILLNESS (First symptom) OR INJURY (Accident) OR PREGNANCY(LMP)

15. IF PATIENT HAS HAD SAME OR SIMILAR ILLNESS. GIVE FIRST DATE MM DD YY

16. DATES PATIENT UNABLE TO WORK IN CURRENT OCCUPATION FROM MM DD YY TO MM DD YY

17. NAME OF REFERRING PHYSICIAN OR OTHER SOURCE

17a. I.D. NUMBER OF REFERRING PHYSICIAN

18. HOSPITALIZATION DATES RELATED TO CURRENT SERVICES FROM MM DD YY TO MM DD YY

19. RESERVED FOR LOCAL USE

20. OUTSIDE LAB? YES NO $ CHARGES

21. DIAGNOSIS OR NATURE OF ILLNESS OR INJURY. (RELATE ITEMS 1,2,3 OR 4 TO ITEM 24E BY LINE)

1. ___ . __
2. ___ . __
3. ___ . __
4. ___ . __

22. MEDICAID RESUBMISSION CODE ORIGINAL REF. NO.

23. PRIOR AUTHORIZATION NUMBER

24. A DATE(S) OF SERVICE From MM DD YY To MM DD YY	B Place of Service	C Type of Service	D PROCEDURES, SERVICES, OR SUPPLIES (Explain Unusual Circumstances) CPT/HCPCS MODIFIER	E DIAGNOSIS CODE	F $ CHARGES	G DAYS OR UNITS	H EPSDT Family Plan	I EMG	J COB	K RESERVED FOR LOCAL USE
1										
2										
3										
4										
5										
6										

25. FEDERAL TAX I.D. NUMBER SSN EIN

26. PATIENT'S ACCOUNT NO.

27. ACCEPT ASSIGNMENT? (For govt. claims, see back) YES NO

28. TOTAL CHARGE $

29. AMOUNT PAID $

30. BALANCE DUE $

31. SIGNATURE OF PHYSICIAN OR SUPPLIER INCLUDING DEGREES OR CREDENTIALS (I certify that the statements on the reverse apply to this bill and are made a part thereof.)

SIGNED DATE

32. NAME AND ADDRESS OF FACILITY WHERE SERVICES WERE RENDERED (If other than home or office)

33. PHYSICIAN'S, SUPPLIER'S BILLING NAME, ADDRESS, ZIP CODE & PHONE #

PIN# GRP#

(APPROVED BY AMA COUNCIL ON MEDICAL SERVICE 8/88) ***PLEASE PRINT OR TYPE*** APPROVED OMB-0938-0008 FORM HCFA-1500 (12-90), FORM RRB-1500, APPROVED OMB-1215-0055 FORM OWCP-1500, APPROVED OMB-0720-0001 (CHAMPUS)

Figure 24-2
HCFA-1500 form.

Administrative Medical Services

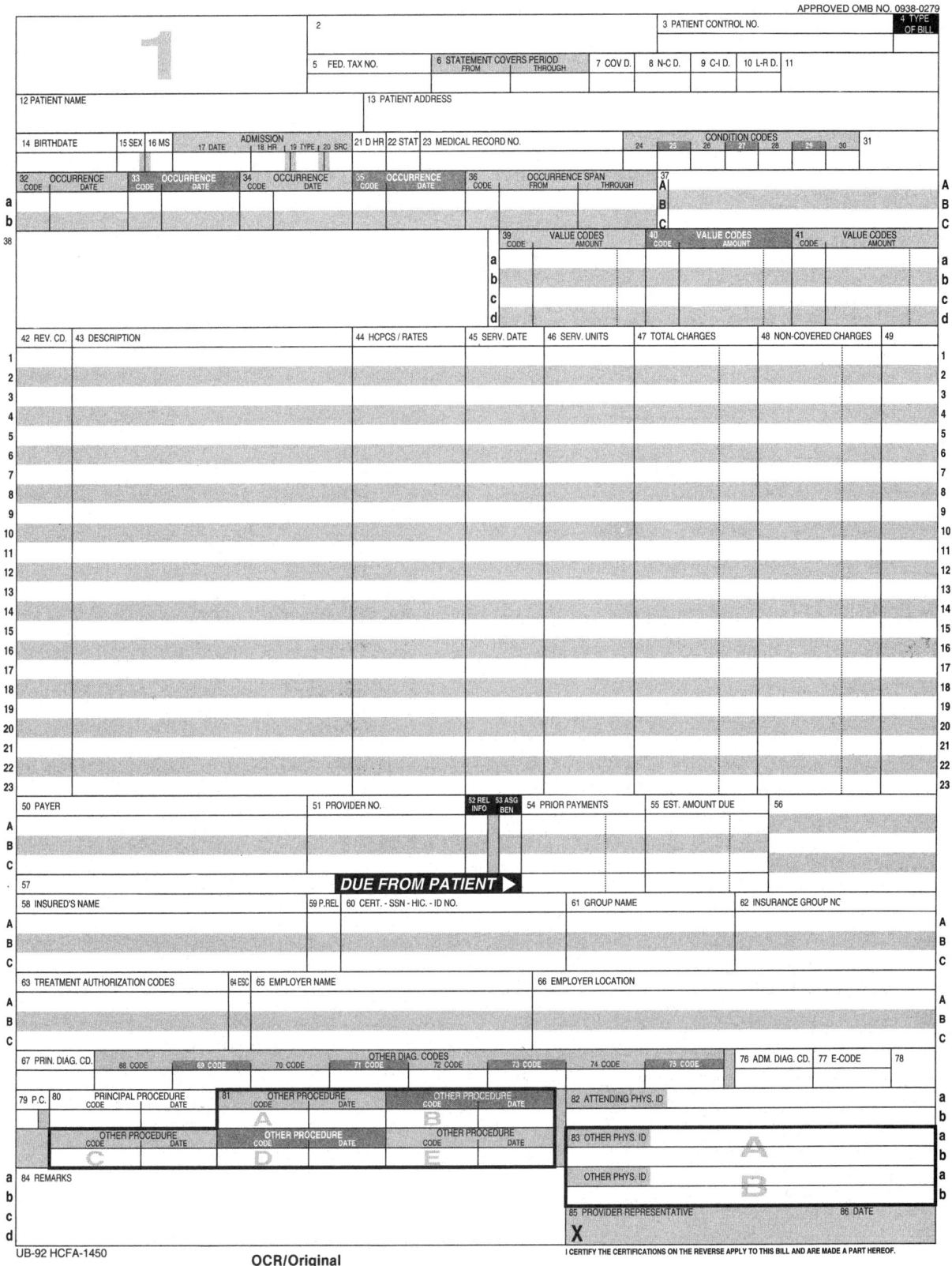

Figure 24-3
UB-92 form.

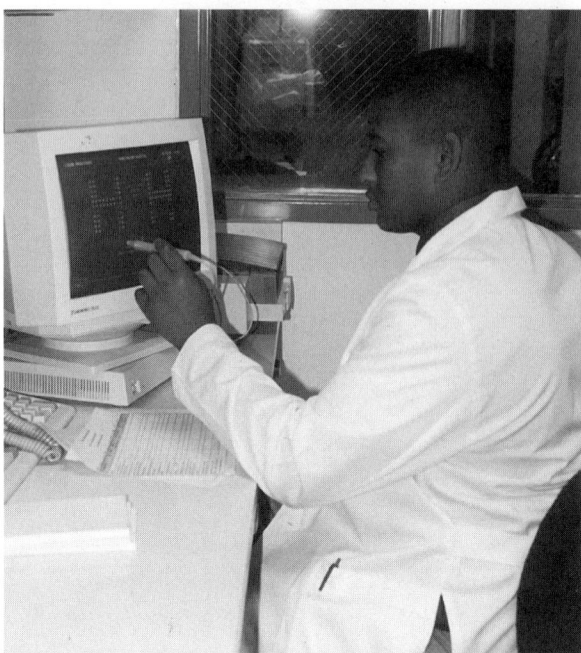

Figure 24-4
Processing electronic claims.

gible for benefits based on a disability may include disabled workers of any age, widows of disabled workers between the ages of 50 and 65, beneficiaries of the disabled aged 18 and older who receive Social Security benefits because of a disability before they turned age 22, people who are blind, and Railroad Retirement annuitants.

If a person has been diagnosed with End Stage Renal Disease (ESRD) and his or her kidneys cease to function, resulting in the need for dialysis treatments, that person may also be eligible for Medicare benefits, since patients in this particular situation are considered to be totally disabled. In order to receive benefits under the Medicare program, patients who suffer from ESRD must follow special rules. If the patient is employed, the employer's group health plan must be the primary payer for the first 18 months after the patient with ESRD becomes eligible for Medicare. This 18-month period begins the month in which a regular course of renal dialysis is initiated or the month in which the patient is hospitalized for a kidney transplantation, whichever is earlier.

Parts of the Medicare Program

There are two parts to the Medicare program. Part A, which is considered to be the basic or hospital plan insurance program, is the part of the program that covers a health care facility that provides services to the patient. If the patient is admitted to the hospital, however, there is a deductible which must be paid, a process which is automatically completed at the time of the patient's first admission to the facility. If the patient is discharged and remains out of the hospital for at least 60 consecutive days, including the day on which he or she is discharged, a new benefit period commences, and another inpatient deductible would be collected if the patient were to be readmitted to the hospital. If a person participating in the Part A program remains confined in the hospital for an extended period, there are additional copayments required. In this case, Medicare deducts the copayment amount from the billed amount and then pays the amount in excess of the copay.

Part B of the Medicare program is often called the supplementary part or medical insurance program. It is this part of Medicare that covers physician services, outpatient and home health services, and additional outpatient services, such as physical and occupational therapy and prosthesis and other durable medical equipment. It is often considered to be supplemental because each person participating in the program is required to pay a specified amount each month in order to receive benefits. The amount which is paid each month can be adjusted to allow for inflation.

After a participant has met his or her yearly deductible, services covered under Part B include manual manipulation of the spine and nonroutine podiatry care, outpatient hospital services and ambulance services, such as physical and occupational therapy, laboratory, and x-ray services, home health care, some nonroutine optometry services, and dental care for surgery of the jaw and related structures and fractures of facial bones. Services not covered under Part B include routine physical examinations and routine foot care, some cosmetic surgeries, immunizations, except in cases of some pneumococcal and hepatitis B infections, and vision or hearing examinations for eyeglasses or hearing aids.

In order to receive benefits automatically under Part A of the Medicare program, an individual must be at least 65 years of age or be receiving a pension under the Railroad Retirement Act. As of July 1, 1973, all persons who reached their 65th birthday and who were not otherwise eligible for receiving benefits under Part A may also enroll by paying the full cost of the coverage, provided that they also enrolled in Part B.

Working with Medicare Providers

Because there are so many members of our society receiving Medicare benefits, payment for services rendered under this government program has become a very large part of the revenue received by many private and public health care facilities. Therefore, if you are employed as a medical coding and

Administrative Medical Services

claims worker, there is a very good possibility that you will encounter these providers.

All providers of services, and medical and health-related equipment suppliers who receive payment benefits under the Medicare program, must meet very stringent licensing requirements of the individual state in which they are located. In addition to meeting certain mandates established by the individual states, in order to be paid for their services, these providers must meet Medicare's requirements, which have been set up by the federal government. Unless a provider has been approved as a "participating facility," it will not be paid under the Medicare program. Sadly, services such as hospital, home health, skilled nursing, hospice care, dialysis, outpatient care, physical and occupational therapy, and laboratory and x-ray services will not be paid by Medicare if insitutions offering these services have not been approved for participation in the Medicare program.

Billing and Receiving Payments under the Medicare Program

Benefit payments under the Medicare program are based on the concept of "reasonable charges," meaning that those charges or amounts that have been approved by the Medicare carrier are based on what is considered reasonable for the geographical area or location in which the provider practices. Unfortunately, however, because of inflation and the way in which Medicare determines its approved benefits to providers, the amount which is finally paid is usually significantly less than the actual amount billed by the provider.

If your position requires that you perform the billing for Medicare participants, one of the first tasks you must complete is to ask the person for his or her Medicare identification card, since this card will provide you with information as to whether the patient is eligible to receive benefits under both Part A and Part B of the program. After you look at the card, it's always a good idea to make a copy of it for your files.

If your employer has agreed to accept the amount Medicare has offered to pay, the next step, as it is in any billing process, is to make sure you record the amount received on the patient's ledger card. After you receive Medicare's Explanation of Benefits (MEOB), you must post each individual payment to each ledger card as well as to the daily journal. And, if the amount you receive is lower than that which your employer has billed, make sure you make an adjustment on the ledger card. Once you have posted the amount, remember that the next entry on the card will be the adjustment to the actual charges. Enter the differences between the billed amount and the amount Medicare approved. This amount should then be entered on the card in the adjustment column. The final step is to subtract the adjusted amount from the remaining balance. Completing this last step will ensure that your Medicare patients are not billed for charges that Medicare has not agreed to pay.

Assignment of Benefits under the Medicare Program

If a provider agrees to accept payments under the Medicare program, that provider should also agree to accept assignment on all Medicare claims. By doing so, the provider will receive his or her payments directly from Medicare instead of having to ask the patient to pay for the services after he or she has received the benefit payment. Remember also that, if your employer has agreed to accept whatever Medicare pays, then the patient is not expected to be responsible for paying any amount in excess of the approved amount. If the patient carries a secondary insurance plan, that carrier is also neither expected to pay nor responsible for paying that excess amount.

Supplements to Medicare Coverage

A supplement is a separate plan which has been written expressly for an individual receiving Medicare benefits. This supplement may have any number of optional benefits chosen by the Medicare participant and generally includes a 20 percent coinsurance for reasonable charges for physicans' services under Part B, coverage of a deductible for hospital services under Part A, and, in some cases, nursing care and prescriptions.

The Medicaid Program

Medicaid, sometimes referred to as Medi-Cal, is not considered a health insurance program, but is part of a program that was first established under Title XIX of the Social Security Act of 1965, which provided for accessibility to medical care and treatment for the needy.

According to law, Medicaid must always be used as a secondary program to private health insurance plans, and if the private plan is required to process a Medicaid request for reimbursement, that insurance carrier must pay Medicaid back the monies it paid the provider for services.

Under the Medicaid program, individuals who are eligible to receive benefits may be entitled to coverage due to medical, family, or fiscal conditions. And, that a participant carries additional, or supplementary, health insurance does not preclude that person from being eligible for coverage under the Medicaid program.

Processing Medicaid Claims

Claims made under the Medicaid program are not processed through the Medicaid program. Rather, Medicaid contracts with an organization who acts as a fiscal intermediary. The amounts which participants may have to provide under a copayment and those amounts which are provided to the health care provider for services rendered are mandated under the Department of Health Services. All monies for reimbursement for hospital inpatient services are also based on each facility's reasonable cost of services as determined from audits and annual limitations on reimbursements.

Under the Medicaid program, no provider is allowed to bill or submit a claim form (Figure 24-5) to the Medicaid beneficiary for any services included under the program's scope of benefits, except to collect money from a patient's private health insurance carrier prior to billing Medicaid. Also, there is a three-year statute of limitations from the date the service was provided until the time of recovery of payment.

Billing and Receiving Payments under the Medicaid Program

All individuals who apply for Medicaid benefits must meet the requirements for eligiblity set forth by the local county's Department of Health and Human Services or the county's welfare department. Therefore, prior to billing a participant for services rendered, you should ask the patient for a copy of his or her Medicaid eligibility card for the month the services are rendered. If the card has adhesive stickers affixed to it, one of the stickers should be attached to the billing form for payment.

Medicaid eligibility cards may be either plastic or paper. The plastic card can be identified through a provider service network (PSN) system. If your employer sees Medicaid participants, he or she will be required to maintain a machine that is used to read the magnetic strip on the back of the card. The card is swiped through the PSN computer network, which will then verify the eligibility of the person seeking services.

If the participant uses the paper eligibility card, it will have adhesive stickers affixed to it. These stickers are used for payment when claims are billed to the Medicaid office. Since Medicaid eligibility is determined on a monthly basis, the paper cards are also issued on a monthly basis. Remember, before the provider agrees to see the participant, you must have a copy of the eligibility card for the month in which the services are to be rendered. Otherwise the provider will not be paid for the services.

Most services provided by health care facilities and private physicians are covered under the Medicaid program. These generally include both inpatient and outpatient hospital care, physician's services, immunizations and home health care services, laboratory and x-ray services, and family planning services. However, just because a service is covered, it does not mean that it will be paid for by Medicaid. In order for that to occur, the provider in many cases must receive prior authorization. This requires that a Treatment Authorization Request, or TAR, be completed, to obtain authorization for specific services. In most cases, when a TAR is required, the form must be completed and sent to the appropriate agency for authorization prior to the services being rendered.

If you have received an authorization for payment under the Medicaid program, you should remember that, like Medicare, the amount which your provider will receive will probably be less than the actual amount of the billed charges. Because Medicaid participants are generally not financially responsible for the balance, an adjustment will probably have to be made and eventually written off. When you receive the explanation of benefits from Medicaid, pull patient's ledger card and record the payments accordingly. Adjustments should then be made on the next corresponding entry line on the ledger card, with the payment amount subtracted from the billed amount and the difference written off.

Processing Medi-Medi Claims

In certain cases, when a participant is eligible for both Medicare and Medicaid benefits, that patient is considered to be a medi-medi patient. This means that, if a patient becomes eligible for benefits under Medicaid, Medicare will automatically forward the claim to the appropriate department for coordination of benefits. Medicare can then cross-reference all claims to check for Medicaid eligibility. Remember, whenever a patient carries Medicare benefits, Medicaid should always be the secondary payer to Medicare.

Processing Workers' Compensation Insurance Claims

Workers' compensation is a benefit program which provides 100 percent medical and disability reimbursement coverage for people who have sustained a job-related injury, illness, or any other condition which may have arisen in the course of employment. By law, an employer is responsible for the benefits due to an employee for any work-related injury or illness.

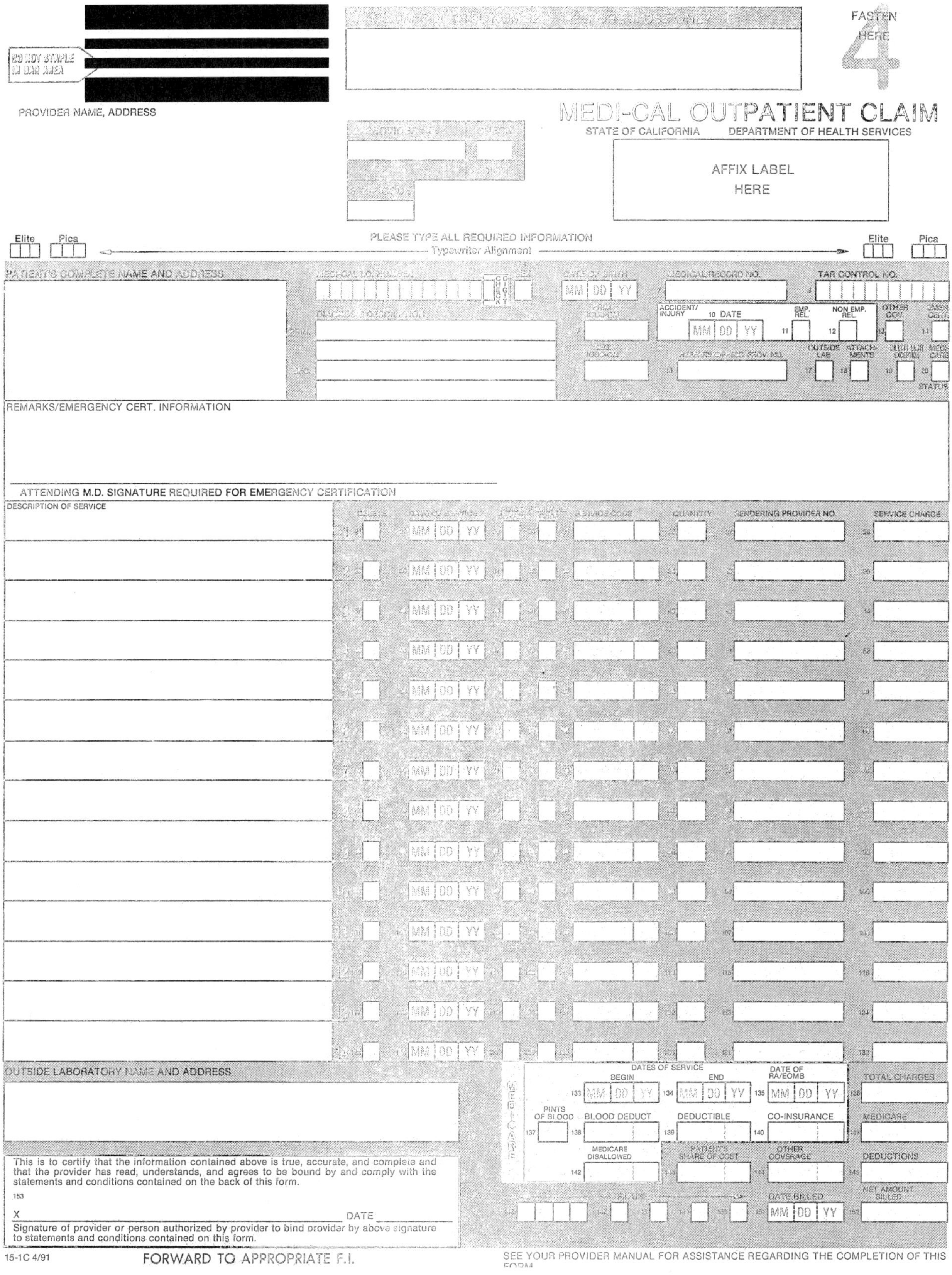

Administrative Medical Services

Figure 24-5
Medicaid claim form.

Each state has its own Workers' Compensation Appeals Board, that is solely responsible for overseeing the rights and benefits of all injured or ill workers in that state. If an employee becomes injured or ill, it is through the Appeals Board that the applicant can file his or her workers' compensation benefit application.

Injuries and medical conditions generally falling under the heading of work-related include those occurring during the employee's working hours. Injury or illness may occur when the employee is working at the specified work location, while the employee is using a company vehicle, while the employee is on a business trip, and while an employee is getting in or out of his or her vehicle in a parking lot. Work-related injuries and illnesses may also involve those conditions that have occurred as a result of an employee's specific occupation.

Types of Workers' Compensation Claims

Claims identified under workers' compensation may be one of three types. The first, called *nondisability claims*, are those processed as minor injuries that generally do not require the patient to be kept away from his or her job. For this type of claim, upon the employee's first visit to the health care provider, the physician is required to complete a "Doctor's First Report of Occupational Injury" form. The form, along with a copy of the bill, should then be submitted to the workers' compensation carrier.

Injuries and medical conditions which require that the employee not return to work until full recovery is made are classified as *temporary disability claims*. When a provider sees a patient in this situation, he or she submits a First Report to the carrier. Ongoing reports should be issued at least every two to three weeks until the patient is discharged to return to work.

An employee who has been processed with a *permanent disability*, which is the third type of workers' compensation claim, is someone who has sustained an injury or illness that prevents him or her from returning to work. Permanent disability benefits always begin after temporary disability has ended. In this situation, the physician prepares a discharge report stating the patient is *permanent and stationary*, meaning that nothing more can be done and that the patient will have the disability for the rest of his or her life. If this occurs, in most instances the case will be reviewed by the Workers Compensation Appeals Board. If it is determined that the disability is permanent, the workers' compensation carrier will offer the employee a *compromise and release*. This often consists of a large lump sum of money, health insurance coverage, or, sometimes, both money and insurance coverage being offered to the disabled employee.

In cases in which the employee has been rated as permanently disabled, a compromise and release will be issued by the insurance carrier for the injuries to the patient. If the physician or provider has been seeing the patient since the onset of the injury, a lien, which is a legal document expressing a claim on the property of another for an owed debt, may be filed for payment of services rendered.

Coordinating Benefits

The process of coordinating health insurance benefits occurs when two or more group health insurance plans provide coverage for the same person. Coordination between the two plans becomes necessary in order to allow for payment of 100 percent of all allowable expenses. The purpose of coordinating benefits is to allow for coverage and usually payment of covered expenses without allowing the covered member(s) to make additional money over and above the total costs for required medical care and services.

Whenever there is a coordination of benefits, there must be a primary plan, that is, the benefit plan that determines and pays its benefits first without regard for the existence of any other coverage; and a secondary plan, which pays after the primary plan has paid its benefits and which takes into consideration the benefits of the primary plan, thereby having the option to reduce its payments so that only 100 percent of allowable expenses are paid.

If the amount of payments made by a plan is more than it should have been under a coordination of benefits provision, the plan may recover the excess from the person(s) it paid or the person(s) it has paid on behalf of other insurers, or plans, or organizations. The amount of the payments made includes the reasonable cash value of any benefits provided in the form of services.

If you are required to bill patients who have more than one form of coverage, you must be sure that you determine who the primary carrier is. Once payment has been received from the primary carrier, you should then post the payment to the patient's ledger card to determine the balance. When billing the secondary carrier, always attach a copy of the primary carrier's explanation of benefits to the claim form being submitted. Remember, the secondary carrier will not process the claim without this information. Upon receipt of the secondary carrier's payment, post the payment to the patient's ledger card. Any balance which remains after all adjustments should then be billed to the patient.

Administrative Medical Services

Summary

In this chapter, we discussed the skills involved in processing medical insurance claims. We talked about how to process government claims, such as those which fall under the Medicare and Medicaid programs, as well as those which are identified as workers' compensation claims. We also discussed the legal issues involved in billing and insurance claims processing.

Review Questions

1. Define the following abbreviations:
    a. UB _____
    b. HCFA _____
    c. HMO _____
    d. TEFRA _____
    e. ESRD _____

2. Briefly explain *Regulation Z* of the Truth and Lending Act.

3. What is the term used to define any intentional misrepresentation of a fact with the intent to deprive a person of his or her property or legal rights?

4. What is the purpose of a *patient information sheet*?

5. Briefly explain the difference between a Release of Information form and an Assignment of Benefits form.

6. Briefly explain the steps involved in processing the following health claims forms:
    a. HCFA-1500 _____
    b. UB-92 _____

7. Briefly explain the steps involved in processing claims for the following two government health benefits programs:
    a. Medicare _____
    b. Medicaid _____

8. What does the phrase *permanent and stationary* mean?

Section VI

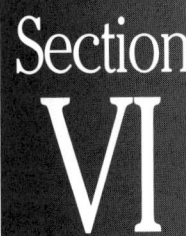

Health Unit Coordinator: Basic Concepts and Applications

25 Introduction to Health Unit Coordinating

26 Transcribing Medical Orders

Introduction to Health Unit Coordinating

Performance Objectives

Upon completion of this chapter, you will be able to:

1. Describe the role of the health unit coordinator as it relates to being a member of the administrative medical services team.
2. Identify the various tasks and responsibilities of the health unit coordinator.
3. Discuss the role communication plays in carrying out health unit coordinator tasks.
4. Explain the purpose and skills involved in managing a nurse's station within a hospital environment.

Terms and Abbreviations

Addressograph a type of imprinting device used to transfer information from a name plate to a label, requisition, or chart form.

AMA abbreviation for *Against Medical Advice*, which is when a patient discharges him- or herself and leaves the hospital against the advice of his or her physician.

Communication the process by which verbal, nonverbal, or written messages are sent from a sender to a receiver.

Emergency admission a hospital admission that is unplanned and is often due to a sudden onset of illness, injury, or accident.

Kardex an up-to-date record, often kept on a 5 × 8" card, which provides information on all orders currently in effect on a patient. It is sometimes referred to as a *patient care plan*.

MAR abbreviation for *medication administration record*, which is a form used to keep an up-to-date record of all medication orders and drugs currently being provided to a patient.

NAHUC abbreviation for the National Association of Health Unit Coordinators, which is a national organization responsible for setting the guidelines for the professional health unit coordinator and for administering a national certifying examination for qualified members of the profession.

Requisition a formal, written request for specific services, supplies, or equipment.

Routine admission a hospital admission which has been planned in advance.

Vital signs the measurements necessary and vital to sustain life, which include blood pressure, temperature, pulse, and respiration.

If you are interested in working in an administrative position that requires a great deal of skill and knowledge in the dissemination of important clinical and clerical information, as well as the expertise and tactfulness necessary to work in a supervisory or leadership role, then you may want to consider a career as a Health Unit Coordinator (Figure 25-1).

Sometimes referred to as a Unit Secretary or Ward Clerk, in most hospitals the role of the health unit coordinator is considered to be as important and necessary as that of the nursing staff. For it is this person who is usually solely responsible for running and managing the nurses station, greeting and communicating with patients, staff, and visitors, and reading,

Figure 25-1
The health unit coordinator.

interpreting, and ultimately transcribing doctor's orders necessary for the diagnosing, care, and treatment of the hospitalized patient.

The Role of the Health Unit Coordinator

Prior to 1940, many of the responsibilities and tasks required of today's health unit coordinators and ward and unit secretaries were performed by registered nurses. However, that all changed during World War II, when it was discovered that nurses were needed more to carry out the everyday hands-on duties involved in caring for the sick and war-torn injured soldiers. This was particularly important, since there was a vast shortage of nurses. Clerical tasks, which were once completed by the nurse, were eventually transferred to ward secretaries. These duties included answering the telephone and intercom system, assembling medical charts, and filing reports. Eventually, as the war began to wind down, the role of the health unit coordinator began to expand and inevitably came to include the important task of reading, interpreting, and transcribing doctor's orders.

While many of the tasks required of the post World War II ward secretary still remain some 50 years after the war ended, the role of today's health unit coordinator has evolved to become one of the most important administrative medical support service providers employed in today's modern hospitals and health care facilities. Today's health unit coordinator is expected to be knowledgeable in hospital organization and the function of all hospital departments, be skilled in various types of hospital communication systems, have a working knowledge and understanding of medical terminology and basic anatomy and physiology, and possess the skills required to read, interpret, and transcribe medical orders necessary for the diagnosis, care, and treatment of the patient. This person must also possess those qualities and attributes necessary to be an effective leader and supervisor.

As a member of the health care team, your role as a health unit coordinator is extremely important. While your job will often require that you be responsible for managing the nurse's station, your position will be supervised by the station's charge nurse or unit coordinator. In addition to having overall responsibility for reading, interpreting, and transcribing physician's orders, your tasks will involve using your interpersonal skills, since the health unit coordinator is generally absorbed in daily contact with physicians, patients, visitors, and other health care and hospital personnel (Figure 25-2).

As a health unit coordinator, many of the tasks mandated by your position will require that you also be responsible for the performance of many clerical duties, such as answering the telephone when visitors, department managers, or physicians call, filing and copying reports, preparing requisitions, and updating the kardex and patient care plans. You will often be the person responsible for greeting the doctor when he or she enters a nursing unit, so you must learn to be tactful and courteous, and your appearance must be impecable and professional. In many cases, you will also be responsible for preparing patient's charts for admissions, transfers, and dis-

Administrative Medical Services

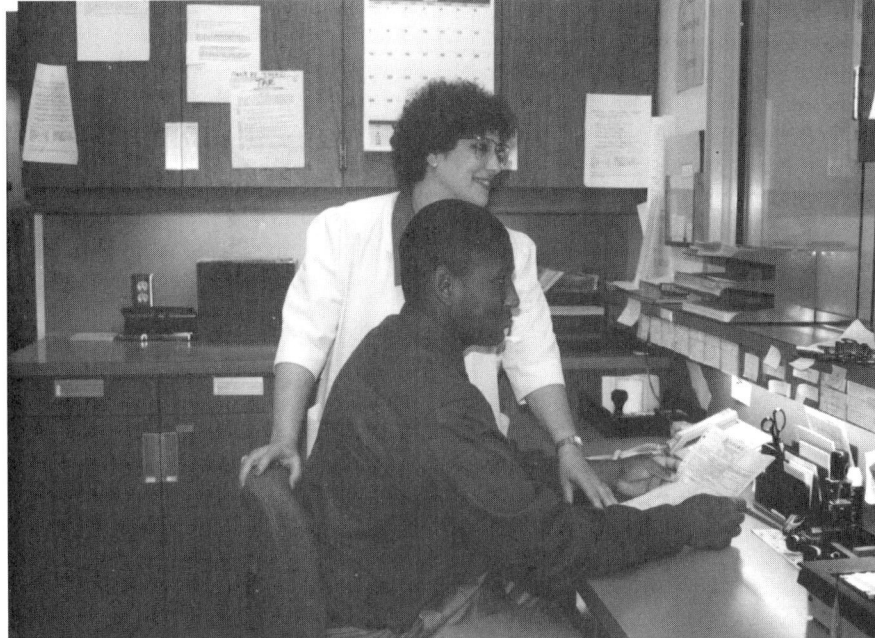

Figure 25-2
Working with other members of the health care team.

charges, as well as maintaining them during the patient's hospitalization and disassembling them after the patient has left the hospital.

In addition to the many tasks required of the health unit coordinator, as with many other positions in the health care industry, you will be expected to keep up-to-date and knowledgeable on new developments in medicine and health care. You will also be expected to read procedure manuals and become familiar with hospital policies and regulations. Your role will also include keeping up on new trends and changes in your field. You'll also be expected to attend inservice training sessions and join professional organizations, such as the National Association of Health Unit Clerks/Coordinators. Remember, as a professional member of the health care team, your role and the responsibilities assigned to your position are extremely important to assure the overall care and treatment of patients admitted to your facility.

Carrying Out Basic Procedures

As we have already stated, the role of the health unit coordinator involves everything from being a good communicator and answering the telephone, greeting patients, staff members, and doctors to being responsible for admitting, transferring, and discharging patients and reading, interpreting, and transcribing physician, nurse, and diagnositic and therapeutic medical orders.

Communications at the Nurse's Station

Earlier in this book, we discussed the skills involved in communication. We noted that the art of communication, whether it is through the use of body language, writing, or speaking, always involves a process by which a message is sent from a sender to a receiver.

As a health unit coordinator, many of your communication skills will involve both verbal and written communications. There are three types of mechanical communication systems used most frequently by the health unit coordinator. These include the *telephone*, the *intercom system*, and the *hospital paging system*. Since much of your time will be spent answering the telephone, you should always have a pad and pencil ready to take down information. Always answer the telephone promptly, and never, never be rude to a caller. Make sure you identify your unit, as well as your name and title, and if the caller tells you he or she wants to give you an order, be courteous, yet firm, as you inform that person that you are not allowed to take orders over the telephone. And remember, when you have completed the call, always say "thank you."

A communication tool which you will be expected to use is the hospital intercom system. An intercom is a mechanical device used by most health care facilities for direct communication between a patient and staff members or between two staff members. Many of today's modern hospitals have a patient's hospital room call light connected directly to the nursing unit intercom system.

Whenever you are required to answer the intercom system, make sure you answer it promptly, in a clear, pleasant, and courteous voice. Listen carefully to what the person is saying, and if it is a patient who is calling you on the system, always acknowledge that person's request by telling him or her that you will notify the nurse.

If you are working in a large health care facility, you should know that there are three types of paging systems that most facilities use. The first, called a *voice system*, involves the hospital operator speaking into a microphone connected to all departments and corridors throughout the facility. When using this system, you must make sure you first dial the operator. You may then ask that person to page the person you are trying to contact. This page is often done either by announcing that person's name or through a coding system.

Another type of paging system used quite frequently by health care facilities is called a *number paging system*. This system is often chosen over others because it is less noisy and can provide a quick response by all members of the facility through an assigned number. There are two major drawbacks to this system. The first is that it requires members of the staff to train themselves to watch for their individually assigned numbers on panels strategically located in corridors throughout the facility. The second, and perhaps the biggest disadvantage of this system, is that it is often inaudible and faint during an emergency or crisis situation.

The third paging system often used by individuals employed by a health care facility is called a *beeper* or *pocket pagefinder system*. It is less frequently used by large health care facilities because it requires members of the staff to carry an electronic "beeper." When this system is used, the operator can either signal the person by using the beeper tone or can actually write or verbally direct the message on the pagefinder, so long as the pager has the capability to respond to either a written or verbal message. If it does not, the person being paged simply acknowledges the call by responding with a telephone call.

There are several other types of electronic communication systems which you may or may not be responsible for using. These include an *imprinter* or *addressograph machine*, which is a device used to transfer information from a name plate to a label, requisition, or form, and a *pneumatic tube system*, which is used to transmit messages, forms, and requisitions from one location or department to another.

Processing Written Communications

There are many different types of written communications the health unit coordinator is reponsible for processing on a daily basis. These include desk sheets, requisitions, kardex forms and cards, and patient chart forms.

Desk sheets are individual, preprinted hospital forms used for a specific purpose. Some of those more frequently encountered in the hospital environment include the following:

- *Work assignment sheet*   This lists the names and room numbers of all patients assigned to a specific unit.
- *Diet sheet*   This lists the names and room numbers of all patients on the unit, indicating which patients must have their food trays held, transferred, or eliminated and which are sent to the dietary department during each shift.
- *Surgery schedule sheet*   This is usually delivered to the nursing unit by the surgery department, and lists all patients who are scheduled for surgery on the following day.
- *Patient's status report sheet*   This lists the names and room numbers of all patients on the unit and is often used by the nursing staff and supervisors to identify each patient by individual diagnosis.
- *Laboratory sheet*   This lists all the patients on the unit who are required to have a lab specimen collected.
- *Census report sheet*   This lists all the names of the patients who are scheduled to be admitted, transferred, and discharged, as well as those who have died. It also provides the exact number of patients on the unit for an individual shift.
- *Transport sheet* or *transportation schedule*   This lists the names of the patients who are scheduled to be transported from the unit during the shift and to what location.

Another type of preprinted hospital communication for which the health unit coordinator is usually responsible is the requisition form. Most hospitals use individual department requisition forms, which can be filled out by the health unit coordinator when a specific task or procedure is required of that department. While there are many different types of requisition forms that may be used throughout the hospital, most are designed to provide more than one copy so that the unit and individual departments within the facility can maintain a copy for their records.

Whenever you are required to process or prepare a requisition, the steps outlined below will assist you to complete the task in an accurate and efficient manner:

- Locate the appropriate requisition. This is easily accomplished by making sure all requisitions are separated according to their use and kept in a location where they are easily accessible.

Administrative Medical Services

- Imprint the requisition form, using the patient's identification card, through the addressograph machine.
- Date and initial the form. Never predate the form or use someone else's initials.
- After you have double-checked the patient's chart for what needs to be ordered or requested, write in or check off the appropriate information on the requisition form.
- Once you have completed the form, remove the copy that is to stay on the nursing unit or is to be kept in the patient's chart.
- Unless otherwise directed by hospital policy, clip the completed form to the patient's kardex or in some other designated location.
- Send the requisition to the appropriate department.

Many large hospitals and health care facilities still use kardexes and patient care plans to assist the nursing staff and other members of the health care delivery team in carrying out the needs and treatment of their patients. A *kardex* is a record, often kept in a metal file holder, that is updated on a daily basis and is used to identify and list all orders currently in effect on an individual patient. In addition to providing the nursing staff with important information such as identification of a patient's problems and nursing goals and approaches for meeting those problems, the kardex usually includes a complete summary of all tests, treatments, and surgical procedures performed on a patient, as well as additional information, such as a list of allergies, special diagnostic tests, consultations, and other data that might be needed for planning the patient's discharge. The kardex may also be used when a member of the staff needs to refer to specific data on a particular patient or to make unit assignments and report at the end of the shift.

The kardex, for the most part, is the only document used in the health care facility for which you may use a pencil to record information. This is because most of the information transcribed and written onto the kardex is changed, sometimes on a daily basis. The only information that may not be recorded in pencil is the patient's name, his or her hospital or medical record number, the name of the attending physician, and the date the patient was admitted.

Perhaps the most important written communication you will be responsible for is the individual patient medical records or charts. These charts have many uses in the health care environment. They serve as a guide in planning the patient's care by providing specific documentation, communication, and written comments and interactions among the medical staff, nurses, and other health care providers involved in the patient's care and treatment. They also play a big part in providing specific data and pertinent information that may be required and necessary for the professional members of the staff in carrying out research projects and in meeting continuing education requirements.

Consisting of a series of standard and supplemental chart forms, the medical record, or chart, is considered a legal document. As such, it may be subpoenaed by a court of law. Therefore, each must be properly maintained and kept available so long as the patient remains on the nursing unit.

Chart forms considered to be a standard part of the patient's medical record include those maintained on all patients as part of the hospital's accreditation practices and legal requirements. They include the following:

- *Admission data form or face sheet* This is kept either at the very front or at the very back of the chart, and is used to provide detailed business information regarding the patient's name, address, occupation, next of kin, religious preference, marital status, and ability to pay.
- *Consent to treatment form* This is used by the facility as a release of liability in providing treatment, and, as such, must be signed by the patient at the time of his or her admission to the hospital.
- *Personal belongings list* This is used to provide the facility with a list and identification of those personal articles and clothing brought into the facility by the patient at the time he or she was first admitted.
- *History and physical form* This is used by the physician while examining the patient, and later as part of the doctor's dictated medical history on the patient.
- *Physician's or doctor's order forms* These are used by the physician to write all medical orders regarding the care and treatment of the patient.
- *Physician's progress notes* These are used as a running documentation by the physician as to how the patient is progressing during his or her hospitalization.
- *Physician's discharge summary* This is used to provide the physician with a space for summarizing the diagnosis, care, treatment, progress, and condition of the patient at the time of his or her discharge from the facility.
- *Nurse's notes* These are sometimes called the patient's progress record and are used by the nursing staff to provide communication among the nursing staff, the doctor, and other members of the health care delivery team, to document the patient's progress, treatments, problems, medications, and any other pertinent informa-

tion the nurses believe may be important to the patient's care and treatment.
- *Nurses' accountability record*   This is used by some health care facilities to provide the nursing staff with a checklist on which information can be recorded as to the patient's activities of daily living (ADL) status.
- *Patient care plan*   If not used as part of the kardex, this may be placed in the patient's chart and used by the nursing staff to identify the patient's problems and nursing goals and approaches.
- *Medication administration record (MAR)*   This is used to keep a daily record of all medication orders currently in effect and to identify those which are to be administered or stopped.
- *Graphic form*   This is used to keep a daily running account of the patient's vital signs, height and weight, intake and output, sugar and acetone results, and insulin coverage.
- *Laboratory report forms*   These are used as a space to provide laboratory results.
- *Charge ticket summary*   This is used by some health care facilities as a place to affix copies of requisitions sent to various departments for services provided to the patient.

Chart forms identified as supplemental include those medical forms and reports that may be added to the patient's chart, based on the patient having experienced a medical condition, procedure, treatment, or any other occurrence considered to be out of the ordinary. These forms usually include the following:

- *Surgical checklist*   This is used by the nursing staff to provide them with a list of all tasks that need to be performed on a patient prior to the patient being taken to surgery.
- *Special consent form*   This is signed by the patient and used by the facility whenever the patient is required to undergo surgery or some other diagnostic study or procedure.
- *Anesthesia record*   This provides the anesthesiologist or anesthetist with a space in which he or she can document observations of the patient while under anesthesia.
- *Postanesthesia record (PAR)*   This is used in the recovery room by the nursing staff to record the observations of the patient during the time he or she is recovering from anesthesia.
- *Emergency room report*   This is used by the nurses in the emergency room to record all observations and treatments performed on the patient during the time he or she was being treated in the emergency room.
- *Special vital signs sheet*   This is used to record the patient's vital signs when they are required to be taken more often than they are routinely taken.
- *Consultation form*   This is used by a physician who has been asked to consult on a patient's case and the type of care or treatment which is being provided. It eventually will be sent to the medical transcription department to be transcribed and become part of the patient's medical record.
- *Respiratory therapy record*   This is used by the respiratory therapist or respiratory caregiver to record all patient observations, care, and treatments provided by the respiratory therapy department, and to record the patient's progress as a result of receiving such care and treatment.
- *Physical therapy record*   This is used by the physical therapist or physical therapy caregiver to record all patient observations, care, and treatments provided by the physical therapy department and to record the patient's progress as a result of such care and treatment.

Whether you work for a health care facility that uses standard chart forms or one which uses both standard and supplemental medical record forms, you should understand that the majority of these hospitals and health care facilities house their patient's medical records and charts in standard metal or plastic, three-ring or clipped binders. These are easily accessible to members of the nursing staff and the physicians and are often located on a circular, movable file or neatly shelved in an area visible to the nurse's station.

Managing the Nursing Station

In many health care facilities throughout the country, another important function of the health unit coordinator that uses both verbal and written communication skills is managing the administrative and operational day-to-day activities of the nursing unit and its adjacent work stations. This generally involves being responsible for such things as ordering medical and nursing supplies and equipment, generating nursing unit reports and patient medical records, and making sure that separate work stations, such as treatment and utility rooms, dictation and medication rooms, the kitchen, and the nurse's lounge, are all kept neat and tidy and readily available for use.

The easiest, and, in most cases, the most efficient way to order supplies and equipment on the unit is to make sure that your station keeps a daily running inventory control list of both disposable and reusable supplies and equipment. Then, when it comes time to order replaceable items, you can readily check the

Administrative Medical Services

lists to make sure you are not ordering too much or too little of what is needed.

Supplies on the unit are generally separated into three categories: those used in the care and treatment of patients; those used in business applications; and those used as reference and resource materials. The lists below identify some of the more frequently used supplies and equipment found at the nurse's station:

Supplies and equipment used in the care and treatment of patients include:

- Disposable supplies, such as needles, syringes, bandages, and dressings.
- Alcohol wipes.
- Minor surgical removal kits.
- Stethoscopes.
- Sphygmomanometer (blood pressure cuff apparatus).
- Instruments used in physical examinations (percussion hammer, ophthalmoscope, otoscope).
- Flashlights.
- Thermometers.
- Disposable examination and isolation gowns.

Supplies and equipment used in business and clerical operations include:

- Pens (usually black and red).
- Message pads.
- Census reports.
- Desk assignment sheets.
- Medical chart forms.
- Medication cards (if used by the facility).
- File folders.
- Addressograph (imprinter) cards and card holders.
- Computer supplies (if a computer is used at the station).
- Requisition forms.
- Telephone directories.

Reference and resource supplies and materials include:

- Pharmacology and medication reference manuals, such as the PDR and hospital formulary.
- General nursing and medical reference textbooks.
- Diagnostic manuals, such as those used for laboratory tests and x-rays.
- Occupational and personnel policy manuals, such as a nursing procedure manual and a health unit coordinator manual.
- Hospital policy manual, which identifies the policies and regulations used by a specific health care facility.

In addition to being responsible for running the clerical operations and ordering most of the supplies and equipment used on the unit, the health unit coordinator is also frequently accountable for completing tasks such as making sure that time sheets and other types of personnel information have been properly posted on the employee bulletin board, keeping track of patients and their charts, and keeping a staff activity log. The best way to accomplish these goals includes setting attainable priorities and then sticking to them, never allowing the unit to run out of supplies, completing one task before moving on to another, making notes to yourself so you won't forget to complete certain tasks, looking for new and more effective ways to make your work, and the work of others, more manageable and efficient, and making out a daily work schedule and sticking to that schedule.

Health Unit Coordinator Tasks and Procedure Skills

In addition to being an effective communicator and carrying out those tasks required of the day-to-day clerical and administrative operations of the nursing station, the health unit coordinator is also responsible for skills and procedures necessary for hospital operations such as preparing the unit and the medical chart for patient admissions, discharges, transfers, and deaths, and preoperative and postoperative procedures for patients scheduled for surgery.

Patient Admissions, Discharges, and Transfers

Patients coming into the hospital are usually admitted in one of two ways. They either come in as a routine admission, in which case their arrival to your facility has already been prearranged by the hospital admissions office or an out-patient clinic, or they are admitted as an emergency through the emergency room.

Unless otherwise prearranged, all routine patient admissions are scheduled through the hospital's admitting department. An admitting clerk prepares the patient's medical chart by filling in the summary or face sheet, assigning a patient medical record or hospital number, and having the patient sign hospital forms, such as the release of liability form and the assignment of benefits form. In many facilities, the admitting department also prepares the patient addressograph plate, as well as the patient's identification bracelet, which is often made of plastic and is attached to the patient's wrist before he or she is brought to the nursing unit. Once the admitting department has completed the admission process, a hospital volunteer or a member of the admitting staff usually escorts the patient to his or her room.

When a patient has been admitted through the emergency room, the admitting department is notified so that a bed can be assigned as soon as possible. In most cases, the admitting clerk does not talk with the patient during evaluation in the emergency room. However, in some instances, a member of the admitting department staff may come to the E.R. or go directly to the patient's room to talk with the patient and complete the admitting process. In these situations, the admitting clerk generally makes every effort to have the patient or his or her representative at least sign the *Conditions of Admissions* or *Release of Liability* form, so that patient care can commence as soon as possible without the hospital being held liable for any problems which may arise from its provision.

Processing the Admission

The role of the health unit coordinator during the admission process involves both clerical and technical tasks. Performing clerical skills usually includes preparing a new patient chart by stamping all chart forms with the patient's identification plate (Figure 25-3). The forms are then assembled in the correct order, with old charts on the patient located by calling medical records and requesting that they be sent to the nursing unit, entering the patient's admitting vital signs and other pertinent information on the patient's graphic chart and in the vital signs book, if the facility uses one, entering the patient's name onto the daily census report, and transcribing all physician orders which may have been written on the patient either during a preadmission or routine admission or during an emergency admission.

Additional clerical responsibilities of the health unit coordinator may include greeting patients and their relatives or visitors at the nurse's station and then directing them to the patient's room, notifying members of the nursing staff of the new patient's arrival to the unit, and checking to make sure that the patient's identification bracelet matches the information found in the chart or, in some cases, processing and applying the bracelet directly onto the patient's wrist.

In most health care institutions, one of the most important responsibilities of the health unit coordinator is reading, interpreting, and transcribing medical or physician's orders. During the admission process, these orders are often referred to as *admission orders* and usually cover the physician's requirements for the patient's as to his or her diet, activity, diagnostic tests, treatments, and medications. If the admission is a routine or prearranged admission, the physician may use "routine" or "standing orders," which are usually preprinted orders kept at the nursing station. If the doctor wishes to change or delete something from these orders, he or she can simply call the unit or write the change in when arriving to see the patient. As with all physician orders, routine and standing orders must be signed off by the doctor.

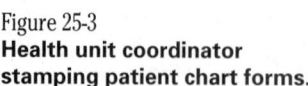

Figure 25-3
Health unit coordinator stamping patient chart forms.

Administrative Medical Services

Processing Patient Discharges

Patients may be discharged for one of four reasons: because they have been cured or relieved of their medical condition, because they are being sent to another institution or facility, because they choose to leave the facility against the advice of their physician, or because they have died while hospitalized.

If the physician has written an order for a routine discharge, you should:

- Make sure that the doctor has written an order for the patient's discharge.
- Make sure that the appropriate staff members know of the discharge so that they can make sure that all treatments have been completed prior to the patient leaving.
- Notify the appropriate hospital departments of the patient's discharge.
- Make sure that the patient has been given proper instructions as to picking up valuables, checking with the business office before leaving the facility, and making a follow-up appointment with the doctor.
- See to it that the patient's discharge is processed in a timely manner.

Once the patient has been discharged from the facility, you should:

- Check to make sure that none of the patient's private belongings have been left in the patient's room.
- Notify the kitchen or dietary department of the patient's discharge.
- Record the discharge on the daily census record form and discharge book if one is used.
- Disassemble and rearrange the medical chart in the appropriate sequence for discharge, and, if required, send it to the medical records department.
- Notify the housekeeping or environmental services department of the discharge so that they may disinfect and prepare the room for a new patient.

If a patient is being discharged to another institution, such as a skilled, extended, or intermediate nursing care facility, the process for discharge is generally the same as it is for a routine discharge, except that some facilities also fill out a patient transfer form.

Patients who decide to leave the hospital without a doctor's order, are classified as discharged *against medical advice (AMA)*. If this occurs, you should:

- Notify the charge nurse, who will then contact the patient's doctor, of the patient's decision to leave.
- Make sure that the patient has signed a special release form, which must also be witnessed, thus releasing the facility of any liability once the patient leaves the hospital.
- Notify the various departments within the hospital, and complete the appropriate clerical tasks, as though it were a routine discharge.

Patients who have died while hospitalized are also considered discharged. As such, it is required that you follow certain procedures to process their discharge. These may include:

- Locating, either by paging or by telephone, the patient's physician.
- Notifying the hospital's nursing services department, the registration or information desk, business office, and admitting departments of the patient's death.
- Entering the patient's name onto the daily census record as a death.
- Making sure that a death certificate is available for the physician to sign when he or she arrives on the unit.
- If an autopsy is to be completed, make sure that a stamped consent form is ready and available for use.
- Requisitioning, if not already available on the unit, a postmortem pack.
- Preparing identification tags which may be used on the body.
- Collecting the patient's personal belongings and placing or sending them to a designated area or location.
- Preparing and disassembling the patient's medical chart in a discharge sequence.

Processing Patient Transfers

As we have previously noted, when a patient is to be transferred to another institution, such as a skilled nursing facility or an intermediate care facility, the process is generally the same as it is for a routine discharge. The only difference is that, in this case, the health unit coordinator also completes a patient discharge form, with one copy being kept for the patient's chart and the other being sent along with the patient. When a patient is being transferred to another room or location within the same facility, however, the process is a little different. For example, if a patient is to be transferred from one room to another within the same nursing unit, the health unit coordinator:

- Makes sure that the change of room is reflected on all of the patient's medical chart forms.
- Makes the appropriate changes on the unit's daily census record and log books.

- Makes the appropriate changes on the patient's kardex and nursing care plan.
- Notifies the patient's relatives and visitors of the change of room.
- Notifies the appropriate hospital departments of the patient's transfer.

When a patient is being transferred from another unit or department to your unit you should:

- Make sure that the room is ready in time to receive the patient.
- Request the patient's medical chart, addressograph plate, and physician's order sheet from the staff member bringing the patient to your unit.
- Notify the nurse in charge and the other appropriate staff members of the patient's arrival at your unit.
- Add the patient's name to the unit's daily census record and identify it as a transfer.
- Add the patient's name to the vital signs book and the dietary sheet.
- Notify the dietary department as to the patient's transfer.
- Receive and transcribe any medical orders that have accompanied the patient.

Processing Surgical Patients

Patients are sometimes admitted into the hospital to undergo some type of surgical procedure. The procedure may be elective, meaning that it is being done completely at the patient's request or option; it may be a required procedure, usually requiring its completion within a set or specific period of time; it could be classified as an urgent case, in which case surgery is generally required within one or two days; or it could be considered an emergency, for which surgical intervention is necessary immediately. Some surgeries may also be classified as being outpatient procedures, in which case overnight hospitalization is rarely required.

When a patient is admitted to the nursing unit to undergo surgery, the health unit coordinator is usually expected to follow the correct procedures for admitting the patient to the unit. Once this has been completed, the health unit coordinator should follow the hospital's procedure for processing a preoperative patient, by:

- Processing, transcribing, and notifying the appropriate personnel of any preoperative orders such as surgical preps, dietary orders, consent forms, preoperative medications, activity restrictions, or need for special equipment and/or supplies.
- Making sure that the surgical checklist is stamped and available for the nurse who will go through it with the patient.
- Making sure the proper consent forms have been signed by the patient, according to the hospital's policy; if the patient is undergoing a procedure that will render him or her sterile, a sterilization form must also be signed by the patient.
- Ensuring that all medical records and laboratory reports have been completed and put into the patient's medical chart.
- Placing the patient's addressograph plate in its holder in the patient's chart.

In most cases, whenever a surgical patient returns to the nursing unit, all medical orders written prior to surgery are considered to be null and void. If this occurs, it will be up to the nurse to contact the patient's doctor to obtain new postoperative orders. Once these orders have been received, your responsibility will be to read them, interpret what they are saying, and then transcribe them so that they can be properly carried out.

Summary

In this chapter, we discussed the role of the health unit coordinator as it relates to members of the administrative medical services team: As such we identified and described the basic tasks required of this position. We also talked about this role encompassing such areas as communication skills, managing the nurse's station, and completing those tasks involved in processing patient admissions, discharges, transfers, deaths, and surgeries.

Review Questions

1. Define the following terms and abbreviations:
    a. AMA _____
    b. MAR _____
    c. communication _____
    d. requisition _____

2. Briefly explain the difference between the following:

a. routine admission _____

b. emergency admission _____

3. What is the name of the national organization most responsible for setting the guidelines and ethical standards for the professional health unit coordinator?

4. List at least three tasks required of the health unit coordinator:

   a. _____
   b. _____
   c. _____

5. Briefly explain the difference among the following types of paging systems:

   a. voice system _____
   b. number system _____
   c. pagefinder system _____

6. List at least five examples of desk sheets used at the nursing station:

   a. _____
   b. _____
   c. _____
   d. _____
   e. _____

7. What is another term used for a kardex?

8. What type of information would be found on an admission data form or face sheet?

9. List at least three examples of reference and resource supplies or materials that might be found at the nursing station:

   a. _____
   b. _____
   c. _____

10. Briefly explain the process involved in the following health unit coordinator procedures:

    a. hospital routine admission
    b. transfer of a patient from one unit to another in the same facility
    c. routine discharge

26

Transcribing Medical Orders

Performance Objectives

Upon completion of this chapter, you will be able to:

1. Describe the role of the health unit coordinator as it relates to processing and maintaining the patient's medical chart and transcribing physician's orders.
2. Identify the purpose and function of transcribing medical orders, and briefly explain the steps and process involved in transcribing them.
3. Briefly define the study of pharmacology and be able to relate it to transcribing medication orders.
4. Define and briefly explain the purpose and function of nursing, diagnostic, and therapeutic orders, and describe the process by which each is transcribed.

Terms and Abbreviations

FDA abbreviation for the *Food and Drug Administration*, which is the branch of the federal government responsible for approving all drugs administered throughout the United States.
MAR abbreviation for *Medication Administration Record*, which is a form kept on each patient that lists the types and dosages of all medications currently in effect.
PDR abbreviation for *Physician's Desk Reference*, which is a drug reference manual published once a year that can be used by members of the health care staff to help identify types of drugs, their usefulness and dosages, side effects, adverse reactions, and routes of administration.

PRN abbreviation used when something is to be carried out as necessary.
STAT abbreviation used when something is to be carried out immediately.
Treatment order an order that always contains a mention of the treatment itself, as well as an indication or time sequence or limitation as to when it is to be administered.
Urinary catheterization the process by which urine is removed from the bladder through the insertion of a nonretention, straight, or indwelling catheter.

If you have decided to work in administrative medical services as a health unit coordinator, one of your most important responsibilities, as we have already stated in Chapter 25, will involve the processing and preparation of the patient's medical chart and the use of knowledge and technical skills involved in reading, interpreting, and transcribing the various reports, records, forms, and physician's orders that make up that chart.

Throughout this text, we discussed the importance of the patient's medical record, or chart, and the function it plays in providing both the health care provider and the health care institution with a legal document that, in effect, tells a story of what is wrong with the patient and what care and treatment were provided to the patient during his or her visit or confinement in the health care facility. We also identified and discussed the various forms that make up the patient's chart, such as those used by the physician for writing orders, progress notes, and data related to a patient's medical history and physical examination. We also discussed forms kept by the nursing staff that

Administrative Medical Services

generally include the nurse's notes, graphics sheets, and medication administration records. We also discussed some of the more frequently used supplemental chart forms, such as consultation reports, special therapy records, and chart forms used when a patient undergoes a surgical procedure.

Since we have already explained the purpose and function of the medical chart, and have identified and defined the various forms comprising it, why discuss it again? Why? Because the medical chart is perhaps considered the single most important document used by all members of the health care delivery team. And maintaining, processing, interpreting, and transcribing those documents that make up the chart are usually the most important responsibilities of the health unit coordinator.

Maintaining the Medical Chart

Because the medical chart is a document used and moved about by many different people, making sure it is properly maintained and returned to where it belongs on the nursing unit often becomes a shared responsibility of the health unit coordinator and the nursing staff. Maintaining its confidentiality, and the information and data contained within the chart, is also an important and mutually shared responsibility of all members of the health care staff.

Maintaining the medical chart also involves making routine checks to ensure its readiness and availability when needed. This often involves the health unit coordinator making periodic checks of all charts to note whether or not all chart forms have been properly stamped with the patient's addressograph plate, making sure that each individual chart has the appropriate forms needed, so that a physician or nurse will not "run out of space" without having a new form ready to continue writing, ensuring that all charts are checked for their proper sequential order, and looking through all charts to safeguard their accuracy and completeness.

Transcribing Data from the Medical Chart

Now that you have a good understanding of the purpose and function of the patient's medical chart, as well as what types of documents comprise it, it's important to learn about what we do with the data and information that are written on those chart forms, particularly those written by the physician, since this is the person solely responsible for identifying what will eventually go into the chart.

Preliminary Information: Reading and Interpreting the Physician's Orders

Before you can begin to start transcribing orders from the patient's chart, there is some preliminary information which you must have.

To begin with, it's important for you to realize that all medical chart transcription starts with understanding what is expected of you. Just as a physician would not expect a nurse to perform a treatment for which that person had not been trained, he or she would not expect the health unit coordinator to transcribe medical orders that were not understood. Therefore, one of the most important preliminary steps of medical chart transcription is knowing what you are going to do before you begin doing it! And the only way this can be accomplished is to read the orders. If you have any doubt as to what they say, or what is needed, then by all means, ask questions.

The first thing that you should understand is that all orders written by the physician can be categorized according to the individual patient's needs. These categories include orders for medications, intravenous fluids, treatments, diagnostic studies, activities, and diet. When the doctor writes an order, he or she may write it as a STAT order, which is an order that must be carried out immediately. A standing or routine order is one that remains in effect and unchanged until the physician decides to change it. A PRN order is one carried out on an as needed basis. And, one-time, or limited, orders are those needed for a specific test or examination, as in the case of a specific laboratory test or x-ray examination.

Another crucial step involved in chart transcription has to do with being able to interpret what the orders say. The best and most accurate way in which this can be achieved is to know which guidelines you can follow to gain the best results. These guidelines include:

- Being familiar with the chart forms your facility uses so that you will be able to determine which ones are needed for what purposes.
- Being knowledgeable in medical terminology, abbreviations, and the use of special symbols used in the health care industry, so that you will be able to interpret what the physician has written.
- Reading all orders before you begin to transcribe them, so that you can prioritize them according to which ones need to be carried out first.
- Transferring all pertinent information from the physician's orders to the patient's kardex or nursing care plan, so that all members of the health care team responsible for caring for the patient are aware of special orders or treatments that must be carried out.

- Knowing which requisitions and forms are required for ordering all supplies, equipment, and medications, and then filling them out and sending them to the appropriate departments.
- Making sure you ask the nurse to check all orders for completeness after you have transcribed them and before they are initiated.

Understanding Basic Pharmacology and Transcribing Medication Orders

Drugs and medications have many purposes and can be used individually and in combination with one another. They help our bodies to combat certain disease-producing microorganisms; they help to provide replacements for fluids and other substances when they are lost due to illness; and they assist the body by improving both involuntary and voluntary body functions. Drugs can be found in plants and animals and can be manufactured in the laboratory. They can be defined by a *chemical name*, which identifies the drug's exact chemical composition, or by a *generic*, *trade*, or *brand name*, all of which are provided by the manufactuer who developed and manufactured the drug.

All drugs used in the United States must be approved by the Food and Drug Administration (FDA), which is a branch of the federal government responsible for regulating the composition of drugs and their usage. These drugs are classified according to their effects on the body, and, as such, are identified in several reference books used in most health care facilities. These books include the following:

- *National Formulary (NF)* Identifies manufacturing and dispensing standards of drugs by the manufacturer.
- *Hospital Formulary* Identifies drugs used by a specific hospital.
- *United States Pharmacopeia (USP)* Used by the FDA to identify all drugs, their chemical breakdown, and information relating to their standard preparation and dispensing.
- *Physician's Desk Reference (PDR)* Used by all health care personnel to provide an in-depth listing of all drugs by their chemical, generic, and trade names, and provides important information about the drug such as side effects, adverse reactions, dosage, and routes of administration.

In addition to drugs being classified as to their chemical, generic, or trade names, all drugs and medications are identified as to their use and effectiveness in the body. For example, some drugs are used to cleanse the skin and prevent the spread of infection. These are called *antiseptics*. Other drugs are used to stop the growth of microorganisms. These are referred to as *disinfectants*. Some of the more frequently ordered drugs you will probably encounter during your career as a health unit coordinator include the following:

- *Stimulants*, *depressants*, *analgesics*, *narcotics* and *hypnotics*, and *sedatives*, which affect the brain and the nervous system.
- *Antibiotics*, which are used to stop or inhibit the spread and growth of disease-producing bacteria and fungi in the body.
- Drugs that affect the gastrointestinal system, such as *laxatives*, *antiemetics* and *antacids*.
- Drugs that affect the heart and cardiovascular system, such as *vasoconstrictors*, *diuretics*, and *anticoagulants*.
- Drugs that affect the endocrine system, such as *insulin* and *hormonal replacements*.

Identifying Common Medication Orders

As you begin to transcribe medication orders, you will soon notice that there are some that are ordered more frequently than others. Some of these include:

- *Analgesics* These are used to relieve pain and discomfort and include *aspirin*, *talwin*, *tylenol*, *advicl*, *darvon*, and *fioinal*.
- *Narcotics* These are also used to relieve pain and discomfort, but are a great deal stonger than analgesics and, therefore, are given less frequently. Some of the most common include *morphine*, *percodan*, *demerol*, *dilaudid*, and *codeine*.
- *Antibiotics* These are used to inhibit the spread of infections. Some of the most common include *ampicillin*, *keflex*, *penicillin*, *erythromycin*, and *gentamycin*.
- *Sedatives* or *hypnotics* and *tranquilizers* These are used to induce sleep and produce a calming effect. Some of the most common include *dalmane*, *halcion*, and *nembutal* (hypnotics) and *compazine*, *valium*, and *mellaril* (tranquilizers).
- *Diuretics* These are used to induce the production of urine. Ssome of the most common include *lasix*, *diuril*, *dyazide*, and *esidrix*.
- *Antiarrhythmics*, *antianginals*, and *cardiotonics* These are used to strengthen and increase heart action. Some of the most common include *lanoxin* and *digoxin* (cardiotonics), *lidocaine* and *inderal* (antiarrhythmics), and *nitroglycerin* (antianginal).
- *Hypoglycemics* and *antidiabetic preparations* These are used to replace insulin and increase the production of glycogen in the body. Some of the most common include *regular*, *NPH*, and *lente*

insulin (injectable hypoglycemics), and *diabinese*, *orinase*, and *micronase* (oral hypoglycemics).

Transcribing the Order

Whenever the physician writes a medication order, he or she is responsible for providing precise and specific information regarding that medication. This information includes both the name and the dosage prescribed, the route of administration, and the frequency with which the medication is to be given. Some doctors may also write a qualifying statement as to how or why the medication should be administered. For example, if the physician wants to write an order for a specific pain medication, he or she might write the order as follows: *Demerol 75 mg IM q 3h prn for abdominal pain*. Interpreted, this order means "the patient is to be given a dosage of 75 mg of the drug demerol, administered intramuscularly, every 3 hours, as needed, for abdominal pain."

Once you have read and interpreted the order, you are ready to transfer it to where it will be seen by the nurse and, eventually, delivered to the patient. This includes writing on the patient's medication administration record (MAR) (Figure 26-1), the kardex or patient care plan, and, if utilized by your facility, a medication ticket. After you have completed the transfer process, you are now ready to order the medications from the pharmacy. But before you do so, remember to have the nurse double-check the order for accuracy and completeness. You may then fill out the proper requisition form and send it to the pharmacy. Following the completion of all these tasks, the medication has been properly "taken off" the chart and transcribed.

Transcribing Orders Related to Nursing Care and Treatments

Orders that relate to the direct care and treatments of patients generally include those tasks and hands-on procedures involved in meeting the daily personal and medical needs of the hospitalized patient. They often incorporate treatments used to assist the patient in measuring such bodily functions as the bowel and bladder, circulatory and respiratory needs, level of consciousness, degree of acitivity while confined to the hospital, and replacement and removal of body fluids through some type of specialized procedure. Some of the more common nursing orders encompassing these areas include the following:

- *Suctioning procedures*   These deal with the removal of fluids by way of a special catheter or tube, and may include gastric suctioning (removal of fluids from the stomach), tracheal suctioning (removal of fluids or secretions from the trachea), throat or nasal suctioning (removal of secretions from the throat or nose), and hemovac suctioning (removal of secretions, fluids, or blood by way of a mechanical device attached directly to a drainage tube, usually inserted after surgery).
- *Enemas, harris flushes, and irrigations*   These involve the insertion of fluids into and removal of fluids from the large intestine to rid the body of fecal materials or flatus (gas).
- *Taking vital signs and height and weight*   These include measuring and recording the patient's blood pressure, pulse, temperature, respiration, and height and weight measurements.
- *Applying hot or cold packs*   These are generally used to decrease edema (swelling), soreness, or pain, and can be either moist (soaks, compresses, wet packs, or baths) or dry (electric applications such as Aquamatic K-pack or heat lamp, thermal blankets or ice packs, and ice collars).
- *Administration of intravenous (IV) or parenteral fluids*   These involve the introduction of specific fluids, such as dextrose, normal saline, electrolytes, total parenteral nutrition (TPN), and blood and blood by-products.
- *Insertion of catheter*   A catheter is inserted into the urinary bladder to assist the patient to void (urinate), and can be done with a nonretention (straight) catheter for one time usage, or with a retention or indwelling (Foley) catheter when the catheter must remain in place for a period of time.
- *Special urinary measurements and procedures*   These include measuring the intake and output of fluids taken in and produced by the patient, and testing the urine for sugar, acetone, or ketones.
- *Special comfort devices*   These may be ordered to assist the patient in meeting his or her needs for added comfort or safety, and often include such items as egg crate mattresses, walkers, soft restraints, sheepskin or lambskin wool, and bed cradle or bed board.

Another type of order frequently identified as a nursing order or a therapeutic order is the request by the patient's physician for a specific type of diet or nutritional component. In the health care facility, there are many different diets available, and the type of diet ordered depends greatly on the health and medical condition of the patient. Following is a list of therapeutic diets offered by most hospitals and health care facilities:

HUNTINGTON MEMORIAL HOSPITAL
MEDICATION RECORD

PAGE: ______ OF ______

Injection Site Codes:

① Rt. Deltoid	⑥ Lt. Ventrogluteal	
② Lt. Deltoid	⑦ Rt. Anterolateral Thigh	**For Pediatric Use Only**
③ Rt. Gluteus Medius	⑧ Lt. Anterolateral Thigh	
④ Lt. Gluteus Medius	⑨ Rt. Abdomen	⑪ Rt. Anterior Thigh
⑤ Rt. Ventrogluteal	⑩ Lt. Abdomen	⑫ Lt. Anterior Thigh

TRANSFER TIME: ______________

ALLERGIES

MEDICATION DATE Rx#	GENERIC NAME DESCRIPTION COMMENTS	(TRADE NAME) ROUTE FREQ.	2300 - 0700 (11-7)	0700 - 1500 (7-3)	1500 - 2300 (3-11)

MEDICATION ADMINISTRATION KEY

(09) - MEDICATION NOT GIVEN (EXPLANATION ON REVERSE SIDE)

~~09~~ - MEDICATION GIVEN

854,002
REV. 1/87

SIGNATURES (11-7)	SIGNATURES (7-3)	SIGNATURES (3-11)

Figure 26-1
Medication administration record. (Courtesy of Huntington Memorial Hospital, Pasadena, CA.)

Administrative Medical Services

- *Regular diet* This type of diet, which provides the patient with all the essentials needed for good nutrition, is generally ordered for patients who do not require any restrictions to their diet.
- *Light, soft, or mechanical soft diet* These diets all contain foods that are soft in consistency and, therefore, easy for the patient to digest. Therefore, they are most frequently ordered for patients who have undergone surgery, before the patient is ready to begin eating a regular or "normal" diet.
- *Clear and full liquid diet* Liquid diets are just what they imply: liquid. A clear liquid diet consists of liquids which can be seen through, and often include such things as clear broth, tea, and ginger ale. Full liquids often include those already given in a clear liquid, as well as others, such as fruit juices, soups, ice cream, and custard.
- *Bland diet* A bland diet consists of foods that are easy to digest and mild in flavor, and are often given to patients who suffer from digestive problems.
- *Diabetic (ADA) diet* A diabetic or ADA (American Diabetic Association) diet is one which provides the patient with the exact balance of nutrients needed to maintain his or her diabetic condition. These diets are usually low in sodium, fat, and carbohydrates.
- *Low residue diets* A diet which is low in residue means that the foods it contains have very little bulk and are, therefore, easy to digest. These types of diets are generally ordered for patients who suffer from rectal and colon diseases.
- *Low calorie and low fat diets* These diets are low in calories or low in fat content and are often prescribed for patients who must lose weight. Low calorie diets are usually low in butter, cream, fats, and deserts, while low fat diets generally contain limited quantities of fats.
- *Low sodium and low cholesterol diets* These diets are often prescribed to cardiac patients because of restrictions in the amount of sodium and cholesterol-producing foods. A low sodium diet contains absolutely no salt, while the low cholesterol diet limits the consumption of eggs, meats, and whole milk.

Reading, Interpreting, and Transcribing the Order

Once the physician has completed writing the orders, the procedure for transcribing them is very much the same as it is for transcribing medication orders. You must first read through all of the orders to make sure you understand them and prioritize those which require immediate attention. The next step is to make sure you transfer the orders onto the kardex or the nursing care plan and the individual desk sheets or procedure or treatment logs. Also, if your facility uses special treatment or procedure tickets, you must transfer the order onto these as well.

After you have completed the transfer process, the next phase of transcribing the orders is to fill out the appropriate requisition forms needed for the individual procedures. However, before these forms are sent off to the various hospital departments, you must remember to have the nurse check the orders to ensure their completeness and accuracy. Once the nurse has made sure that all of the orders are complete, you may send the requisitions to their appropriate departments.

Transcribing Orders Related to Diagnostic Treatments

Medical orders which fall under the classification of *diagnostic* include many of the tests and specialized procedures necessary and required by the physician for an accurate diagnosis of the patient's medical condition or illness.

Treatments classified as diagnostic involve both basic and advanced laboratory and radiology tests, procedures, and examinations of body organs and specimens. Some of the more common and frequently encountered of these are such laboratory studies as a complete blood count (CBC); blood chemistry examinations, such as fasting blood sugar (FBS), blood urea nitrogen (BUN), glucose tolerance testing (GTT); and specialized microbial or pathology testing, such as testing of the cerebral spinal fluid, testing the blood for antibodies for a potential blood donor recipient, and bone marrow testing through aspiration for possible transplantation. Some of the more frequently ordered radiology examinations include chest x-rays and x-ray examinations using contrast media, such as intravenous pyelogram (IVP) studies; upper and lower gastrointestinal studies (GI series); barium enemas (BE); computerized tomography (CT) scans; and magnetic resonance imaging (MRI) studies.

Diagnostic orders are not always limited to the laboratory or radiology department. These may also include other departments within the hospital, such as respiratory therapy, cardiology, physical and occupational therapy, speech pathology, and gastroenterology. These are all units within the health care facility, the primary goal of which is to provide the physician with the results of studies and tests that will ultimately lead to a diagnosis. Studies often completed in these departments include electrocardiograms (ECG), which provide the physician with a graph or tracing of the heart's electrical action; electroencephalograms (EEG), a procedure used to provide a tracing of the brain's electrical patterns; various types of respiratory or inhalation therapy treatments, such as intermittent positive pressure breathing (IPPB); positive and expiratory pressure (PEEP) therapy, which provides the physician with a means of measuring the amount and duration of oxygen being taken in and carbon dioxide being pushed out of the lungs; and sonography, imaging, and tomography, which use a computer to outline and identify such medical conditions as tumors, fractures, and other catastrophic diseases which might otherwise not be distinguishable using conventional x-ray therapy.

Reading, Interpreting, and Transcribing the Order

Like those treatments and procedures classified as nursing orders, those which are identified as diagnostic in nature must also be read and interpreted before you can transcribe them. After you have read through all of the orders, you must prioritize those which require immediate attention. As with others listed under priority, these types of orders are generally preceeded by the word STAT, meaning that the order must be transcribed and completed immediately. The next step is to transfer the order onto the patient's kardex or nursing care plan, the unit desk sheets or special procedures logs, and any other type of treatment ticket.

Once you have finished transferring the orders to their respective and appropriate sheets, logs, and tickets, you are ready to begin filling out the requisition forms which may be needed to order a specific treatment. However, as with both the medication orders and the nursing care orders, before you send the requisition form you must have the nurse check over all orders which have been transcribed to ensure accuracy and completeness. After the nurse has checked the transcribed order, you may send the requisitions to the approrpriate departments.

Cancelling, Renewing, and Stopping Orders

Whenever the physician writes an order to cancel, renew, or stop an order, the responsibility for completing this task is often left to the health unit coordinator. As with any other medical order, a cancellation, renewal, or stopage order, must also be transcribed. However, the main difference between these orders and those which we have already discussed (medication, nursing, and diagnostic orders) is that, once you have read through and interpreted an order to the end, you must notify the nurse and the appropriate departments immediately. If, for example, the doctor writes an order to discontinue a specific medication, you would not only notify the nurse responsible for administering the medication but you would also notify the pharmacy. The pharmacy would in turn either come to your unit and remove the existing medications or would ask that the medication be sent back to the pharmacy (Figure 26-2).

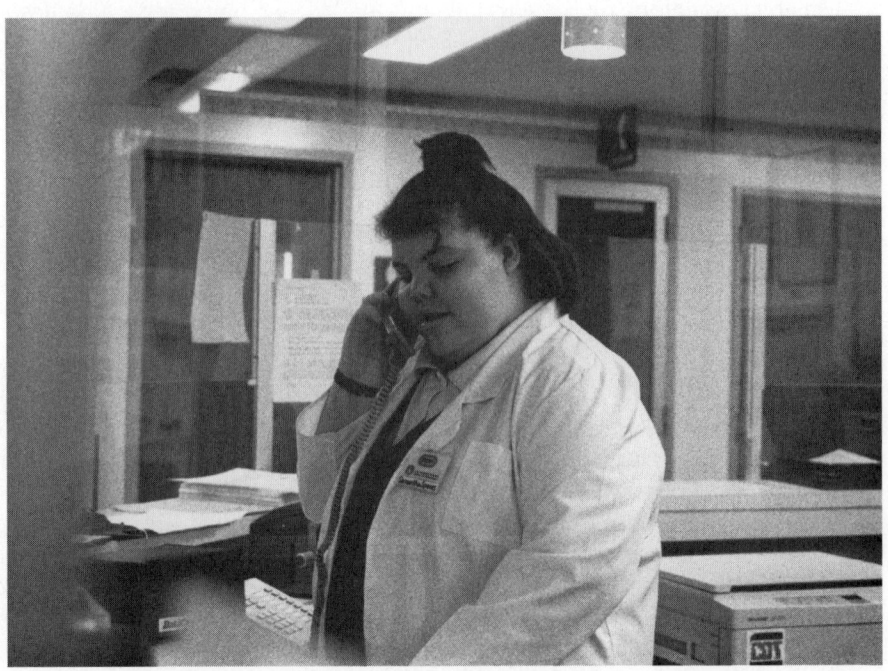

Figure 26-2
Notifying other departments of a cancelled order is part of the health unit coordinator's responsibility.

Administrative Medical Services

Summary

In this chapter, we discussed the role of the health unit coordinator as it relates to the processing and maintenance of the patient's medical chart and to the transcribing of those medical orders which comprise the chart. We also discussed both the purpose and function of medical orders, including those identified as medication orders, nursing, care orders, treatment orders, and diagnostic orders. We also briefly explained the various steps involved in transcribing each type.

Review Questions

1. Define the following abbreviations:
   a. STAT _____
   b. PRN _____
   c. FDA _____
   d. PDR _____
   e. MAR _____

2. Whose responsibility is it to maintain the patient's medical chart?

3. What are the three steps involved in transcribing all medical orders?
   a. _____
   b. _____
   c. _____

4. Briefly explain the difference among the following names given to drugs:
   a. chemical name _____
   b. generic name _____
   c. brand name _____

5. Give at least three pharmacology reference books which might be found at the nursing unit:
   a. _____
   b. _____
   c. _____

6. Transcribe the orders listed below:
   a. Demerol 50 mg IM q 4h prn for back pain
   b. Sse hs until clear
   c. May have lo Na tray this p.m., then NPO after MN

7. Define the following abbreviations:
   a. CBC _____   e. ECG _____
   b. GTT _____   f. PEEP _____
   c. BUN _____   g. IPPB _____
   d. FBS _____   h. EEG _____

8. Give at least one example of a diagnostic order.

9. Briefly define the difference between the following:
   a. CT scan _____
   b. MRI _____

10. List at least three types of x-rays which the physician might order:
    a. _____
    b. _____
    c. _____

Section VII

Job Seeking Skills and Employment Opportunities in Administrative Medical Services

27
Job Seeking Skills and Employment Opportunities

27

Job Seeking Skills and Employment Opportunities

Performance Objectives

Upon completion of this chapter, you will be able to:

1. Identify where most jobs will occur by the year 2000.
2. Describe the steps and process involved in beginning a job search.
3. Explain how to fill out a job application form.
4. Discuss the purpose of a cover letter.
5. Define the function of a resume and identify its basic components.
6. Identify the three types of resumes most often used in securing a position in the field of administrative medical services and briefly explain when each should be used.
7. List the proper guidelines for preparing a resume.
8. Discuss the steps involved in preparing for a job interview.
9. Discuss the process involved in conducting oneself in a professional manner during the job interview process.
10. Briefly explain the follow-up process that should occur after an interview has taken place.

Terms and Abbreviations

Chronological resume   a type of resume which provides an employer with specific dates and employers listed in chronological order.
Combination resume   a type of resume which emphasizes the applicant's work skills and lists previous employers with dates of employment.
Cover letter   a personal introduction to a prospective employer that is always enclosed with a resume.
DNA   abbreviation for *does not apply*.
Employment agency   an agency or organization whose sole purpose is to assist applicants in attaining employment and employers in filling a position.
Functional resume   a type of resume which highlights an applicant's qualifications and marketable skills.
Interview   the process by which an applicant speaks with a potential employer regarding employment.
N/A   abbreviation for *not applicable*.
Resume   a one- to two-page document which is used to provide perspective employers with information regarding a future employee's educational and employment history.

According to the U. S. Bureau of Labor Statistics, by the year 2000 over half of the jobs available in the United States will occur in professional, clerical, and service-oriented occupations. All information generated from these statistics point to health care being the industry offering the highest number of jobs and the brightest prospects. Physicians, nurses, administrative medical assistants, medical records clerks, medical insurance coders, medical transcriptionists, and health unit coordinators will be in great demand throughout rural and urban communities, especially when a national health insurance program goes into effect. Key factors contributing to increased employment opportunities for administrative healthcare workers and other allied health care professionals, include the need to staff medical offices for the increasing numbers of medical doctors being graduated from medical schools, public awareness for the need to provide quality health care, and an increase in the volume of paperwork emanating from insurance companies and from state and federal regulatory agencies.

Research studies seem to support the fact that although the job-seeker has all the hands-on technical and administrative skills and theoretical knowledge to do a job, he or she often lacks the basic skills required to *get* the job. Therefore, before you begin actively seeking employment in a health care facility, there are specific steps you should take in order to prepare to find and *obtain* the best position possible. The first question you should consider, however, is why do employers reject job applicants?

There are many reasons why an employer rejects a job applicant. Some of the most frequent responses employers give include applicants showing little interest or poor reasons for desiring a position. Also, an application showing a past history of "job hopping" and the inability to verify previous employment and/or educational background tends to make a prospective employer hesitant to hire an individual. Demonstrating a lack of maturity or an inability to communicate effectively during the interview process are two additional reasons employers sometimes give for not being interested in an applicant. Dressing unacceptably or showing a lack of professionalism and good manners during the interview are also reasons. And finally, providing the employer with a poorly completed job application form or resume or demonstrating a lack of job-related skills are more good reasons why employers tend to shy away from hiring someone.

Initiating the Job Hunt

When you are ready to begin your search for a position, you should try to follow through on all possible job leads. An individual who recently completed his or her education might contact the school placement office and place his or her name on file. You might also attend meetings, seminars, and workshops that are open to the public, which are run by professional organizations involved in administrative health care occupations. Those announcements are usually placed in a special section of your local newspaper. Because many jobs are never formally advertised, you should try to spread the word that you are "looking" to classmates, teachers, family, and acquaintances who work for cardiovascular medical practices, out-patient centers, individual physicians, or health care facilities. If you own a computer, you may also want to go "online" or hook up with an E-mail listing, since many online services such as America OnLine and Prodigy offer users bulletin boards of job listings and classified ads.

You should understand that the hiring for jobs, especially those in health care, is often most effective when it is completed through a network of contacts or grapevine, called a *"hidden job market."* This process involves an employer passing the word of a specific job opening to his or her friends and colleagues, who in turn spread the word and refer qualified persons they know. Approximately 75 percent of all health-related jobs are secured through personal contacts, the hidden job market, and the greatest and most abundant source, acquaintances.

As you conduct your search for employment, try to visit public and private employment agencies, fill out application forms, and take any skills tests that may be a requirement for a specific position. Many of the employment and personnel agencies work as referral services and make their money by placing applicants in jobs. Often when a job applicant calls an employment agency about an advertised job opportunity, chances are that it may have already been filled. Even so, you can expect the agency to suggest that you still visit the office so that you can talk about other openings.

Before agreeing to an appointment at an employment or personnel agency, it is always in your best interest to first find out as much as possible about other job listings, since you may already have learned about them from another source or agency. That's because many positions are listed with more than one agency. Some agencies may require you to sign a written binding contract before they can assist you in your job search. Before you agree to this, make sure you read it very carefully to determine whether you or the employer is required to pay the agency a fee for finding you the job.

Another good source of leads is the help wanted section of your local newspaper. The advertisement explains how to make the initial contact, usually by telephone or written communication. Help wanted advertisements appearing in newspapers generally cannot mention the age, sex, and race requirements of the job; however, some ads, such as situation want ads, do allow this.

Another good way to begin your job search is to visit the personnel or human resources departments of local hospitals, clinics, and health care facilities. Many of these facilities also keep questionnaires and application forms on hand for potential employees to review. If you do read through these forms, it's important to study each question carefully before indicating an answer. This is because in some agencies and facilities, the object of the question may be to see if you can follow directions, rather than just answer the question. If you do fill out one of the application forms, always try to furnish as much information as possible, but if an item does not apply or cannot be answered, you should indicate that the question has not been overlooked by inserting *"no," "none," "NA (not applicable)," or "DNA (does not apply)."*

Filling Out the Job Application Form

When filling out a job application form, you should always read the entire form before beginning to fill it out, since some instructions may be on the last line of the page. Make sure you read the fine print and take note of any special instructions, such as *"please print"* or *"put last name first,"* since doing so shows the employer your ability to follow instructions. Also, unless you are told to do otherwise, always make sure you fill the application out in ink. If your handwriting is illegible, you should block print and never abbreviate words when there is ample space to write them out. Try to keep a completed application form with you that has been checked for accuracy. You may use it as a "guide" to copy information onto each master application form. If the form asks for your experience using equipment, make sure you list each specific type of equipment. When listing your previous employment, make sure you are exact with dates of previous employment. This is very important, since most employers will require a verification of your employment history. If the form asks the reason why you left a position, this may be left blank for discussion during the interview. And finally, make sure you sign the application upon its completion.

Occasionally, a blind letter to an office where a possible job opportunity is discovered through a friend or where you would especially like to work may result in a positive response.

Seeking Employment in Another Location

If you are planning to seek employment in a geographical area that may be unfamiliar to you, advance planning is highly desirable. Nothing is more discouraging to an unemployed person than a prolonged job search in unfamiliar surroundings. Dun and Bradstreet, Inc's. *Million Dollar Directory* and the *Middle Market Directory* are two national publications which can provide you with information on potential employers in many cities throughout the United States. Your local *Yellow Pages* directory, which is provided by most telephone companies, can also provide you with a comprehensive listing of professional offices and hospitals. Chambers of Commerce publications also include membership directories, names of major professional employers in specific areas, and brochures describing regional facts. It's even a good idea to subscribe to the principal newspaper of a city where employment is desired, since this is one way of *tuning in* to the local job market of that area, and responding to their advertisements.

The Cover Letter

A *cover letter* is like a personal introduction to a prospective employer (Figure 27-1). It should be typed and addressed to a specific person. If you are applying for a position that you saw advertised in the newspaper and that ad listed only a telephone number or a post office box, you should call and ask for the name of the personnel supervisor or the person that is most likely to be doing the interviewing. This is done so that the cover letter can include that person's name. If you are unable to ascertain the name of the person to whom the letter is to be addressed, it is acceptable to use the salutation "Dear Sir or Madam," or "To Whom It May Concern." In some cases, you may even use a less formal salutation, such as "Hello" or "Good Morning."

If your intention is to attract the reader's interest, you should use the first 20 words of the letter to state the reason you are applying for the position. This part of the letter should also provide the reader with a summary of your qualifications and a statement as to how you learned about the position. If you choose, you may even provide a short accounting of your career objectives. The concluding sentence of your cover letter should pave the way for an interview appointment by calling attention to a telephone number. Remember that the cover letter should be a preview of your writing skill and should therefore be free of any errors in spelling, punctuation, and grammar. A personal touch can also be added to make the letter stand out.

Preparing the Resume

Most human resources supervisors agree that the sole purpose of providing a perspective employer with a resume is to sell your job qualifications in a format that is as brief and as attractive as possible, in order to obtain an interview. A well-written resume should be able to summarize your educational and vocational history, with an analysis of problems effectively

May 3, 1995

Personnel Dept.
Free Memorial Hospital
601 Graceland Avenue
Anywhere, USA 55555

Dear Sir or Madam:

Please allow me a few moments from what I am sure is a very busy schedule, to respond to the ad you are currently running in the Anywhere Times, in which your company is seeking to fill the position of Director of Health Unit Coordinators for your facility.

As you will note from my enclosed resume, I have an extensive background as a health unit coordinator, a medical billing and coding specialist, and a medical records clerk. In addition to being highly skilled in these areas, my expertise has also afforded me with a strong background in public relations and advertising, both of which are disciplines that I believe to be of great benefit to the position you are seeking to fill.

If you believe, as I do, that my background and expertise may be what you are looking for in order to fill the very challenging position of Director of Health Unit Coordinators, my hope is that you will consider me for this very rewarding position. I eagerly look forward to hearing from you so that we may further discuss the benefits I could bring to your very prestigious hospital.

Sincerely,

Susie Jones

Figure 27-1
Example of a cover letter.

confronted and solved, emphasizing organizational, administrative, and clinical skills (Figure 27-2).

The ideal resume is a one-page document, written in language similar to a telegram. It should contain as many action words as possible and should avoid the personal pronoun. If it is impossible to confine the resume to one page, it can be typed on a 10 × 14-inch page for reduction to standard letter size. Brevity is also important as long as it is possible to include all relevant information.

All data on the resume should be originally typed or printed on high-quality, wrinkle-free bond paper, single spaced internally, with double spacing between unrelated items, and with balanced spacing on all four margins. Headings should stand out, and can be typed using a different style of type from the rest of the resume.

Types of Resumes

There are three formats which are generally accepted for job seekers in the health care industry. These inclue the *chronological resume*, the *functional resume*, and the *combination resume*.

The chronological resume (Figure 27-3), is the most widely used because of its familiarity to employers. It provides the employer with an interview script and is the easiest for the applicant to prepare. Specific dates and employers are listed in chronological order, with either experience or education first, depending on the background area to be emphasized.

The major advantage of using a chronological format is that it stresses a steady employment record. However, the two greatest disadvantages of this resume are that any lack of experience is quickly revealed, and while a steady working history may be exposed, absence of job skills also seem to stand out. To counteract these weaknesses, you may want to enclose photocopies of any certificates or licenses which emphasize specific skills.

For most applicants who have recently graduated from high school or college, the functional resume (Figure 27-4) seems to be the best choice for highlighting the person's qualifications and marketable skills. It does not list specific job titles or descriptions with dates, rather it emphasizes growth and development in skills.

The combination resume emphasizes the applicant's work skills and identifies specific employers and dates in which the employee worked. A major advantage of this format is that it shows at a glance an applicant's entire working and educational history, particularly stressing the most relevant skills and employment experience, while minimizing less significant experience. A major weakness of this format, however, is that it tends to run on and on.

Whether you select the chronological format, the functional format, or the combination format for your resume, personal data such as your name, address, and telephone number should be stated in as brief a manner as possible. According to a decision made by the Civil Rights Act of 1964 and enforced by the Equal

Administrative Medical Services

HELAINE J. HOWARD
5619 Newstreet Avenue
Thistown, VT 55555   (888) 999-9999

EMPLOYMENT HISTORY

9/87 - present   VALLEY MEDICAL & DENTAL CLINIC
Thistown Vermont
Personnel Director

Position requires management and supervision of all departments, including sales and marketing, patient admissions, billing and coding, and insurance processing and collections. Also responsible for managing clinic which sees more than 100 medical and dental patients per day, and for assisting all employees in learning how to complete tasks required of their positions. Duties also include implementation of human resources tasks, such as presenting seminars, in-services, and training workshops to new and existing employees, and processing documents for all departments which are related to personnel matters. Position requires supervision and management over a staff of 25 employees.

8/85 - 9/87   Newberry Medical Group
Granada Hills, California
Medical Office Manager

Completed all tasks required in medical billing, coding, insurance processing, completion of medical records, appointment scheduling, personnel matters, and assisting patients and members of the staff whenever necessary.

1/84 - 8/85   Dr. M. D. Smith
Hacienda Heights, California
Front Office Assistant

Responsibilities included being responsible for the front office of a family practice physician. Duties included answering the telephone, scheduling patient appointments, filing and processing medical reports, processing payments and insurance forms, coding medicare and medicaid insurance forms, and assisting the physician with matters which required correspondence with members of the community.

6/77 - 1/84   Jones Medical Group
Westernville, California
Medical Receptionist

Responsibilities included answering the telephone and scheduling appointments, filing medical reports, and typing letters and other correspondence for a medium-sized medical group.

References Furnished Upon Request

Figure 27-2
Example of a resume.

Employment Opportunity Commission, no employer is allowed to ask you about information regarding your height, weight, date of birth, marital status, social security number, or presence of any physical disability, therefore, you may omit any information related to these topics from your resume.

Guidelines for Preparing the Resume

When preparing your resume, you will be much more successful in obtaining the position you desire if you follow some very simple guidelines. These include:

- Being as brief as possible; get to the purpose of your resume, and remember, you will always get better results if you do not "bore" your prospective employer with all the details of your skills on your resume; save something for your interview!
- Explaining previous job titles and duties, and mentioning awards, special interests, credentials, certifications, and licenses.
- Emphasizing the positive points about your employment history, while being very careful not to identify your weaknesses.
- Enclosing photocopies of certificates, licenses, or diplomas, only if they will be of value to the interviewer for recall purposes and indication of skills.
- Indicating that references are available on request.
- Always typing your resume, the cover letter, and the envelope.
- Never mentioning salary requirements, unless you are requested to do so.

SUSIE SMITH
12345 Green Street
Youknow, CA 11111
(213) 555–5555

EMPLOYMENT HISTORY

1981 to Present: **Medical Billing Services, Inc.,** Van Nuys, Pennsylvania

As the owner of my own *medical billing* company, I have been involved in all aspects of medical billing, coding and insurance processing. These areas of billing and collection procedures involved HMOs, Medicare and Medicaid, private insurance companies, and workers' compensation. Most recently, I was involved in two start-up projects. In the first, from 1992 to April of 1993, I was responsible for creating, designing, developing, and implementing a new billing program for a large medical practice in Northern California. Within six months, the practice grew to over 300 patients, and grossed over $2 million in revenue. From May of 1993 until May of 1994, I was involved in the conception, development, and implementation of a brand new company which specialized in collections for the health care industry. This company currently trains other health care facilities in billing and collection procedures.

1979 to 1981: **Doctor's Medical Group,** Los Angeles, California

As a part-time *medical coding and billing clerk* for a large physicians-operated medical group, I was responsible for doing all the medical coding and billing. My responsibilities in this role also included performing and teaching others how to code insurance forms for billing, process different types of insurance forms, bill patients for services rendered, and perform collections on non-paid accounts.

1976 to 1979: **Concorde Billing Services,** Kansas City, Missouri

As the *Medical Receptionist,* I was responsible for greeting patients and other people coming into the office, answering telephones, scheduling appointments, completing correspondence for executives, and performing other clerical tasks required of my position.

EDUCATION

B.A.　　　　　　　　Health Sciences Management, Anywhere University, Jonesville, CA 1976
A.A.　　　　　　　　Business Management, Los Angeles Valley College, Greenville, CA 1971

References Furnished Upon Request

Figure 27-3
The chronological resume.

- Only listing volunteer activities if they are related to the health care industry or the position for which you are seeking employment.

If the bulk of your resume is devoted to educational qualifications and skills, you should provide details of courses that relate to clinical and organizational skills. Awards may also be mentioned, as well as the grade averages. The names of high schools and any colleges attended may also be included. And any diplomas from vocational schools or degrees achieved, with the most recent listed first, should also be identified.

Work experience should include the name and address of the employer, the job title, the length of time employed, and a brief description of duties. Most recent employment is always listed first. Remember, prospective employers usually verify the applicant's employment history, so it is important that you are able to state the exact employment dates of any of your previous employers. Also, unless required, past salaries should never be included in the resume or cover letter. Miscellaneous information, se-

Administrative Medical Services

JANE DOE
5311 Corteen Place, #28
No. Hollywood, CA 91607
(818) 761–2345

7/90–5/95	DR. JOHN SMITH, M.D. Los Angeles, California **Office Manager**	
	Overall responsibility for management of high-volume medical chiropractic office. Duties included answering the telephone and scheduling appointments, billing, banking, and collections, insurance claim processing, maintenance and management of all medical and financial records, correspondence, generation of new patients into practice, public and patient relations, and administration of all clerical and office tasks as required for practice.	
4/87–5/90	TWIN-TOWERS RADIOLOGY GROUP Los Angeles, California **Patient Accounts Representative**	Figure 27-4 **The functional resume.**
	Responsibilities included billing, insurance claim processing, financial record-keeping, answering the telephone and appointment scheduling, banking and collections, and maintenance and management of patients' medical records for high-volume radiology medical group.	
1/78–12/86	MEDICAL MANAGEMENT COMPANY Philadelphia, Pennsylvania **Medical Coding Specialist**	
	Responsibilities included accounts payable and accounts receivable, medical coding, yearly taxes, dealing with insurance companies, and all other tasks related to bookkeeping and coding practices for large medical management group.	

References Furnished Upon Request

---

lectively chosen, belongs at the end of the resume. This information might include volunteer work, honors, awards, certificates, special language skills, and leadership positions in the health care industry or professional health care organizations and associations.

Because prospective employers usually ask for names of people who can provide references, you should contact friends and former employers *before* an interview in order to ask them if they will serve as references. This courtesy allows the person time to prepare a response before receiving an unexpected telephone call or communication from the prospective employer.

Taking Part in the Interview Process

The interview process is perhaps the most important aspect in securing a position, chiefly because it often determines whether an applicant will be offered employment. Research studies have shown that when two applicants have similar skills and education, the choice of which candidate to hire is based almost entirely on the physical appearance and attitude of the person at the time of the interview. The importance of good grooming in projecting a favorable first impression cannot be overemphasized. Your clothing should always be immaculate. And only wear conservative clothing that fits the image of the job. You may choose to use coordinated accessories, but only if they show good taste. Remember, a picture is worth a thousand words, and looks often speak much louder and may reveal inward feelings. When you know you look your best, you will be more confident and relaxed at the time of your interview.

What Should You Bring?

A portfolio of information related to the current job opening, carried to the interview, suggests that you are serious about obtaining the position and are well organized. Such information often includes letters of recommendation, school diplomas or degrees, transcripts, certificates, names and addresses of references, a copy of your resume, your social security card, and items related to your prior education and training.

It is important to find out as much as possible about the position you are applying for before the interview. And because the interview appointment is the only time you may have with the interviewer or prospective employer, this is the time to show that person firsthand your ability to be a professional. Remember, during the interview process, it is the employer who desires a firsthand evaluation of your maturity, manners, personality, and verbal skills. You will find that being knowledgable about the requirements of the position and only responding to questions that are asked of you will help you to feel much more relaxed during the interview process.

The Actual Interview

If you are contacted by the employer to be interviewed, you should always arrive on time, usually at least 15 minutes before your scheduled appointment time. Once the interviewer comes to greet you, always allow the first handshake to be initiated by the interviewer, and remember to address him or her by name. It's also a good idea to wait to be seated until the interviewer sits down, or directs you to be seated. Listen attentively to each question and ask for clarification if there can be more than one interpretation. Always use your common sense, and think before answering a question. There is no need to hurry through your responses.

Never answer questions about your personal life unless the answer demonstrates a reflection of your ability to perform the job, and always be discreet in your references to former employers. Make sure your responses are short, but avoid one-word answers, such as "yes" or "no." And never, never use slang or swear words!

A confident prospective employee always demonstrates a genuine interest in the position for which he or she is applying by asking questions about job specifications, continuing education policies, medical benefits, and to whom he or she will be responsible. You should also be prepared to respond to a question concerning salary expectations. If you are unable to answer a question, be honest and state that you are unable to respond. And most important, always speak confidently to the interviewer without letting your eyes wander. You must appear relaxed, yet interested in the position.

There may be some questions that may be asked by the interviewer that legally do not have to be answered unless the information is job related. Jobs are to be offered on the basis of qualifications, so any questions related to your marital status, religious preference, club memberships, height, weight, dependents, or age do not have to be answered. Inquiries dealing with credit ratings, home ownership, family planning, and pregnancy are also illegal.

The way in which the interview process ends is as important as how it begins. Most prospective employers usually terminate the interview by standing. This should be your cue to stand and prepare to leave. As you did in the beginning, allow the interviewer to initiate the shaking of hands at the conclusion of the interview process. Before departing, you may want to ask the interviewer when you might expect to hear if the job is to be offered, so that you will not miss the call. The interview process is completed by saying a simple "thank you" to both the interviewer and the receptionist.

After the Interview, What's Next?

In today's employment market, an offer of employment is seldom initiated during the interview process. Therefore, if there has been no contact from the employer within a day or two, it is time for you to write a follow-up note of thanks (Figure 27-5). If the interviewer has provided you with a telephone number, you may also choose to place a call, in order to express your continued interest in the position and to keep from being forgotten. The object of these "gentle reminders" is to briefly restate your assets, emphasizing your strong feelings toward obtaining the position, while at the same time, reinforcing the time you may be reached by telephone. Even if the position is not offered, you will learn from each interview until you are able to secure the right job.

Once you have secured a position, it will be well for you to remember that, no matter what level you are hired at, you are still a beginner in that position, with much to learn. It is a good idea to borrow the office or department's procedure manual to preview the facility's routines. Arriving for work the first day with a notebook to take notes saves the subsequent embarrassment of having to ask how to carry out tasks after they have been explained, and taking notes also reinforces procedures and demonstrates your genuine desire to do the best job possible.

Summary

In this chapter, we discussed the basic skills involved in obtaining a position in the health care industry. We discussed the various steps and processes involved in beginning the job search, how to properly fill out a job application form, and how to create a cover letter, resume, and thank you note. We also explained how to prepare for and take part in the actual interview process. Finally, we discussed the follow-up process required of the prospective employee, once the interview has taken place.

Administrative Medical Services

JANE DOE
5311 Corteen Place, Apt. #28
No. Hollywood, CA 91607
(818) 761-2345

May 15, 1995

Dr. John Smith
9001 Green St., Suite 306
Anywhere, USA 12345

Dear Dr. Smith:

Please allow me to take this opportunity to personally thank you for the time you gave me during my recent interview.

As I am sure you are aware, I am most interested in working for you and would consider it a privilege if you were to select me for a position in your practice. If I am fortunate enough to be offered a job, I can promise you someone that is a hard worker, and who is extremely dedicated to working "above and beyond the call of duty" for her employer.

Once again, I would like to thank you for your consideration. I eagerly look forward to hearing from you, so that we may further discuss the possibility of me working as a member of your team.

Most sincerely,

Jane Doe

Figure 27-5
The thank you note.

Review Questions

1. Identify at least three reasons an employer might reject a job applicant.
    a. _____
    b. _____
    c. _____

2. Differentiate between a *chronological resume*, a *functional resume*, and a *combination resume*.

3. Give at least one example of how an applicant may show interest in a job interview.

4. Why is it important to read the "fine print" of an employment agency contract before signing it?

5. Identify at least one disadvantage of using a chronological resume.

6. If three days have passed since the applicant has had an interview, and he or she has not heard from the prospective employer, what two follow-up steps should be taken to help secure the position?

7. List three personal items that legally do not have to be included on a resume or answered during an interview.

8. What steps can be taken by the administrative medical services worker to help him or her in securing a position?

Appendix A

Medical Abbreviations and Symbols

a	before	AKA	above-the-knee amputation
@	at	ALS	amyotrophic lateral sclerosis
aa	of each	alt	alternate
AAA	abdominal aortic aneurysm	a.m.	morning
AB	abortion	AMA	against medical advice
abd	abdomen	amp	ampule
ABG	arterial blood gases	amt	amount
abn	abnormal	A&O	alert and oriented
ac	before meals	AODM	adult onset diabetes mellitus
accom	accommodation	A&P	auscultation and percussion; anterior and posterior
ACH	adrenocortical hormone		
ACTH	adrenocorticotropic hormone	AP	anterior-posterior
AD	right ear	AP&Lat	anterior-posterior and lateral
ADA	American Dietetic Association	ART	accredited records technician
ADH	antidiuretic hormone	AS	left ear
ADL	activities of daily living	ASA	aspirin
ad lib	as desired	ASAP	as soon as possible
adm	admission	ASCVD	arteriosclerotic cardiovascular disease
AF	acid fast	ASCVRD	arteriosclerotic cardiovascular renal disease
AFB	acid-fast bacillus ratio		
AgNo2	silver nitrate	ASHD	arteriosclerotic heart disease
A/G ratio	albumin globulin ratio	as tol	as tolerated
AK	above knee	astig	astigmatism

255

at fib	atrial fibrillation	**conv**	convalescent
AU	both ears	**COPD**	chronic obstructive pulmonary disease
aud	auditory	**CORT**	certified operating room technician
aux	auxillary	**CPT**	chest physical therapy
AV	atrioventricular	**CRNA**	certified registered nurse anesthetist
Ax	axillary	**CRP**	C-reactive protein
Bact	bacteria	**C&S**	culture and sensitivity
BacT	bacteriology	**C-sect**	cesarean section
BBB	bundle branch block	**CSD**	central supply room
BE	barium enema	**CSF**	cerebrospinal fluid
bet	between	**CT**	computerized tomography
bid	twice a day	**CVA**	cerebrovascular accident (stroke)
bilat	bilateral	**CVP**	central venous pressure
BKA	below-the-knee amputation	**CXR**	chest x-ray
bld	blood	**cysto**	cystoscopy
BM	bowel movement	**D**	dorsal
BP	blood pressure	**DAT**	diet as tolerated
BPH	benign prostatic hypertrophy	**D&C**	dilatation and curretage
BR	bathroom	**DC**	discontinue
BRP	bathroom privileges	**DDS**	doctor of dental surgery
BS	blood sugar	**decub**	decubitus
BSC	bedside commode	**del**	delivery
BUN	blood urea nitrogen	**dept**	department
Bx	biopsy	**dict**	dictation
c̄	with	**diff**	differential
C	centigrade	**Dig**	digoxin; digitalis
C1, C2, C3	cervical vertebrae	**dil**	dilute
Ca	cancer; calcium	**Dir**	director
cal	calorie	**disc**	discontinue
Cap	capsule	**disch**	discharge
cardio	cardiology	**DJD**	degenerative joint disease
CAT	computerized axial tomography	**SM**	diabetes mellitus
Cath	Catholic	**DO**	doctor of osteopathy
cath	catheter	**DOA**	dead on arrival
cauc	caucasian	**DON**	director of nursing
caut	cauterization	**doz**	dozen
CBC	complete blood count	**DP**	discharge plan
CBI	continuous bladder irrigation	**DPM**	doctor of podiatric medicine
CBR	complete bed rest	**DPT**	diptheria, pertussis, tetanus
cc	cubic centimeter	**dr**	dram
CC	chief complaint	**Dr**	doctor
CCU	coronary care unit	**DRGs**	diagnosis related groups
CHF	congestive heart failure	**D/S**	dextrose in saline
CHO	cholesterol	**dsg**	dressing
CHUC	certified health unit coordinator	**DT**	delirium tremens
Cl	chloride	**DUB**	dysfunctional uterine bleeding
CL	chlorine	**D/W**	dextrose in water
clysis	hypodermoclysis	**Dx**	diagnosis
cm	centimeter	**dysp**	dyspnea
CNA	certified nurse assistant	**EBL**	estimated blood loss
CNS	central nervous system	**ECF**	extended care facility
c/o	complains of	**ECG or EKG**	electrocardiogram
CO$_2$	carbon dioxide	**Echo**	echocardiogram
coag	coagulation	**Echo EG**	echoencephalogram
compd	compound	*E. coli*	*escherichia coli*
cond	condition	**ECT**	electroconvulsive therapy

EDC	expected day of confinement	**GTT**	glucose tolerance test
EEG	electroencephalogram	**gtts**	drops
EENT	eyes, ears, nose, and throat	**GU**	genitourinary
e.g.	for example	**Gyn**	gynecology
elix	elixir	**HB**	heart block
EMG	electromyogram	**HCG**	human chorionic gonadotropin
EMS	electrical muscle stimulation	**Hct**	hematocrit
EMT	emergency medical technician	**Hgb**	hemoglobin
ENT	ears, nose, and throat	**H&H**	hemoglobin and hematocrit
eq	equivalent	**HHC**	home health care
equip	equipment	**H&P**	history and physical
ER	emergency room	**H$_2$O**	water
ERCP	endoscopic retrograde cholangio-pancreatography	**hosp**	hospital
ESR	erythrocyte sedimentation rate	**hr**	hour
est	estimate	**hs**	bedtime
et	and	**hyper**	above
etiol	etiology	**hypo**	under or below
ETOH	ethanol	**hyst**	hysterectomy
EUA	examination under anesthesia	**Hx**	history
evac	evacuate	**ICF**	intermediate care facility
ex	example	**ICU**	intensive care unit
exam	examination	**I&D**	incision and drainage
exp lap	exploratory laparotomy	**ID**	identification
ext	external	**i.e.**	that is
F	Fahrenheit	**IM**	intramuscular
FB	foreign body	**IMC**	intensive medical care
FBS	fasting blood sugar	**in**	inch
Fe	iron	**inf**	infusion
FeSO$_4$	ferrous sulfate	**ing**	inguinal
FHT	fetal heart tones	**inj**	injection
fib	fibrillation	**insuff**	insufficiency
fl	fluid	**int**	internal
fl dr	fluid dram	**invol**	involuntary
fl oz	fluid ounce	**I&O**	intake and output
FP	flat plate	**IOP**	interoccular pressure
FS	frozen section	**IPPB**	intermittent positive pressure breathing
FSH	follicle stimulating hormone	**irrig**	irrigat
ft	foot	**ISC**	intensive surgical care
FTI	free thyroxin index	**IU**	International Units
FUO	fever of unknown origin	**IUD**	intrauterine device
FWB	full weight bearing	**IV**	intravenous
Fx	fracture	**IVC**	intravenous cholangiogram
g	gram	**IVP**	intravenous pyelogram
gal	gallon	**JCAHO**	Joint Commission for Accreditation Health Care Organizations
GB	gallbladder	**K**	potassium
GC	gonorrhea	**KCl**	potassium chloride
G&E	gastroscopy and esophagoscopy	**Kg**	kilogram
GED	gastroscopy, esophagascopy, duodenoscopy	**KJ**	knee jerk
GH	growth hormone	**KUB**	kidneys, ureters, bladder
GI	gastrointestinal	**KVO**	keep vein open
glauc	glaucoma	**l**	liter
gr	grain	**L**	left
grav	gravida	**lab**	laboratory
GSW	gunshot wound	**lac**	laceration
		lact	lactation

lam	laminectomy	**MSW**	master of science in social work
lap	laparotomy	**myop**	myopia
lat	lateral	**Na**	sodium
LBBB	left bundle branch block	**NaCl**	sodium chloride
LBP	low back pain	**NaHCo$_3$**	sodium bicarbonate
LDH	lactic dehydrogenase	**NAHUC**	National Association of Health Unit Co-ordinators
LE	lower extremity		
lg	large	**N/C**	no complaints
LH	luetenizing hormone	**neg**	necessary
lig	ligament	**neuro**	neurology
liq	liquid	**NF**	*National Formulary*
LLE	left lower extremity	**NG**	nasogastric
LLL	left lower leg	**NG-tube**	nasogastric tube
LLQ	left lower quadrant	**NH**	nursing home
l/M	liters per minute	**NICU**	neonatal intensive care unit
LMP	last menstrual period	**NKA**	no known allergies
LOA	leave of absence	**no**	number
LOC	level of consciousness	**N$_2$O**	nitrous oxide
LPN	licensed practical nurse	**noct**	night
L-S	lumbosacral	**norm**	normal
lt	left	**NPO**	nothing by mouth
LUE	left upper quadrant	**NS**	normal saline
LV	left ventricle	**Nsg**	nursing
LVH	left ventricular hypertrophy	**NSR**	normal sinus rhythm
LVN	licensed vocational nurse	**NTG**	nitroglycerin
L&W	living and well	**N/V; N&V**	nausea and vomiting
lymphs	lymphocytes	**NWB**	non-weight bearing
lytes	electrolytes	**o**	oral
M	male	**O$_2$**	oxygen
macro	macrocytic	**Ob**	obstetrics
MAO	manoaminooxidase	**Ob-Gyn**	obstetrics and gynecology
MAR	medication administration record	**OBS**	organic brain syndrome
mcg	microgram	**OD**	right eye; overdose
MCH	mean cell hemoglobin	**off**	office
MCL	midclavicular line	**oint**	ointment
MCV	mean cell volume	**OJ**	orange juice
MD	doctor of medicine	**OOB**	out of bed
MDR	minimum daily requirement	**op**	operation
med	medical	**OPD**	outpatient department
meds	medications	**ophth**	ophthalmology
mEq	milliequivalents	**OR**	operating room; open reduction
Met Ca	metastatic cancer	**ORIF**	open reduction with internal fixation
mg	milligram	**ORT**	operating room technician
MgSo$_4$	magnesium sulfate	**ortho**	orthopedic
MI	myocardial infarction	**os**	mouth
min	minute	**OS**	left eye
misc	miscellaneous	**OT**	occupational therapy
mL	milliliter	**OU**	both eyes
mm	millimeter	**oz**	ounce
mn	midnight	**p**	pulse
mod	moderate	**p̄**	after
MOM	Milk of Magnesia	**P&A**	percussion and auscultation
MRA	medical records administrator	**PA**	posterior-anterior
MRI	magnetic resonance imaging	**PAC**	premature atrial contraction
MS	morphine sulfate	**PA & Lat**	posterior-anterior and lateral
MSN	master of science in nursing	**palp**	palpitation

Administrative Medical Services

Pap	Papanicolaou (smear)	qhs	every hour of sleep
PAT	paroxysmal atrial tachycardia	qid	four times a day
path	pathology	QNS	quantity not sufficient
PBI	protein bound iodine	qod	every other day
pc	after meals	qs	quantity sufficient
PE	physical examination	qt	quart
Ped	pediatric	quad	quadrant
PEEP	positive end expiratory pressure	quant	quantity
per	by	R	right
PERRLA	pupils equal, round, reactive to light and accommodation	Ra	radium
		RA	rheumatoid arthritis
PFT	pulmonary function test	RBBB	right bundle branch block
PG	pregnant	RBS	random blood sugar
PH	past history	re	regarding
Pharm	pharmacy	rec	record
phys	physician	ref	reference
Physio	physiotherapy	ref phys	referring physician
PID	pelvic inflammatory disease	reg	registered
PKU	phenolketonuria	rehab	rehabilitation
p.m.	afternoon	resp	respiration
PMH	past medical history	ret cath	retention catheter
PMS	premenstrual syndrome	retro	retrograde
pneu	pneumonia	RFR	refraction
po	by mouth	Rh	rhesus blood factor
PO	phone order	RHD	rheumatic heart disease
postop	postoperative	RHF	right heart failure
PPBS	postprandial blood sugar	R/L	Ringer's lactate
PPN	partial parenteral nutrition	RLE	right lower extremity
PRBC	packed red blood cells	RLL	right lower lobe
preop	preoperative	RLQ	right lower quadrant
prep	prepare	RN	registered nurse
PRN	as needed	R/O	rule out
Proct(o)	proctology	Roent	roentgenology
prog	prognosis	ROM	range of motion
Prot	Prostestant	RPh	registered pharmacist
pro-time	prothrombin time	R&R	rate and rhythm
PSRO	Professional Standards Review Organization	RR	recovery room
		RRA	registered records administrator
psych	psychiatric	Rt	right
pt	patient	RT	respiratory therapy
PT	physical therapy	RUE	right upper extremity
PTA	physical therapy assistant	RUL	right upper lobe
PTT	partial thromboplastin time	RUQ	right upper quadrant
PVC	premature ventricular contraction	Rx	prescription
PVD	peripheral vascular disease	$\bar{s}$	without
PWB	partial weight bearing	S&A	sugar and acetone
Px	prognosis	SA	sinoatrial
PZI	protamin zinc insulin	SB	stillborn
q	every	SC	subcutaneous
qd	every day	sec	second
qh	every hour	sed rate	sedimentation rate
q2h	every 2 hours	SGOT	serum glutamic pyruvic transaminase
q4h	every 4 hours	sib	siblings
q6h	every 6 hours	sig	instructions
q8h	every 8 hours	SL	sublingual
q12h	every 12 hours	SNF	skilled nursing facility

SO	significant other	TP	total protein
SOB	short of breath	TPN	total parenteral nutrition
Soc Sec	social security	TPR	temperature, pulse, respiration
Soc Serv	social services	tr	tincture
sol	solution	T/R	timed release
SOP	standard operating procedure	trach	trachea
SOS	once only, if necessary	TSH	thyroid stimulating hormone
sp	specific	tsp	teaspoon
spans	spansule	TURP	transurethral resection of prostate
spec	specimen	TV	tidal volume
Sp Gr	specific gravity	TWB	total weight bearing
ss	one half	TWE	tap water enema
S&S	signs and symptoms	Tx	traction
SSE	soap suds enema	U	units
Staph	staphylococcus	UA	urinalysis
STAT	immediately	UCD	usual childhood diseases
STD	sexually transmitted disease	UCG	urinary chorionic gonadotropin pregnancy test
Strep	streptococcus		
Sub Q	subcutaneous	UGI	upper gastrointestinal
suff	sufficient	ULQ	upper left quadrant
supp	suppository	umb	umbilical
surg	surgical	UMS	urethral meatal stricture
susp	suspension	ung	ointment
SW	social worker	UO	undetermined origin
Sx	symptom(s)	U/O	urinary output
syph	syphilis	URI	upper respiratory infection
syr	syrup	urol	urology
T	temperature	URQ	upper right quadrant
T1; T12	thoracic vertebrae or nerves	USP	*United States Pharmacopeia*
T3	triiodothyronine	Ut Dict	as directed
T4	thryroxine	UTI	urinary tract infection
T&A	tonsillectomy and adenoidectomy	VA	Veterans Administration; visual acuity
tab	tablet	vag	vaginal
tach	tachycardia	vasc	vascular
TAH	total abdominal hysterectomy	VD	venereal disease
TB	tuberculosis	VDRL	Venereal Disease Research Laboratory
tbsp	tablespoon	vent	ventricular
TCDB	turn, cough, deep breath	vent fib	ventricular fibrillation
TCI	transient cerebral ischemia	VER	visual evoked response
TDP	tentative discharge plan	vert	vertical
tech	technician	VF	visual fields
TENS	transcutaneous electrical nerve stimulator	V fib	ventricular fibrillation
		vib	vibration
THR	total hip replacement	Vit	vitamin
TIA	transient ischemic attack	VNA	Visiting Nurse Association
tid	three times a day	VO	verbal order
tinct	tincture	vol	volume
TJR	total joint replacement	VP	venous pressure
TKO	to keep open	VS	vital signs
TL	team leader	WBC	white blood count
TLC	tender loving care	WBTT	weight bearing to tolerance
TMJ	temperomandibular joint	Wt	weight
TO	telephone order	ZSR	zeta sedimentation rate

Appendix B

Root Words, Prefixes, and Suffixes

Root Words

abdomin/o	abdominal
acro/o	extremity
aden/o	gland
adip/o	fat
aer/o	air
alveol/o	air sac
an/o	anus
angi/o	vessel
aort/o	aorta
append/o	appendix
aqua	water
arteri/o	artery
arthr/o	joint
articul/o	joint
atri/o	atrium
audi/o	to hear
axill/o	armpit
blephar/o	eyelid
brachi/o	arm
bronch/o	bronchial tube
bucc/o	cheek
calcul/o	stone
cardi/o	heart
caud/o	tail
cephal/o	head
cerebell/o	cerebellum
cerebr/o	cerebrum
cerumin/o	wax
cervic/o	cervix, neck
cheil/o	lip
chole	bile, gall
chondr/o	cartilage
clavic/o	clavicle
col/o	colon
colp/o	vagina
corona	crown
cost/o	rib
crani/o	skull
crypt/o	hidden
cutis	skin
cyan/o	blue
cyst/o	bladder
cyt/o	cell
dacry/o	tear
dent/o	tooth
derm/o	skin

261

duct	tube	**or/o**	mouth
dura	hard	**orchi/o**	testicle
dyps/o	thirst; drink	**oste/o**	bone
encephal/o	brain	**ot/o**	ear
endocrin/o	endocrine glands	**ovari/o**	ovary
enter/o	intestines	**para**	to give birth
episi/o	perineum	**patell/o**	patella
erythr/o	red	**path**	disease
esophag/o	esophagus	**perine/o**	perineum
femor/o	femur	**phag/o**	swallow
fib/o	fibrous	**pharyng/o**	pharynx
gastr/o	stomach	**phil**	like
gingiv/o	gum	**phleb/o**	vein
gluc/o	glucose	**pneum/o**	lung; air
glyc/o	glycogen	**poly**	many
gravid/o	pregnant	**primi**	first
gynec/o	female reproductive system; woman	**proct/o**	rectum
hepat/o	liver	**pseud/o**	false
herni/o	hernia	**psych/o**	mind
hist/o	tissue	**pty/o**	saliva
hydr/o	water	**pulm/o**	lung
hyster/o	uterus	**pyel/o**	kidney pelvis
ile/o	ilium; part of the small intestine	**py/o**	pus
immun/o	immune system	**ren/o**	kidney
inguin/o	groin	**reticul/o**	network
intestin/o	intestine	**rhin/o**	nose
irid/o	iris	**rubr/i**	red
jejun/o	jejunem	**salping/o**	fallopian tube
kal/o	potassium	**sarc/o**	flesh
lact/o	milk	**scapul/o**	scapula
lapar/o	abdomen	**scler/o**	hard
larynx/o	larynx	**sepsis**	decay
leuk/o	white	**sept/o**	wall
lingu/o	tongue	**sphygmos**	pulse
lith/o	stone	**spir/o**	breathe
lumbar	lower back	**splen/o**	spleen
lymph/o	lymphatic system	**spondyl/o**	spine; vertebra
mast/o	breast	**stern/o**	sternum
maxillo/o	upper jaw	**steth/o**	chest
meat/o	meatus	**tend/o**	tend
melano	black	**thorac/o**	chest
men/o	month flow, menses	**thrombo/o**	clot
mening/o	meninges	**thyr/o**	thyroid
multi	many	**toxic/o**	toxis
my/o	muscle	**trache/o**	trachea
myel/o	bone marrow; spinal cord	**trich**	hair
nas/o	nose	**tympan/o**	drum
nat/o	birth	**ur/o**	urine
nephr/o	kidney	**ureter/o**	ureter
neur/o	nerve	**urethr/o**	urethra
noct	night	**uter/o**	uterus
ocul/o	eye	**valv/o**	valve
odont/o	teeth	**varic/o**	enlarged and twisted vein
onc/o	cancer; tumor	**vas/o**	vessel
oophor/o	ovary	**ven/o**	vein
ophthalm/o	eye	**ventriculo/o**	ventricle

venul/o	venule
vertebr/o	vertebrae
vesic/o	bladder
viscer/o	large internal organ; intestine
vitre/o	glassy

Prefixes

a	without
ab	away from
ad	toward
aero	air
ambi	both
an	without
andro	male
ante	before
anti	against
auto	self
bi	two
brady	slow
circum	around
con	together
contra	against
cryo	cold
de	down
di	two
dia	through
dis	apart
dys	painful; difficult
ecto	outer
en	in
endo	within
epi	upon
eu	good
ex	away from
exo	outside
extra	outside
hemato	blood
hemi	half
hemo	blood
hydro	water
hyper	above
hypo	below
in	in
infra	beneath; below
inter	between
intra	within
intro	into
macro	large
mal	bad
mega	large
meso	middle
meta	change
micro	small
mono	one
multi	many
neo	new
noct	night
null	none
oligo	few
pan	all
para	beside
per	through
peri	around
poly	many
post	after
pre	before
primi	first
pro	before
pseudo	false
pyo	pus
quadri	four
re	again
retro	backward; behind
semi	half
sub	under
super	above
supra	above
syn	together
tachy	fast
tetra	four
trans	across
tri	three
ultra	excessive or beyond

Suffixes

ac	pertaining to
al	pertaining to; of
alg	pain
ar	pertaining to
asis	condition
cele	protrusion
centesis	puncture
cise	to cut out
cyte	cell
dyn	pain
eal	pertaining to
ectomy	to cut out
emia	blood
er	one who
esthesia	pain
genesis	beginning
genic	producing
graph	instrument used to record
graphy	recording
ia	condition; pathological state
ism	condition
itis	inflammation
ive	performs

logist	one who studies and practices	**ptosis**	downward displacement
logy	study of	**ptysis**	to spit
lysis	breakdown	**rrhage**	excessive flow
malacia	softening	**rrhaphy**	surgical repair
megaly	enlargment	**rrhea**	discharge or flow
metry	measurement	**rrhexis**	rupture
oid	resembling	**sclerosis**	hardening
ole	small	**scope**	instrument for viewing
oma	tumor	**scopy**	examination of
opsy	to view	**sis**	condition
osis	condition of	**stasis**	stopping or controlling
path	disease	**stat**	stopping or controlling
penia	abnormal reduction	**stomy**	to make an opening
pexy	surgical fixation	**thermy**	heat
phagia	swallowing	**tic**	pertaining to
phasia	speaking	**tion**	state of; result
phobia	exaggerated fear	**tome**	instrument to cut
phone	voice	**tomy**	incision
plasty	surgical repair	**trophy**	development; nourishment
plegia	paralysis	**ule**	very small
pnea	breathing	**uria**	urine
poiesis	formation	**y**	condition

Appendix C

Billing Forms

HCFA 1500

Field No.	Description
1	Medicare, Medicaid, CHAMPUS, CHAMPVA, Feca Black Lung or Other. Check the appropriate box for which you are submitting this claim for payment
1A	Insured's ID number. Social security number of the insured
2	Patient's name. Same
3	Patient's birth date and sex. Write all dates as month/day/year; check box of appropriate sex
4	Insured's name. Subscriber's name
5	Patient's address and phone number. Same
6	Patient's relationship to insured. Same
7	Insured's address and phone number. Same
8	Patient's status. Check the appropriate boxes
9	Other insured's name. Other insured whose coverage may be responsible, in whole or in part, for the payment of this claim
9A	Other insured's policy or group number. Same
9B	Other insured's date of birth and sex. Same
9C	Employer's name or school name. Employer or school name of other insured party
9D	Insurance plan name of program name. Name of the insurance company and/or the group plan for the other insured
10A	Was condition related to employment? If yes, then there is workers' compensation involved. If no, then no workers' compensation was involved. Circle whether employment is current or previous

Field No.	Description
10B	Was condition related to an auto accident? If yes, check for an injury date (block 14) and an injury diagnosis (block 23). The state in which the accident occurred should also be indicated. If no, then the claim may not be for an injury
10C	Was condition related to other accident? If yes, check for an injury date (block 14) and an injury diagnosis (block 23). If no, then the claim may not be for an injury
10D	Reserved for local use. Same
11	Insured's policy group of FECA number. Subscriber's group number. This information refers to primary insured listed in 1A above
11A	Insured's date of birth. Same
11B	Employer's name or school name. Refers to employer or school name of insured party
11C	Insurance plan name or program name. Refers to the name of insurance company and/or the group plan
11D	Is there another health benefit plan? Check the appropriate box. If yes, then items 9A through 9D must be completed
12	Patient's or authorized person's signature. Patient's release of information for medical services
13	Assignment of benefits. This box should be signed by the patient in order to allow the insurer to pay provider for services rendered directly instead of paying the patient and waiting for patient to pay provider
14	Date of illness, injury, accident, or pregnancy. All claims for injury must have an injury or accident date. If the patient's condition is pregnancy, then put down the date of the last menstrual period
15	If patient has had same or similar illness, give first date. Same
16	Dates patient unable to work in current occupation. Same
17	Name of referring physician or other source. Fill in if patient was referred to current physician by another physician, a hospital, or a clinic. Also list the name of the referring party
17A	I.D. number of referring physician. Same
18	Hospitalization dates relating to current services. Same
19	Reserved for local use
20	Outside lab. Was laboratory work performed outside of your office? If it was, check box yes and show total amount of charges
21	Diagnosis or nature of illness or injury. The diagnosis states why the patient went to see the provider. Use both an ICD-9 code and a description
22	Medicaid resubmission code. Leave this blank
23	Prior authorization number. Refers to authorization number for services that were approved prior to being provided
24A	Date of service. This is the date service was rendered by the provider. Use complete date
24B	Place of service. This is the location where the services were performed
24C	Type of service. Leave blank
24D	Procedure code. Use the 5-digit procedure code found in the CPT or HCPC manuals
24D	Modifier code. Use the 2-digit modifier from the CPT
24E	Diagnosis code. This is used in conjunction with field 21. The number placed in field 24E refers to diagnosis 1, 2, 3, and 4 in block 21
24F	Charges. This is the charge per line of service
24G	Days or units. This is the number of times a service was performed
24H	EPSDT family plan. Leave this blank
24I	EMG. If service was performed in the hospital emergency room, this should match the service code listed in item 24B
24J	COB. This is the coordination of benefits. List any other insurance policies or plans which may be responsible for payment on this claim. Indicate by a Y for yes or an N or no
24K	Reserved for local use. Leave this blank
25	Federal tax ID number. If the provider is a physician or an individual, his or

Field No.	Description
	her social security number should be used. If provider is a facility, use employer identification number
26	Patient's account number. Same
27	Accept assignment for government claims. This refers to medicare or CHAMPUS. Do not use to assign payment on this claim to the provider. Instead, use item 13 only for your assignment of payment
28	Total charge. This is the total charge of the claim
29	Amount paid. This is the amount paid by the patient or the subscriber
30	Balance due. This is the difference between the total charge and the amount paid by the patient or subscriber, if any
31	Signature of physician or supplier of service. Must be signed by the provider requesting the payment
32	Name and address of facility where services rendered. If the same as item number 33, leave blank
33	Physician's/supplier's billing name, address, zip code, and phone number. This is the name, address, and phone number of the provider of service. This is the address that payments will be addressed and sent to if assignment of benefits has been signed in field 13

HCFA 1500 ITEM 24B: PLACE OF SERVICE

This is a numerical code that is used to indicate the place where the service was rendered.

00–10	Unassigned
11	Office (location other than a hospital, SNF, military treatment facility, community health center, state or local public health clinic, or ICF, where the health professional routinely provides health care services, diagnosis and treatment of illness or injury on a walk-in or ambulatory basis)
12	Home (location other than a hospital or other health care facility where the patient receives care in a private residence
13–20	Unassigned
21	Inpatient hospital, other than psychiatric, which provides diagnostic, therapeutic, and rehabilitative services by or under the supervision of a licensed physician to patients admitted for a number and variety of medical conditions
22	Outpatient hospital, provides diagnostic, therapeutic, and rehabilitation services to sick and injured persons not requiring hospitalization; a patient who is not admitted to a hospital is defined as an outpatient
23	Emergency room, hospital. Patients in the emergency room are considered to be facility outpatients, therefore, remember to complete box number 24
24	Ambulatory surgical center, free-standing facility, other than a doctor's office, where surgical and diagnostic procedures are performed on an ambulatory basis
25	Birthing center, a facility other than a doctor's office or hospital's maternity department that provides a setting for labor, delivery, and immediate postpartum care and immediate care of the newborn
26	Military treatment facility, a facility operated by one or more of the uniformed services
27–30	Unassigned
31	Skilled nursing facility, a facility that primarily provides inpatient skilled nursing care and related services to patients who require medical, nursing, or rehabilitative care
32	Nursing facility, a facility providing skilled nursing care and related services for the rehabilitation of injured, disabled, or sick persons or, on a regular basis, health-related care services above the level of custodial care to mentally retarded persons
33	Custodial care facility, a facility providing room, board, and personal assistance services, usually on a long-term basis
34	Hospice, a facility other than a patient's home, which provides palliative and supportive care for terminally ill patients and their families
35–40	Unassigned
41	Ambulance (land), a vehicle designed,

Field No.	Description
	equipped, and staffed for lifesaving and transporting the sick or injured
42	Ambulance (air or water), a vehicle specifically designed, equipped, and staffed for life-saving and transporting the sick or injured
43–50	Unassigned
51	Inpatient psychiatric facility, a facility providing inpatient psychiatric services for the diagnosis and treatment of mental illness
52	Psychiatric facility partial hospitalization, a facility for the diagnosis and treatment of mental illness providing a planned therapeutic program for patients who do not require inpatient or full-time hospitalization
53	Community mental health center, a facility providing comprehensive mental health care and services on an ambulatory basis
54	Intermediate care facility/mentally retarded, a facility providing care and other health-related services for persons above the level of custodial care of mentally retarded persons
55	Residential substance abuse treatment facility, a facility providing treatment for substance (alcohol and drug) abuse to live-in residents who do not require acute medical care
56	Psychiatric residential treatment center, a facility or part of a facility in which psychiatric care is provided on a 24-hour basis
57–60	Unassigned
61	Comprehensive inpatient rehabilitation facility, a facility providing comprehensive rehabilitation services under the supervision of a physician to inpatients with physical disabilities
62	Comprehensive outpatient rehabilitation facility, a facility providing comprehensive rehabilitation services under the supervision of a physician to inpatients with physical disabilities
63–64	Unassigned
65	End-stage renal disease treatment facility, a facility other than a hospital that provides dialysis treatment, maintenance, and/or training to patients or care-givers on an ambulatory or home-care basis
66–70	Unassigned
71	State or local public health clinic, a facility maintained by either state or local health departments that provides ambulatory primary medical care under the direction of a physician
72	Rural health clinic, a certified facility located in a rural medically underserved area that provides ambulatory primary medical care under the direction of a physician
73–80	Unassigned
81	Independent laboratory, a laboratory that has been certified to perform diagnostic and/or clinical tests independent of an institution or a physician's office
82–98	Unassigned
99	Other unlisted facility, a facility other than those listed and identified

UB-92 BILLING FORM

Field No.	Description
1	Provider name, address, and telephone number. This is the name, address and telephone number of the hospital or clinic where the services were provided
2	Reserved (untitled). This is where all the unlabeled fields are reserved for state or national use
3	Patient control number. This refers to the patient's account number
4	Type of bill. This is a 3-digit code that is used to provide information regarding the type of bill being submitted
5	Federal tax number. This is the provider's identification or social security number
6	Statement covers period. These are the dates of services this billing statement represents
7	Covered days. These are the number of days services were covered by the primary payer
8	Non-covered days (inpatient only). These are the days services were not covered by the primary payer

Administrative Medical Services

Field No.	Description
9	Coinsurance days. These are the number of days for which the patient is responsible for paying a portion of the costs of services
10	Lifetime reserve days. Under medicare, each beneficiary has a lifetime reserve of 60 additional days of inpatient hospital services after using 90 days of inpatient services during a particular period of an illness
11	Reserved for state assignment
12	Patient's name. Same
13	Patient's address. Same
14	Birthdate. This is the patient's date of birth
15	Sex. This is the patient's sex
16	Marital status. This represents the patient's marital status. Mark as S (single), M (married), X (legally separated), D (divorced), W (widowed), or U (unknown)
17	Date of admission. This represents the date in which the patient was admitted to the hospital
18	Hour of admission. This is the hour in which the patient was admitted to the hospital based on a 24-hour clock
19	Type of admission. This is the numerical code which denotes the priority of the admission
20	Source of admission. This is the numerical code denoting the source of this admission
21	Discharge hour. This is the hour in which the patient was discharged from the hospital, based upon a 24-hour clock
22	Patient status. This is the numerical code used to denote the status of the patient as of the time of the statement through date
23	Medical record number. This is the number that has been assigned by the provider by the medical record
24–30	Condition codes. These are codes that are used to identify conditions relative to the claim that may affect the payer processing
31	Reserved for national assignment
32–35	Occurrence codes. These are the codes and the associated dates that define a significant event relative to this bill that might effect the payer processing
36	Occurrence span. These are the codes and the related dates that identify a specific event that relates to the payment of the claim
37	Internal control number. This is the control number that has been assigned to the original bill by the payer or the payer's intermediary
38	Responsible party name and address. This is the name and address of a person who is ultimately responsible for insuring payment of the bill
39–41	Value codes and amounts. These are codes and related dollar amounts that identify specific data regarding the monetary nature that is necessary for the processing of this claim
42	Revenue code. This is a code that is referenced as the type of services that were provided
43	Revenue description. This is a description of the type of services that were provided
44	HCPCS/Rates. This is the accommodation rate for inpatient bills, or the CPT or HCPCS code for ancillary or outpatient services
45	Service date. This is the date in which services were provided if this is a series bill in which the date of service differs from the from/through date on the bill
46	Units of service. This is the quantitative measure of services, days, miles, pints of blood, units, or treatments
47	Total charges. This is the amount of the total charges for that line of services
48	Non-covered charges. This is the amount per line of service that was not covered by the primary payer
49	Reserved for national assignment
50	Payer identification. This is the name of the insurer(s) covered by the patient who may be ultimately responsible for payment on this bill.
51	Provider name. This is the number that has been assigned to the provider by the listed payer
52	Release information. Y for yes and N

Field No.	Description
	for no as to whether the patient signed a release of information form
53	Assignment of benefits. Y for yes and N for no as to whether the patient signed an assignment of benefits form
54	Prior payments. This is the amount that has been paid toward this bill prior to the current billing date
55	Estimated amount due. This is the amount that has been estimated by the provider to be due from the indicated payer
56	Reserved for state assignment
57	Reserved for national assignment
58	Insured's name. This is the name of the person listed on the insurance forms (subscriber's name)
59	Patient's relationship to insured. This is a numerical code designation that indicates the relationship between the patient and the insured
60	Subscriber's certificate number. This is the policy number under which the insured is covered if it is an individual policy
61	Insured group name. This is the name of the group or company that holds the insured's policy
62	Insurance group number. This is the group number that denotes the group policy or plan under which the insured is covered
63	Treatment authorization code. This is a number that indicates that the treatment described by this bill has been authorized by the payer
64	Employment status code. This is a code that denotes whether the employee is currently working part or full-time, retired, or is in active military service
65	Employer name. This is the name of the insured person's employer
66	Employer location. This is the address of the employer of the insured or responsible party
67	Principal diagnosis code. This is the ICD-9-CM code used for the diagnosis of the patient's condition
68–75	Other diagnosis codes. These are the ICD-9-CM, V, and E codes that may be used for additional diagnosis of the patient's condition
76	Admitting diagnosis. This is the ICD-9 code used at the time of the admission to the hospital
77	External cause of injury code (E code). This is the ICD-9 code used for an external cause of injury, poisoning, or adverse effect
78	Reserved for state assignment
79	Procedure coding method used. This is an indicator code used to identify the coding method used for procedure coding on the claim (1-3 reserved for state assignment, 4 - CPT-4, 5 HCPCS, 6-8 reserved for national assignment, and 9 ICD-9-CM)
80	Principal procedure codes and date. This is the CPT code for the principal procedure that was rendered. For medicare, the ICD-9-CM codes must be entered here and on item number 81
81	Other procedure codes and dates. These are the CPT codes used for additional procedures that were provided and the dates those procedures were performed
82	Attending physician ID. This is the name and license number of the physician who was primarily responsible for the patient during this hospitalization
83	Other physician ID. This is the name and license number of the secondary physician, the assistant surgeon, and any other physician who provided services to the patient during this hospitalization
84	Remarks. This is used for any pertinent data for which there is no other specific place on the form
85	Provider representative signature. This is the signature of the provider's representative
86	Date bill submitted. This is the date on which this bill was signed and submitted for payment

UB-92 ITEM NO. 4

Type of Bill

The following code structure should be used to classify the type of bill used for this hospitalization. Each

Administrative Medical Services

individual claim should have its own 3-digit code entered in the space provided that corresponds with the following information.

First Digit	Type of Facility
1	Hospital
2	Skilled nursing facility
3	Home health
4	Christian Science (hospital)
5	Christian Science (extended care facility)
6	Intermediate care facility
7	*Clinic
8	*Special facility
9	Reserved for national use

*If the type of facility is a clinic, then the "Bill classifications for clinics only" must be used; if the type of facility is a special facility, then the "Bill classifications for special facilities only" must be used.

Second Digit	Bill Classification (used for all except clinics)
1	Inpatient, including Medicare Part A only
2	Inpatient, including Medicare Part B only
3	Outpatient
4	Other (for hospital referenced diagnostic procedures or home health under the plan of treatment)
5–7	Reserved for national use only
8	Swing beds
9	Reserved for national use only

Bill Classification (used for clinics only)

1	Rural health
2	Hospital based or independent renal dialysis center
3	Free standing
4	Outpatient rehabilitation facility
5	Comprehensive outpatient rehabilitation facility
6–8	Reserved for national use only
9	Other

Bill Classification (used for special facilities only)

1	Hospice (non-hospital based only)
2	Hospice (hospital based only)
3	Ambulatory surgical center
4	Free standing birthing center
5–8	Reserved for national use only
9	Other

Third Digit	Frequency of Billing
0	Non-payment/zero claim
1	Admit through discharge claim
2	Interim first claim. Pertains to the first claim in a series for the same course of action
3	Interim continuing claim. Pertains to a prior claim has been submitted for this course of treatment or for hospital confinement and a subsequent bill is also expected to be issued
4	Interim last claim. Pertains to prior claims having already been submitted for this course of treatment or for hospital confinement and this is expected to be the last bill issued
5	Late charge(s). Pertains to a prior claim or complete set of claims having already been submitted to the provider and late charge(s) being added to the prior billing(s)
6	Adjustment of prior claim. This is an adjustment(s) being made that ultimately alters the prior claim with the addition of an explanation and a credit or additional charge(s) added to the claim
7	Replacement of prior claim. This pertains to a claim replacing a prior claim and the prior claim then being considered null and void
8	Void/Cancel prior claim. This pertains to the prior bill being voided, and subsequently cancelled
9	Reserved for national assignment

UB-92 ITEM 19

Type of Admission

These pertain to 1-digit codes which are used to indicate the priority of this admission according to the following structure.

Code No.	Priority
1	Emergency. Used when the patient requires immediate medical intervention as a result of a severe, life-threatening, or potentially disabling medical condition, in which the

Code No.	Priority
	patient must be admitted through the hospital emergency room
2	Urgent. Used when the patient requires immediate attention for the care and/or treatment of a physical or mental disorder in which the patient is admitted to the first available and suitable accommodation
3	Elective. Used when the patient's medical condition permits adequate time to schedule the availability of a suitable accommodation
4	Newborn. Used in conjunction with a special Source of Admission Code (item 20), when a baby is born within the healthcare facility
5–8	Reserved for national assignment
9	Information not available

UB-92 ITEM 20

Source of Admission

These pertain to 1-digit codes used to indicate the source of this admission according to the following structure.

Code No.	Emergency, Elective, or Other Types of Admission
1	Physician referral
2	Clinic referral
3	HMO referral
4	Transfer from a hospital
5	Transfer from a skilled nursing facility
6	Transfer from another health care facility
7	Emergency room
8	Court or law enforcement
9	Information not available
A–Z	Reserved for national assignment

Newborn Admission

1	Normal delivery. Delivery with no complications
2	Premature delivery. Baby delivered with time and/or weight factors qualifying it for premature status
3	Sick baby. Baby delivered with medical complications, other than those related to prematurity
4	Extramural birth. Baby born in a non-sterile environment
5–8	Reserved for national assignment
9	Information not available

UB-92 ITEM 92

Patient Status

These are 2-digit codes that refer to the status of the patient at the time of the last date that was covered by this billing statement.

Code No.	Status
01	Discharged to home or self care
02	Discharged or transferred to another short-term general inpatient hospital
03	Discharged or transferred to a skilled nursing facility
04	Discharged or transferred to an intermediate care facility
05	Discharged or transferred to another type of inpatient or outpatient institution
06	Discharged or transferred to home under the care of an organized home health care service agency
07	Left or discontinued care against medical advice
08	Discharged or transferred to home under the care of a home IV provider
09	Admitted as an inpatient to this hospital
10–19	Discharged to be defined at the state level
20	Expired
21–29	Expiration to be defined at the state level
30	Patient is still expected to return for outpatient services
31–39	Patient is still defined at the state level
40*	Expired at home
41*	Expired in a health care facility such as a hospital, free-standing clinic, or hospice
42*	Expired, location unknown

*These codes are for use only on Medicare claims for hospice care.

Administrative Medical Services

UB-92 ITEMS 24 through 30

Condition Codes

These are 2-digit codes that are used to identify specific conditions relating to this bill that may affect the payer processing. There are no specific dates associated with these codes as there are in items 32 through 36.

Code No.	Insurance Code
01	Military service related
02	Condition is employment related
03	Patient is covered by insurance not reflected here
04	Patient is enrolled in an HMO
05	Lien has been filed on this claim
06	Patient is in end-stage renal disease (ESRD) and the first 18 months of the entitlement has been covered by the employer group health insurance
07	Treatment of nonterminal condition for hospital patient
08	Beneficiary would not provide information regarding other insurance coverage
09	Neither the patient nor his/her spouse is employed
10	Patient and his/her spouse is employed but no employer group plan exists
11	Disabled beneficiary but no large group health plan
12–16	Payer codes

Special Condition

17	Reserved for national assignment
18	Maiden name has been retained
19	Child retains the mother's name
20	Beneficiary has requested billing
21	Billing for denial notice
22–25	Reserved for national assignment
26	VA-eligible patient chooses to receive services in a Medicare-certified facility
27	Patient has been referred to a sole community hospital for a diagnostic laboratory test
28	Patient and/or his/her spouse's employee health plan is secondary to Medicare
29	Disabled beneficiary and/or his/her family member's large group health plan is secondary to Medicare
30	Reserved for national assignment

Student Status

(Used only when the patient is a dependent child over the age of 18)

31	Patient is a full-time day student
32	Patient is a cooperative or work-study student
33	Patient is a full-time night student
34	Patient is a part-time student
35	Reserved for national assignment

Accommodation

36	General care patient in a special unit
37	Ward accommodation at the patient's request
38	Semi-private room is not available
39	Private room is medically necessary
40	Same-day transfer
41	Partial hospitalization services are necessary
42–45	Reserved for national assignment

CHAMPUS and SNF

46	Non-availability certificate on file
47	Reserved for CHAMPUS
48	Psychiatric residential treatment centers for children and adolescents
49–54	Reserved for national assignment
55	SNF bed is not available
56	Medical appropriateness. Patient's SNF admission was delayed more than 30 days after discharge from hospital because condition made it inappropriate to begin active care within that period
57	SNF readmission
58–59	Reserved for national assignment

Prospective Payment

60	Day outlier
61	Cost outlier
62	Payer code indicating that the claim was paid under a DRG
63–65	Payer codes set aside for payer use
66	Provider does not wish cost outlier payment

Renal Dialysis Setting

70	Self-administered erythropoietin (EPO)
71	Full care in the dialysis unit
72	Self-care in the dialysis unit

Code No.	Renal Dialysis Setting
73	Self-care training
74	Home
75	Home 100 percent reimbursement for dialysis
76	Back-up in facility dialysis
77	Provider accepts or is obligated to accept payment by a primary payer as payment in full
78	New coverage has not been implemented by the HMO
79	CORF services (physical therapy, occupational therapy, or speech pathology) have been provided offsite

	PPO Approval Indicator Service
C0	Reserved for national assignment
C1	Approved as billed
C2	Automatic approval as billed based on a focused review
C3	Partial approval
C4	Admission and services have been denied
C5	Postponement review applicable
C6	Admission has been pre-authorized
C7	Extended authorization has been approved
C8–C9	Reserved for national assignment

UB-92 ITEMS 32 through 35

Occurrence Codes

These codes and their associated date define a significant event that relates to this bill and that may affect the payer's processing.

Code No.	Occurrence (relating to accidents)
01	Auto accident
02	No fault insurance involved
03	Accident involving a tort liability
04	Accident that was employment related
05	Other accident
06	Crime victim involved
07–08	Reserved for national assignment

	Occurrence (arising from medical condition)
09	Start of infertility treatment cycle
10	Last menstrual period
11	Onset of symptoms and/or illness
12	Date of onset for a chronically dependent person
13–16	Reserved for national assignment

	Occurrence (related to insurance)
17	Date outpatient occupational therapy plan established or was last reviewed
18	Date of patient's or beneficiary's retirement
19	Date of spouse's retirement
20	Guarantee in which payment began
21	Utilization review (U/R) notice was received
22	Date in which active care ended
23	Reserved for national assignment
24	Date insurance was denied
25	Date benefits were terminated by primary payer
26	Date skilled nursing facility bed become available
27	Date home health plan was established or was last reviewed
28	Date CORF (outpatient plan) was established or last reviewed
29	Date outpatient physical therapy plan was established or last reviewed
30	Date outpatient speech pathology plan was established or last reviewed
31	Date beneficiary was notified of intent to bill accommodations
32	Date beneficiary was notified of intent to bill for procedures or treatment
33	First day in which Medicare coordination of benefits for ESRD beneficiaries were instituted by the employee's group health plan
34	Date of election of extended care facilities
35	Date treatment began for physical therapy
36	Date patient was discharged as an inpatient for covered transplant procedure
37	Date patient was discharged as an inpatient for noncovered transplant procedure
38–39	Reserved for national assignment
40	Schedule date of admission as an inpatient in the hospital
41	Date of first test for a pre-admission into the hospital

Administrative Medical Services

Code No.	Service-Related
42	Date of discharge
43	Scheduled date of canceled surgery
44	Date occupational therapy treatment began
45	Date speech therapy treatment began
46	Date cardiac rehabilitative treatment began
47–49	Payer codes. These are reserved for use by the payer
50–69	Reserved for state assignment
70–99	These are occurrence span codes and dates and include the following:
A1	Birth date-insured A. This is the birth date of the insured person who is covered by the insurance and who is considered as the primary payor
A2	Effective date-insured A policy. This is the date the insurance coverage under the primary payer first began
A3	Benefits exhausted-payer A. This is the last date for which benefits are available
A4–A9	Reserved for national assignment
B0	Reserved for national assignment
B1	Birth date-insured B. This is the date of birth of the insured person who is covered by the insurance and who is considered to be the secondary payor
B2	Effective date-insured B policy. This is the date the insurance coverage under the secondary payer first began
B3	Benefits exhausted-payer B. This is the last date for which benefits became available.
B4–B9	Reserved for national assignment
C0	Reserved for national assignment
C1	Birth date-insured C. This is the birth date of the insured person who is covered by the insurance and who is considered to be the tertiary payor
C2	Effective date-insured C policy. This is the date the insurance coverage for the tertiary payer first began
C3	Benefits exhausted-payer C. This is the last date for which the benefits were available
C4–I9	Reserved for national assignment
J0–L9	Reserved for state assignment
M0–Z9	Occurence span codes and dates

Occurrence Span

These are codes and their related dates that identify a specific event that relates to the payment of the claim.

70	Qualifying stay dates
70	Payer code-nonutilization dates
71	Prior stay dates
72	First/last visit
73	Benefit eligibility period for primary payer
74	Noncovered level of care
75	SNF level of care
76	Patient liability
77	Provider liability period
78	SNF prior stay dates
79	Payer code
80–99	Reserved for state assignment
M0	PRO/UR approved stay dates
M1–W9	Reserved for national assignment
Z0–Z9	Reserved for state assignment

UB-92 ITEMS 39 through 41

Value Codes and Amounts

These are codes and their related dollar amounts that are used to identify specific data of a monetary nature that is necessary to process the claim.

Code No.	Value and Amount
01	Most common semi-private rate
02	Hospital has no semi-private rooms
03	Reserved for national assignment
04	Inpatient professional component
05	Outpatient professional component included in charges and is also billed separately to the carrier
06	Medicare blood deductible

Code No.	Value and Amount
07	Reserved for national assignment
08	Medicare lifetime reserve amount
09	Medicare co-insurance amount
10	Lifetime reserve amount during the second calendar year
11	Co-insurance amount during the second calendar year
12	Working aged beneficiary/spouse with employer group health plan
13	ESRD beneficiary in a medicare coordination period with an employer group health plan
14	No fault auto/other
15	Workers' compensation
16	PHS or other federal agency
17–20	Payer codes. These are used for payer only
	Medicaid-Specific Codes
21–24	Reserved for state assignment
25–29	Reserved for national assignment for Medicaid
	Code Structure
30	Pre-admission testing
31	Patient liability amount
32–36	Reserved for national assignment
37	Pints of blood that have been furnished
38	Blood deductible pints
39	Pints of blood that have been replaced
40	New coverage that has not been implemented by an HMO
41	Black Lung
42	VA
43	Disabled beneficiary under the age of 65 with a large group health plan
44	Amount provider agreed to accept from primary payer, which amount is less than charges but higher than payment received for which a Medicare secondary payment is due
45	Accident hour
46	Number of grace days
47	Amount of liability insurance
48	Amount for hemoglobin reading
49	Amount for hematocrit reading
50	Amount for physical therapy visits
51	Amount for occupational therapy visits
52	Amount for speech therapy visits
53	Amount for cardiac rehabilitation visits
54–55	Reserved for national assignment
56	Skilled nursing-home visit hours
57	Home health aide-home visit hours
58	Amount for arterial blood gas (PO_2/PA_2)
59	Amount for oxygen saturation (O_2 SAT/Oximetry)
60	HHA branch MSA. This is the branch location of the Metropolitan Statistical Area
61–67	Reserved for national assignment
68	EPO-drug. This is the number of units of erythropoietin administered and/or supplied during the billing period. Amount is reported in whole units
69	Reserved for national assignment
70–72	Payer codes
73	Reserved for national assignment
75–79	Payer codes
80	Most common ward rate
81–99	Reserved for state assignment
A0	Reserved for national assignment
A1	Deductible payer A
A2	Coinsurance payer A
A3	Estimated responsibility for payer A
A4–A9	Reserved for national assignment
B0	Reserved for national assignment
B1	Deductible for payer B
B2	Coinsurance payer B
B3	Estimated responsibility for payer B
B4–B9	Reserved for national assignment
C0	Reserved for national assignment
C1	Deductible for payer C
C2	Coinsurance for payer C
C3	Estimated responsibility for payer C
C4–C9	Reserved for national assignment

Administrative Medical Services

Code Structure

D0–D2	Reserved for national assignment
D3	Estimated responsibility for patient
D4–W9	Reserved for national assignment
X0–Z9	Reserved for state assignment

UB-92 ITEM 42

Hospital Revenue Codes

These codes are used to identify a specific accommodation, ancillary service, or billing calculation.

Code No.	Revenue
001	Total Charges: used to reflect the total of all charges on this bill
01X	Reserved for internal payer use
02X–06X	Reserved for national assignment
07X–09X	Reserved for state use
10X	All inclusive rate
	0 All inclusive room and board plus ancillary (ALL INCL R&B/ANC)
	1 All inclusive room and board (ALL INCL R&B)
11X	Room and board private medical or general: routine service charges for single-bed rooms
	0 General classification (ROOM-BOARD/PVT)
	1 Medical/surgical/gyn (MED-SUR-GYN/PVT)
	2 OB (OB/PVT)
	3 Pediatric (PEDS/PVT)
	4 Psychiatric (PSYCH/PVT)
	5 Hospice (HOSPICE/PVT)
	6 Detoxification (DETOX/PVT)
	7 Oncology (ONCOLOGY/PVT)
	8 Rehabilitation (REHAB/PVT)
	9 Other (OTHER/PVT)
12X	Room and board semi-private two-bed medical or general: Routine service charges that have been incurred for accommodations with two beds
	0 General classification (ROOM-BOARD/SEMI)
	1 Medical/surgical/gyn (MED-SUR-GYN/2 BED)
	2 OB (OB/2 Bed)
	3 Pediatric (PEDS/2 Bed)
	4 Psychiatric (PSYCH/2 Bed)
	5 Hospice (HOSPICE/2 Bed)
	6 Detoxification (DETOX/2 Bed)
	7 Oncology (ONCOLOGY/2 Bed)
	8 Rehabilitation (REHAB/2 Bed)
	9 Other (OTHER/2 Bed)
13X	Semi-private, three, and four beds: Routine service charges incurred for accommodations with three or four beds
	0 General classification (ROOM-BOARD/3&4 Bed)
	1 Medical/surgical/gyn (MED-SUR-GYN/3&4 Bed)
	2 OB (OB/3&4 Bed)
	3 Pediatric (PEDS/3&4 Bed)
	4 Psychiatric (PSYCH/3&4 Bed)
	5 Hospice (HOSPICE/3&4 Bed)
	6 Detoxification (DETOX/3&4 Bed)
	7 Oncology (ONCOLOGY/3&4 Bed)
	8 Rehabilitation (REHAB/3&4 Bed)
	9 Other (OTHER/3&4 Bed)
14X	Private or deluxe: Deluxe rooms are accommodations with special amenities substantially in excess of those that have been provided to other patients
	0 General classification (ROOM-BOARD/PVT/DLX)
	1 Medical/surgical/gyn (MED-SUR-GYN/DLX)
	2 OB (OB/DLX)
	3 Pediatric (PEDS/DLX)
	4 Psychiatric (PSYCH/DLX)
	5 Hospice (HOSPICE/DLX)
	6 Detoxification (DETOX/DLX)
	7 Oncology (ONCOLOGY/DLX)
	8 Rehabilitation (REHAB/DLX)
	9 Other (OTHER/DLX)
15X	Room and board, ward (medical or general): Routine service charge for accommodations with five or more beds
	0 General classification (ROOM-BOARD/WARD)
	1 Medical/surgical/gyn (MED-SUR-GYN/WARD)
	2 OB (OB/WARD)

Code No.	Revenue
	3 Pediatric (PEDS/WARD)
	4 Psychiatric (PSYCH/WARD)
	5 Hospice (HOSPICE/WARD)
	6 Detoxification (DETOX/WARD)
	7 Oncology (ONCOLOGY/WARD)
	8 Rehabilitation (REHAB/WARD)
	9 Other (OTHER/WARD)
16X	Other room and board: Any routine service charges for accommodations that cannot be included in the more specific revenue codes
	0 General classification (R&B)
	4 Sterile environment (R&B/STERILE)
	7 Self care (R&B/SELF)
	9 Other (R&B/Other)
17X	Nursery: Charges for nursing care to the newborn and premature infants in the nurseries
	0 General classification (NURSERY)
	1 Newborn (NURSERY/NEWBORN)
	2 Premature (NURSERY/PREMIE)
	5 Neonatal ICU (NURSERY/ICU)
	9 Other (NURSERY/OTHER)
18X	Leave of absence: Charges that are made to hold a room while the patient is temporarily away from the provider
	0 General classification (LEAVE OF ABSENCE OR LOA)
	1 Reserved (RESERVED)
	2 Patient convenience (LOA/PT CONV)
	3 Therapeutic leave (LOA THERAPEUTIC)
	4 ICF/MR-any reason (LOA/ICF/MR)
	5 Nursing home (for hospitalization (LOA/NURS HOME)
19X	Not assigned
20X	Intensive care: Routine service charges for medical or surgical care that is provided to patients who require a more intensive level of care than that which is rendered within the general medical or surgical unit
	0 General classification (INTENSIVE CARE or ICU)
	1 Surgical (ICU/SURGICAL)
	2 Medical (ICU/MEDICAL)
	3 Pediatric (ICU/PEDS)
	4 Psychiatric (ICU/PSYCH)
	6 Post ICU (POST ICU)
	7 Burn care (ICU/BURN CARE)
	8 Trauma (ICU/TRAUMA)
	9 Other intensive care (ICU/OTHER)
21X	Coronary care: Routine service charges for medical or surgical care that has been provided to patients with a coronary illness and who require a more intensive level of care than care which is rendered within the general medical care unit
	0 General classification (CORONARY CARE or CCU)
	1 Myocardial infarction (CCU/MYO INFARC)
	2 Pulmonary care (CCU/PULMONARY)
	3 Heart transplant (CCU/TRANSPLANT)
	4 Post CCU (POST CCU)
	9 Other coronary care (CCU/OTHER)
22X	Special charges: Charges that have been incurred on a daily basis during a patient's stay in the hospital
	0 General classification (SPECIAL CHARGES)
	1 Admission charge (ADMIT CHARGE)
	2 Technical support charge (TECH SUPPT CHG)
	3 U.R. service charge (UR CHARGE)
	4 Late discharge, medically necessary (LATE DISCH/MED NEC)
	9 Other special charges (OTHER SPEC CHG)
23X	Incremental nursing charge rate: The charge for nursing services that are assessed in addition to those for room and board
	0 General classification (NURSING INCREM)
	1 Nursery (NUR INCR/NURSERY)
	2 OB (NUR INCR/OB)
	3 ICU (NUR INCR/ICU)
	4 CCU (NUR INCR/CCU)
	5 Hospice (NUR INCR/HOSPICE)
	9 Other (NUR INCR/OTHER)

Administrative Medical Services

Code No.	Revenue
24X	All inclusive ancillary: This is a flat rate charge for incurred services on either a daily or total-stay basis for ancillary services only

    0 General classification (ALL INCL ANCIL)
    9 Other inclusive ancillary (ALL INCL ANCIL/OTHER)

25X Pharmacy: Charges for medications produced, manufactured, packaged, controlled, assayed, dispensed, and distributed to the patient under the direction of a licensed pharmacist. Also includes blood plasma and other components of blood and IV solutions

    0 General classification (PHARMACY)
    1 Generic drugs (DRUGS/GENERIC)
    2 Non-generic drugs (DRUGS/NONGENERIC)
    3 Take home drugs (DRUGS/TAKEHOME)
    4 Drugs incident to other diagnostic services (DRUGS/INCIDENT OTHER DX)
    5 Drugs incident to radiology (DRUGS/INCIDENT RAD)
    6 Experimental drugs (DRUGS/EXPERIMT)
    7 Nonprescription (DRUGS/NONPSCRPT)
    8 IV solution (IV SOLUTIONS)
    9 Other pharmacy (DRUGS/OTHER)

26X IV therapy: Includes administration of intravenous solution by specially trained personnel to patients requiring such treatment

    0 General classification (IV THERAPY)
    2 Infusion pump (IV THER/INFSN PUMP)
    3 IV therapy-pharmacy services (IV THER/PHARM/SVC)
    4 IV therapy/drug/supply delivery (IV THER/DRUG/SUPPLY DELV)
    9 Other IV therapy (IV THERP/OTHER)

27X Medical-surgical supplies and devices: These are charges for supply items required for the patient's care

    0 General classification (MED-SUR SUPPLIES)
    1 Nonsterile supply (NON-STER SUPPLY)
    2 Sterile supply (STERILE SUPPLY)
    3 Take home supplies (TAKE HOME SUPPLY)
    4 Prosthetic/orthotic devices (PROSTH/ORTH DEV)
    5 Pacemaker (PACEMAKER)
    6 Intraocular lens (INTRA OC LENS)
    7 Oxygen-take home (O_2/TAKEHOME)
    8 Other implants (SUPPLY/IMPLANTS)
    9 Other supplies (SUPPLY/OTHER)

28X Oncology: These are charges for the treatment of tumors and related diseases

    0 General classification (ONCOLOGY)
    9 Other Oncology (ONCOLOGY/OTHER)

29X Durable medical equipment other than renal: These are charges for medical equipment, other than for renal equipment, that can withstand use repeated use

    0 General classification (MED EQUIP/DURAB)
    1 Rental (MED EQUIP/RENT)
    2 Purchase of new DME (MED EQUIP/NEW)
    3 Purchase of used DME (MED EQUIP/USED)
    4 Supplies/drugs for DME effectiveness (MED EQUIP/SUPPLIES/DRUGS)
    9 Other equipment (MED EQUIP/OTHER)

30X Laboratory: These are charges for the performance of diagnostic and routine clinical laboratory tests

    0 General classification (LABORATORY or LAB)
    1 Chemistry (LAB/CHEMISTRY)
    2 Immmunology (LAB/IMMUNOLOGY)
    3 Renal patient (home) (LAB/RENAL HOME)

Code No.	Revenue
	4 Nonroutine dialysis (LAB/NR DIALYSIS)
	5 Hematology (LAB/HEMATOLOGY)
	6 Bacteriology and microbiology (LAB/BACT-MICRO)
	7 Urology (LAB/UROLOGY)
	9 Other laboratory (LAB/OTHER)
31X	Laboratory pathological: These are charges for diagnostic and routine laboratory tests on tissues and cultures
	0 General classification (PATHOLOGY LAB or PATH LAB)
	1 Cytology (PATHOL/CYTOLOGY)
	2 Histology (PATHOL/HYSTOL)
	4 Biopsy (PATHOL/BIOPSY)
	9 Other (PATHOL/OTHER)
32X	Radiology-diagnostic: These are charges for diagnostic radiology services that are provided for the purpose of examination and care of patients, and include taking, processing, examining, and interpreting radiographs and fluorographs
	0 General classification (DX X-RAY)
	1 Angiocardiography (DX X-RAY/ANGIO)
	2 Arthrography (DX X-RAY/ARTH)
	3 Arteriography (DX X-RAY/ARTER)
	4 Chest x-ray (DX X-RAY/CHEST)
	9 Other (DX X-RAY/OTHER)
33X	Radiology-therapeutic: These are charges for therapeutic radiology services and chemotherapy that are required for the patient's care and treatment, and include therapy for injection or ingestion of radioactive substances
	0 General classification (RX X-RAY)
	1 Chemotherapy-injected (CHEMOTHER/INJ)
	2 Chemotherapy-oral (CHEMOTHER/ORAL)
	3 Radiation therapy (RADIATION RX)
	5 Chemotherapy-IV (CHEMOTHERAP-IV)
	9 Other (RX X-RAY/OTHER)
34X	Nuclear medicine: These are charges for procedures and tests performed by a radioisotope laboratory that utilizes radioactive materials required for diagnosis and treatment of patients
	0 General classification (NUCLEAR MEDICINE or NUC MED)
	1 Diagnostic (NUC MED/DX)
	2 Therapeutic (NUC MED/RX)
	9 Other (NUC MED/OTHER)
35X	CT scan: These are charges for computerized tomography scans of the head and other parts of the body
	0 General classification (CT SCAN)
	1 Head scan (CT SCAN/HEAD)
	2 Body scan (CT SCAN/BODY)
	9 Other scan (CT SCAN/OTHER)
36X	Operating room services: These are charges for services provided to patients during the performance of surgical and related procedures during and immediately following surgery
	0 General classification (OR SERVICES)
	1 Minor surgery (OR/MINOR)
	2 Organ transplant-other than (OR/ORGAN TRANS) kidney
	7 Kidney transplant (OR/KIDNEY TRANS)
	9 Other operating services (OR/OTHER)
37X	Anesthesia: These are charges for anesthesia services in the hospital
	0 General classification (ANESTHESIA)
	1 Anesthesia incident to radiology (ANESTHE/INCIDENT RAD)
	2 Anesthesia incident to other diagnostic services (ANESTHE/INCDNT OTHER DX)
	4 Acupuncture (ANESTHE/ACUPUNC)
	9 Other anesthesia (ANESTHE/OTHER)
38X	Blood
	0 General classification (BLOOD)
	1 Packed red cells (BLOOD/PKD RED)
	2 Whole blood (BLOOD/WHOLE)
	3 Plasma (BLOOD/PLASMA)
	4 Platelets (BLOOD/PLATELETS)

Administrative Medical Services

Code No.	Revenue

    5 Leukocytes (BLOOD/LEUKOCYTES)

    6 Other components (BLOOD/OTHER)

    7 Other derivatives (BLOOD/DERIVATIVES)

    9 Other blood (BLOOD/OTHER)

39X Blood storage and processing

    0 General classification (BLOOD/STOR-PROC)

    1 Blood administration (BLOOD/ADMIN)

    9 Other blood storage and processing (BLOOD/OTHER STOR)

40X Other imaging services

    0 General classification (IMAGE SERVICES)

    1 Diagnostic mammography (DIAG MAMMOGRAPHY)

    2 Ultrasound (ULTRASOUND)

    3 Screening mammography (SCRN MAMMOGRAPHY)

    4 Positron emission tomography (PET SCAN)

    9 Other imaging services (OTHER IMAGE SVS)

41X Respiratory therapy services: These are charges for the administration of oxygen and certain potent drugs through inhalation or positive pressure and other forms of rehabilitative therapy through the exchange of oxygen and other gases

    0 General classification (RESPIRATORY SVC)

    2 Inhalation services (INHALATION SVC)

    3 Hyperbaric oxygen therapy (HYPERBARIC O_2)

    9 Other respiratory therapy services (OTHER RESPIR SVS)

42X Physical therapy: These are charges for therapeutic exercises, massage, and utilization of effective properties of light, heat, cold, water, electricity, and assistive devices used in the diagnosis and rehabilitation of patients who have neuromuscular, orthopedic, and other physical disabilities

    0 General classification (PHYSICAL THERP)

    1 Visit charge (PHYS THERP/VISIT)

    2 Hourly charge (PHYS THERP/HOUR)

    3 Group rate (PHYS THERP/GROUP)

    4 Evaluation or re-evaluation (PHYS THER/EVAL)

    9 Other physical therapy (OTHER PHYS THERP)

43X Occupational therapy: These are charges for teaching manual skills and independence in activities of daily living and personal care that are necessary to stimulate mental and emotional activity on the part of the patient

    0 General classification (OCCUPATION THERP)

    1 Visit charge (OCCUP THERP/VISIT)

    2 Hourly charge (OCCUP THERP/HOUR)

    3 Group rate (OCCUP THERP/GROUP)

    4 Evaluation or re-evaluation (OCCUP THERP/EVAL)

    9 Other occupational therapy (OTHER OCCUP THERP)

44X Speech-language pathology: These are charges for services provided to patients with impaired functional communication skills

    0 General classification (SPEECH PATHOL)

    1 Visit charge (SPEECH PATH/VISIT)

    2 Hourly charge (SPEECH PATH/HOUR)

    3 Group rate (SPEECH PATH/GROUP)

    4 Evaluation or re-evaluation (SPEECH PATH/EVAL)

    9 Other speech-language pathology (OTHER SPEECH PATH)

45X Emergency room: These are charges for emergency treatment to those persons who are ill or injured and who require immediate and unscheduled medical or surgical care or treatment

    0 General classification (EMERG ROOM)

    9 Other emergency room (OTHER EMER ROOM)

46X Pulmonary function: These are charges for test that measure inhaled and

Code No.	Revenue
	exhaled gases and the analysis of blood and for tests that are used to evaluate the patient's ability to exchange oxygen and other gases
	0 General classification (PULMONARY FUNC)
	9 Other pulmonary function (OTHER PULMON FUNC)
47X	Audiology: These are charges for the detection and management of communication disabilities centering in whole or in part on the function of hearing
	0 General classification (AUDIOLOGY)
	1 Diagnostic (AUDIOLOGY/DX)
	2 Treatment (AUDIOLOGY/RX)
	9 Other (OTHER AUDIOL)
48X	Cardiology: These are charges for cardiac procedures that are rendered in a separate unit within the hospital, and include, but are not limited to, cardiac catheterization, Swan-Ganz catheterization, coronary angiography, and exercise stress testing
	0 General classification (CARDIOLOGY)
	1 Cardiac cath lab (CARDIAC CATH LAB)
	2 Stress test (STRESS TEST)
	9 Other cardiology (OTHER CARDIOL)
49X	Ambulatory surgical care
	0 General classification (AMBUL SURG)
	9 Other ambulatory surgical care (OTHER AMBL SURG)
50X	Outpatient services: These are charges for services that have been rendered to the patient who is admitted as an inpatient before midnight of the day following the date of service
	0 General classification (OUTPATIENT SVS)
	9 Other outpatient services (OUTPATIENT/OTHER)
51X	Clinic: These are charges for providing diagnostic, preventive, curative, rehabilitative, and educational services on a scheduled basis to ambulatory patients
	0 General classification (CLINIC)
	1 Chronic pain center (CHRONIC PAIN CL)
	2 Dental clinic (DENTAL CLINIC)
	3 Psychiatric clinic (PSYCH CLINIC)
	4 OB-GYN clinic (OB-GYN CLINIC)
	5 Pediatric clinic (PEDS CLINIC)
	9 Other clinic (OTHER CLINIC)
52X	Free-standing clinic
	0 General classification (FREESTAND CLINIC)
	1 Rural health-clinic (RURAL/CLINIC)
	2 Rural health-home (RURAL/HOME)
	3 Family practice (FAMILY PRACTICE)
	9 Other freestanding clinic (OTHER FR/STD CLINIC)
53X	Osteopathic services: These are charges for a structural evaluation of the cranium and the entire cervical, dorsal, and lumbar spine by a doctor of osteopathy
	0 General classification (OSTEOPATH SVS)
	1 Osteopathic therapy (OSTEOPATH RX)
	9 Other osteopathic services (OTHER OSTEOPATH)
54X	Ambulance: These are charges for ambulance service, usually on an unscheduled basis to the ill or injured person who requires immediate medical attention
	0 General classification (AMBULANCE)
	1 Supplies (AMBUL/SUPPLY)
	2 Medical transport (AMBUL/MED TRANS)
	3 Heart mobile (AMBUL/HEART MOBL)
	4 Oxygen (AMBUL/OXY)
	5 Air ambulance (AIR AMBULANCE)
	6 Neonatal ambulance services (AMBUL/NEONAT)
	7 Pharmacy (AMBUL/PHARMACY)
	8 Telephone transmission ECG (AMBUL/TELEPHONIC ECG)
	9 Other ambulance (OTHER AMBULANCE)

Administrative Medical Services

Code No.	Revenue
55X	Skilled nursing: These are charges for nursing services that must be provided under the direct supervision of a licensed nurse to assure the safety of the patient and to achieve the medically desired results
	0 General classification (SKILLED NURSING)
	1 Visit charge (SKILLED NURS/VISIT)
	2 Hourly charge (SKILLED NURS/HOUR)
	9 Other skilled nursing (SKILLED NURS/OTHER)
56X	Medical social services: These are charges for services such as counseling and interviewing patients and interpreting problems of social situations that are rendered to patients on any basis
	0 General classification (MED SOCIAL SVS)
	1 Visit charge (MED SOC SERVS/VISIT)
	2 Hourly charge (MED SOC SERVS/HOUR)
	9 Other medical social services (MED SOCIAL SERVS/OTHER)
57X	Home health aide: These are charges made by a home health agency for personnel that are primarily responsible for the personal care of the patient
	0 General classification (AIDE/HOME HEALTH)
	1 Visit charge (AIDE/HOME HLTH/VISIT)
	2 Hourly charge (AIDE/HOME HLTH/HOUR)
	9 Other home health aide (AIDE/HOME HLTH/OTHER)
58X	Other visits (home health): These are charges by a home health agency for visits other than for physical therapy, occupational therapy, or speech therapy, and must be specifically identified
	0 General classification (VISIT/HOME HEALTH)
	1 Visit charge (VISIT/HOME HLTH/VISIT)
	2 Hourly charge (VISIT/HOME HLTH/HOUR)
	9 Other home health (VISIT/HOME HLTH/OTHER)
59X	Units of service (home health): These are revenue codes used by home health agencies that bill on the basis of units of service
	0 General classification (UNIT/HOME HEALTH)
	9 Home health other units (UNIT/HOME HLTH/OTHER)
60X	Oxygen home health: These are charges by a home health agency for oxygen equipment, supplies, or contents, excluding purchased equipment
	0 General classification (O_2/HOME HEALTH)
	1 Oxygen-stationary equipment, supplies or contents (O_2/STAT EQUIP/SUPPL/CONT)
	2 Oxygen-stationary equipment or supplies under 1 LPM (O_2/STAT EQUIP/UNDER 1 LPM)
	3 Oxygen-stationary equipment or supplies over 4 LPM (O_2/STAT EQUIP/OVER 4 LPM)
	4 Oxygen-portable add-on (O_2/PORTABLE ADD-ON)
61X	MRI: These are charges for magnetic resonance imaging of the brain and other parts of the body
	0 General classification (MRI)
	1 Brain (including brain stem) (MRI-BRAIN)
	2 Spinal cord (including spine) (MRI-SPINE)
	9 Other MRI (MRI-OTHER)
62X	Medical-surgical supplies: These are charges for supplies required for patient care
	1 Supplies incident to radiology (MED-SUR SUPP/INCDNT RAD)
	2 Supplies incident to other diagnostic services (MED-SUR SUPP/INCDNT ODX)
63X	Drugs requiring specific identification
	0 General classification (DRUGS)
	1 Single source drug (DRUG/SNGLE)
	2 Multiple source drug (DRUG/MULT)
	3 Restrictive prescription (DRUG/RSTR)

Code No.	Revenue
	4 Erythropoietin (EPO) less than 10,000 units (DRUG/EPO<10,000 Units)
	5 Erythropoietin (EPO) more than 10,000 units (DRUG/EPO>10,000 Units)
	6 Drugs requiring detailed coding (DRUGS/DETAIL CODE)
64X	Home IV therapy services
	0 General classification (IV THERAPY SVC)
	1 Non-routine nursing, central line (NON RT NURSING/CENTRAL)
	2 IV site care, central line (IV SITE CARE/CENTRAL)
	3 IV start-change peripheral line (IV STRT/CHNG/PERIPHRL)
	4 Non-routine nursing peripheral line (NON RT NURSING/PERIPHRL)
	5 Training patient/caregiver, central line (TRNG PT/CAREGVR/CENTRAL)
	6 Training disabled patient, central line (TRNG DSBLPT/CENTRAL)
	7 Training patient/caregiver, peripheral line (TRNG PT/CAREGVR/PERIPHRL)
	8 Training disabled patient, peripheral line (TRNG DSBLPT/PERIPHRL)
	9 Other IV therapy services (OTHER IV THERAPY SVC)
65X	Hospice service: These are charges for hospice care services for a terminally ill patient
	0 General classification (HOSPICE)
	1 Routine home care (HOSPICE/RTN HOME)
	2 Continuous home care (HOSPICE/CTNS HOME)
	3 Reserved
	4 Reserved
	5 Inpatient respite care (HOSPICE/IP RESPITE)
	6 General inpatient care (HOSPICE/IP NON-RESPITE)
	7 Physician services (HOSPICE/PHYSICIAN)
	9 Other hospice (HOSPICE/OTHER)
66X	Respite care: These are charges for hours of service under the Respite Care Benefit for homemaker or home health aide, personal care services, and nursing care provided by a licensed professional nurse
	0 General classification (RESPITE CARE)
	1 Hourly charge/skilled nursing (RESPITE/SKILLED NURSE)
	2 Hourly charge/home health aide/homemaker (RESPITE/HMEAID/HMEMKR)
67X	Not assigned
68X	Not assigned
69X	Not assigned
70X	Cast room: These are charges for services related to the application, maintenance, and removal of casts
	0 General classification (CAST ROOM)
	9 Other cast room (OTHER CAST ROOM)
71X	Recovery room
	0 General classification (RECOVERY ROOM)
	9 Other recovery room (OTHER RECOV RM)
72X	Labor and delivery
	0 General classification (DELIVROOM/LABOR)
	1 Labor (LABOR)
	2 Delivery (DELIVERY RM)
	3 Circumcision (CIRCUMCISION)
	4 Birthing center (BIRTHING CENTER)
	9 Other labor room/delivery (OTHER/DELIV-LABOR)
73X	ECG services
	0 General classification (EKG/ECG)
	1 Holter monitor (HOLTER MON)
	2 Telemetry (TELEMETRY)
	9 Other EKG/ECG (OTHER EKG-ECG)
74X	EEG services
	0 General classification (EEG)
	9 Other EEG services (OTHER EEG)
75X	Gastrointestinal services
	0 General classification (GASTR-INTS SVS)

Administrative Medical Services

Code No.	Revenue

- 9 Other gastrointestinal services (OTHER GASTRO-INTS)

76X Treatment and observation room: These are charges for the use of a treatment room or other observation room for outpatient observation services
- 0 General classification (TREATMENT-OBSERVATION RM)
- 1 Treatment room (TREATMENT RM)
- 2 Observation room (OBSERVATION RM)
- 9 Other treatment/observation room (OTHER TREAT/OBSERV RM)

77X Not assigned

78X Not assigned

79X Not assigned

80X Inpatient renal dialysis
- 0 General classification (RENAL DIALYSIS)
- 1 Inpatient hemodialysis (DIALY/INPT)
- 2 Inpatient peritoneal (DIALY/INPT/PER)
- 3 Inpatient continuous ambulatory peritoneal dialysis (DIALY/INPT/CAPD)
- 4 Inpatient continuous cycling peritoneal dialysis (DIALY/INPT/CCPD)
- 9 Other inpatient dialysis (DIALY/INPT/OTHER)

81X Organ acquisition: These are charges for the acquisition of a kidney, liver, or heart for use in transplantation. Other organs are included in category 89X
- 0 General classification (ORGAN ACQUISIT)
- 1 Living donor-kidney (KIDNEY/LIVE)
- 2 Cadaver donor-kidney (KIDNEY/CADAVER)
- 3 Unknown donor-kidney (KIDNEY/UNKNOWN)
- 4 Other kidney acquisition (KIDNEY/OTHER)
- 5 Cadaver donor-heart (HEART/CADAVER)
- 6 Other heart acquisition (HEART/OTHER)
- 7 Donor-liver (LIVER ACQUISIT)
- 9 Other organ acquisition (ORGAN/OTHER)

82X Hemodialysis-outpatient or home
- 0 General classification (HEMO/OP OR HOME)
- 1 Hemodialysis/composite or other rate (HEMO/COMPOSITE)
- 2 Home supplies (HEMO/HOME/SUPPL)
- 3 Home equipment (HEMO/HOME/EQUIP)
- 4 Maintenance 100% (HEMO/HOME/100%)
- 5 Support services (HEMO/HOME/SUPSERV)
- 9 Other outpatient hemodialysis (HEMO/HOME/OTHER)

83X Peritoneal dialysis-outpatient or home
- 0 General classification (PERITONEAL/OP OR HOME)
- 1 Peritoneal/composite or other rate (PERTNL/COMPOSITE)
- 2 Home supplies (PERTNL/HOME/SUPPL)
- 3 Home equipment (PERTNL/HOME/EQUIP)
- 4 Maintenance 100% (PERTNL/HOME/100%)
- 5 Support services (PERTNL/HOME/SUPSERV)
- 9 Other outpatient peritoneal (PERTNL/HOME/OTHER)

84X Continuous ambulatory peritoneal dialysis (CAPD)-outpatient or home
- 0 General classification (CAPD/OP OR HOME)
- 1 CAPD/composite or other rate (CAPD/COMPOSITE)
- 2 Home supplies (CAPD/HOME/SUPPL)
- 3 Home equipment (CAPD/HOME/EQUIP)
- 4 Maintenance 100% (CAPD/HOME/100%)
- 5 Support services (CAPD/HOME/SUPSERV)
- 9 Other outpatient CAPD (CAPD/HOME/OTHER)

Code No.	Revenue
85X	Continuous cycling peritoneal dialysis (CCPD)-outpatient or home
	0 General classification (CCPD/OP OR HOME)
	1 CCPD/composite or other rate (CCPD/COMPOSITE)
	2 Home supplies (CCPD/HOME/SUPPL)
	3 Home equipment (CCPD/HOME/EQUIP)
	4 Maintenance 100% (CCPD/HOME/100%)
	5 Support services (CCPD/HOME/SUPSERV)
	9 Other outpatient CCPD (CCPD/HOME/OTHER)
86X	Reserved for dialysis (national assignment)
87X	Reserved for dialysis (national assignment)
88X	Miscellaneous dialysis: These are charges for dialysis services not identified anywhere else
	0 General classification (DAILY/MISC)
	1 Ultrafiltration (DIALY/ULTRAFILT)
	2 Home dialysis aid visit (HOME DIALYSIS AID VISIT)
	9 Miscellaneous dialysis other (DIALY/MISC/OTHER)
89X	Other donor bank: These are charges for the acquisition, storage, and preservation of all human organs except kidneys
	0 General classification (DONOR BANK)
	1 Bone (DONOR BANK/BONE)
	2 Organ (other than kidney) (DONOR BANK/ORGN)
	3 Skin (DONOR BANK/SKIN)
	9 Other donor bank (OTHER DONOR BANK)
90X	Psychiatric and psychological treatments: These are charges for treatment for emotionally disturbed patients, including those admitted for diagnosis and for treatment
	0 General classification (PSYCH TREATMENT)
	1 Electroshock treatment (ELECTRO SHOCK)
	2 Milieu therapy (MILIEU THERAPY)
	3 Play therapy (PLAY THERAPY)
	9 Other (OTHER PSYCH RX)
91X	Psychiatric and psychological services: These are charges for providing nursing care and employee, professional services for emotionally disturbed patients, including those admitted for diagnosis and treatment
	0 General classification (PSYCH SERVICES)
	1 Rehabilitation (PSYCH/REHAB)
	2 Day care (PSYCH/DAYCARE)
	3 Night care (PSYCH/NIGHTCARE)
	4 Individual therapy (PSYCH/INDIV RX)
	5 Group therapy (PSYCH/GROUP RX)
	6 Family therapy (PSYCH/FAMILY RX)
	7 Bio feedback (PSYCH/BIOFEED)
	8 Testing (PSYCH/TESTING)
	9 Other (PSYCH/OTHER)
92X	Other diagnostic services: These are charges for other diagnostic services not otherwise categorized
	0 General classification (OTHER DX SVS)
	1 Peripheral vascular lab (PERI VASCUL LAB)
	2 Electromyogram (EMG)
	3 Pap smear (PAP SMEAR)
	4 Allergy testing (ALLERGY TEST)
	5 Pregnancy testing (PREG TEST)
	9 Other diagnostic service (ADDL DX SVS)
93X	Not assigned
94X	Other therapeutic services: These are charges for other therapeutic services not otherwise categorized
	0 General classification (OTHER RX SVS)
	1 Recreational therapy (RECREATION RX)
	2 Education and training (EDUC/TRAINING)
	3 Cardiac rehabilitation (CARDIAC REHAB)

Administrative Medical Services

Code No.	Revenue
	4 Drug rehabilitation (DRUG REHAB)
	5 Alcohol rehabilitation (ALCOHOL REHAB)
	6 Complex medical equipment-routine (CMPLX MED EQUIP-ROUT)
	7 Complex medical equipment-ancillary (CMPLX MED EQUIP-ANC)
	9 Other therapeutic services (ADDITIONAL RX SVS)
95X	Not assigned
96X	Professional fees: These are charges for medical professionals that the hospitals or third-party payers require to be separately identified
	0 General classification (PRO FEE)
	1 Psychiatric (PRO FEE/PSYCH)
	2 Ophthalmology (PRO FEE/EYE)
	3 Anesthesiologist (MD) (PRO FEE/ANES MD)
	4 Anesthetist (CRNA) (PRO FEE/ANES CRNA)
	9 Other professional fees (OTHER PRO FEE)
97X	Professional fees (continued)
	1 Laboratory (PRO FEE/LAB)
	2 Radiology-diagnostic (PRO FEE/RAD/DX)
	3 Radiology-therapeutic (PRO FEE/RAD/RX)
	4 Radiology-nuclear medicine (PRO FEE/NUC MED)
	5 Operating room (PRO FEE/OR)
	6 Respiratory therapy (PRO FEE/RESPIR)
	7 Physical therapy (PRO FEE/PHYSI)
	8 Occupational therapy (PRO FEE/OCUPA)
	9 Speech pathology (PRO FEE/SPEECH)
98X	Professional fees (continued)
	1 Emergency room (PRO FEE/ER)
	2 Outpatient services (PRO FEE/OUTPT)
	3 Clinic (PRO FEE/CLINIC)
	4 Medical social services (PRO FEE/SOC SVC)
	5 ECG (PRO FEE/ECG)
	6 EEG (PRO FEE/EEG)
	7 Hospital visit (PRO FEE/HOS VIS)
	8 Consultation (PRO FEE/CONSULT)
	9 Private duty nurse (PRO FEE/NURSE)
99X	Patient convenience items: These are charges for items that are generally considered by the third-party payers to be strictly convenience items and, therefore, are not generally covered
	0 General classification (PT CONVENIENCE)
	1 Cafeteria-guest tray (CAFETERIA)
	2 Private linen service (LINEN)
	3 Telephone-telegraph (TELEPHONE)
	4 TV-radio (TV/RADIO)
	5 Nonpatient room rentals (NONPT ROOM)
	6 Late discharge charge (LATE DISCHARGE)
	7 Admission kits (ADMIT KITS)
	8 Beauty shop-barber (BARBER/BEAUTY)
	9 Other patient convenience items (PT CONVENCE/OTH)

UB-92 ITEM 59

Relationship Codes

These are numerical codes that provide the designation of the patient to the insured.

Code No.	Relationship
01	Insured party
02	Spouse of insured
03	Child of insured
04	Natural child of the insured who does not have financial responsibility of the child
05	Step-child of the insured
06	Foster child
07	Ward of the court
08	Employee, who is employed by the insured
09	Unknown
10	Handicapped dependent
11	Organ donor
12	Cadaver donor
13	Grandchild

Code No.	Relationship
14	Niece or nephew
15	Injured plaintiff
16	Sponsored dependent
17	Minor dependent of a minor dependent
18	Parent
19	Grandparent
20–99	Reserved

UB-92 ITEM 64

Employment Status Codes

These are codes that are used to define the employment status of the person identified in Item 63.

Code No.	Employment Status
1	Employed full time
2	Employed part time
3	Not employed
4	Self-employed
5	Retired
6	On active military duty
7–8	Reserved for national assignment
9	Employment status unknown

MEDICAID CODES

Field No.	Description
1	Leave blank
1A	Type providers name and address
2	Type in the provider's 9-digit medicaid provider number
3	Leave blank
4	Enter an "X" in the medicaid box
5	For patients that have medicare and medicaid, enter an "X" in boxes 4 and 5
6	Enter the 5-digit zip code number
7A	Enter the provider's telephone number
7B	Attach a medicaid sticker for the month of services
8	Enter the patient's last name, first name, and middle initial and full address as it appears on the medicaid card
9	Leave blank on medicaid forms. For medi-medi, enter the medicare number as it appears on the medicare ID card
10	Enter the patient's sex. M for male, F for female
11	Complete this field only for services associated with an accident by entering an "X" in the yes or no box in order to show if the patient's condition is related to his or her employment
12	Enter the date of onset of the patient's illness or the date of the last menstrual period
13	Enter the complete 11-digit TAR control number; if a TAR is not required, leave blank
14	Enter the patient's medicaid ID number exactly as it appears on the eligibility label
15	Enter the patient's date of birth in six digits
16	Enter the patient's last name, account number, or record number
17–18	Enter the hospital admission and discharge dates if the services were related to a hospitalization
19	Leave this field blank unless billing for emergency services
20	Enter an "X" if the patient has other insurance coverage; enter the name and address of the carrier
21	Claims must be submitted within two months of the month the services are rendered. Exceptions to this include the following:

Code No.	Description
1	Proof of eligibility was unknown or unavailable
2	Other coverage
3	Delays in obtaining authorization
4	Delay by department or fiscal intermediary in providing certification or in supplying billing
5	Do not use
6	Substantial damage to provider's records was caused by fire, flood, or other natural disaster
7	Employee committed theft, sabotage, or other willful acts

Administrative Medical Services

	Code No.	Description
	8	Court decisions, fair hearing decisions, provider appeals, or other circumstances in which the provider had no control
22		Enter an "X" if attachments are included; if not, leave space blank
23		Leave blank
24		Leave blank if patient is receiving medicare benefits; if not, enter one of the codes listed below:

Code No.	Description
0	Under the age of 65, and does not qualify for medicare benefits
1	Benefits are exhausted
2	Utilization committee has denied or physician noncertification
3	No prior hospitalization
4	Facility has denied claim
5	Provider is not eligible
6	Recipient is not eligible
7	Medicare benefits have been denied or cut short by a medicare intermediary
8	Services are not covered
9	PSRO denial

25	Optional
26	Optional
27	Enter the full 9-digit medicaid provider number of the facility where the services were provided
28	If a claim submitted to a health insurer has not been paid within 90 days, the provider may then bill medicaid using the Billing Limit Exception reason (code 2)
29	Leave blank
30	Leave blank
31	Optional
32	Enter the medicaid number of the referring or prescribing doctor
33	Optional if an entry is made in field number 34
34	Enter all the letters and numbers of the ICD-9 code
35	Optional if an entry is made in field number 36
36	Enter all the letters and numbers of the ICD-9 code, if applicable
37	Optional, except if billing for procedures not listed
38	Leave blank
39	If a mistake was made, delete the entire line by typing an "X" in this space; then enter correct information on next line
40	Enter date of service in 6-digit format
41	Enter the location where the services were rendered, using the codes listed below:

Code No.	Location of Services Rendered
1	Office
2	Home
3	Inpatient hospital
4	Skilled nursing facility
5	Out-patient hospital
6	Independent laboratory
7	Other (describe in section identified as "Remarks")
8	Independent kidney treatment center
9	Clinic
A	Surgery clinic
B	Emergency room
C	Intermediate-care facility
D	Extended-care facility

42	Enter code 1 or 2 if the services are related to family planning
43	Enter the medicaid provider number or state license number of the provider only if it is different from the billing provider number
44	Enter the applicable code and modifiers from the CPT
45	Enter the number of medical visits or surgical lesions
46	Enter the usual and customary fee for services provided
47–116	Follow the same instructions for each claim line
117	Leave blank
118	Leave blank
119	Leave blank

Code No.	Description
120	Leave blank
121	Leave blank
122	Leave blank
123	Leave blank
124	Leave blank
125	Enter amount of payment received from third-party payer identified in field number 28
126	Enter amount paid by other insurance coverage (the amount in field 127); leave blank unless medicare denies or this is a noncovered service
128	Using a 6-digit format, enter date the claim is being submitted to medicaid
129	Enter the difference between the total charges and deductions (field 119 minus field 126 equals field 129)
130–134	Leave blank
135	Leave blank
136	Leave blank
137A	Leave blank
137B	Use this space to identify any services that required additional explanation or for entering the "emergency certification statement"
138	Leave blank
139	The claim form must be signed and dated by the provider of services. Do not use a stamped signature or initials
140	Leave blank
141	Leave blank

Index

Page numbers followed by *t* indicate tables; page numbers in italics indicate figures.

AAMA (American Association of Medical Assistants), 3
AAMT (American Association of Medical Transcriptionists), 3, 107, 108
abbreviation(s), medical, 33–34, 193–196, 255–260
abduction, 36
abstracting, medical, 74, 75, 78–79
   of data, from medical records, 83–85, 145
account due statement(s), 57, 61
accountability record(s), nurses', 228
accreditation, definition of, 74
   standards of, role of medical records staff in, 181, 182
Accredited Record Technician (ART), 3, 146, *146*, 187
accumulation period, 194
acromegaly, 48
adaptive model of health care delivery, 10
address, inside, in letter writing, spacing of, 118
addressograph(s), 223, 226
adduction, 36
adjustment(s), in claim coding, 194
administrative medical service(s), career opportunities in, 3–7. See also *Career opportunity(ies)*.
   history of, 4
   training and education for, 4
administrative services only (ASO), 194
administrator(s), in insurance plans, 194
admission(s), 64–65
   emergency, 223, 229–230
   forms for, 65, *66–70*, 227, 230
   processing of, for medical records, 155
   routine, 223, 229–230
admission agreement(s), 64, 65, *70*
admission data form(s), 227, 230
admission order(s), 230
adrenal gland, 50, *50*
aerobic bacteria, 24
against medical advice (AMA), 223, 231
AHIMA (American Health Information Management Association), 187
alcohol psychosis, insurance coding for, 200, 204
alignment, definition of, 23
alphabetical filing system(s), 74, 76
alphanumeric code(s), definition of, 87
AMA (against medical advice), 223, 231
American Association of Medical Assistants (AAMA), 3

American Association of Medical Transcriptionists (AAMT), 3, 107, 108
American Guild for Patient Account Managers, 3
American Health Information Management Association (AHIMA), 187
anaerobic bacteria, 24
analgesic(s), 236
analysis, of medical records, 78–81, *79–81*, 145, 148, 154, 157, 180
anatomical position(s), 34–35, *35*
anatomical posture(s), 23, 27, *27*, 29, *30*, 35–36, *36*
anemia, 43
anesthesia record(s), 228
anesthesiology report(s), 156, *172–179*
aneurysm(s), 39
anterior, 34
antiarrhythmic(s), 236
antibiotic(s), 236
antiseptic(s), 26
anus, anatomy of, 41, *43*
aorta, 41, *45*
appendicitis, 41
appendix, anatomy of, 41, *43*
ART (Accredited Record Technician), 3, 146, *146*, 187
arteriosclerosis, 43
artery(ies), 41, *45*
arthritis, 39, 41
ascending colon, 41, *43*
asepsis, and infection control, 24–27
   definition of, 23
   surgical, 25–26, *26*
aseptic handwashing, 25–26, *26*
ASO (administrative services only), 194
assembly, of medical records, 78–81, *79–81*, 148, *148*, 156–157, *158–179*
assignment of benefit(s), 194
   in insurance claims, 208, 210
   in Medicare program, 215
asthma, 44
attending physician(s), in medical record analysis, 154, 157, 180
attention line(s), in letter writing, 120
attestation, 181
audit(s), 96, 97
audit claim(s), 194
authorization, for medical record release, 145, 150, *151*

autonomy, hospitalization and, 15
autopsy report(s), 156, *172–179*
average length of stay, calculation of, 84

bacteria, 23, 24
    definition of, 24
bacteriology, definition of, 23
balance, definition of, 23, *28*, 28–29, *30*
base of support, definition of, 23, *28*, 28–29
battery, definition of, 17
beeper(s), 57, 60, 226
behavior, ethical, 18
bending, correct positioning for, *30*
beneficiary(ies), 194
benefit(s), assignment of, 194
    coordination of, 199, 208, 218
billing, for services rendered, 210–211. See also *Insurance claim(s)*.
    forms for, 265–290
birth certificate(s), processing of, 84–85
      by medical clerks, 183–184
bladder, anatomy of, 47, *49*
block style, in letter writing, 120, *121*
blood, composition of, 41–42
blood count, complete, insurance coding for, 200
blood pressure, measurement of, 42–43
blood-forming organ(s), diseases of, insurance coding for, 203–204
body cavity(ies), 34, *34*
body mechanics, 23, *27–28*, 27–29, 30
Bright's disease, 47
bronchi, anatomy of, 44
bronchitis, 44

calculator(s), 62
calculi, urinary, 47
cancer, registry for tracking, 148
capillary(ies), 41, *45*
capitalization, in medical transcription, 114
cardiac arrest, emergency care for, 31
cardiac muscle(s), 40, *42–43*
cardinal number(s), 110
cardiovascular system, diseases of, insurance coding for, 203–204
care, duty of, 17, 19
    standard of, 93, 94
care agreement(s), with release, *67*
career opportunity(ies), 3–7, 245–252
    employment interview for, 251–252
      thank-you note after, *253*
    job application forms for, 247
    search for, initiation of, 246–247
carry-over(s), last quarter, 194
catherization(s), urinary, 234
caudal, 35
cavity(ies), of body, 34, *34*
CBC (complete blood count), insurance coding for, 200
CCS (Certified Coding Specialist), 3, 88
cecum, anatomy of, 41, *43*
census, 83, 84
census report sheet(s), at nurse's station, 226
center of gravity, 23, *28*, 28–29
central processing unit (CPU), 101, *102*
cerebrovascular accident (CVA), insurance coding for, 200
Certified Coding Specialist (CCS), 3, 88
certified mail, 61
certified medical assistant (CMA), 3
certified medical transcriptionist (CMT), requirements for, 108
certified nurse assistant (CNA), 195
certified registered nurse anesthetist (CRNA), 195
CHAMPUS (Civilian Health and Medical Program for Uniformed Services), 87, 88
CHAMPVA (Civilian Health and Medical Program for Veterans Administration), 87, 88
charge(s), processing of, for coding insurance claims, 90
charge ticket summary(ies), 228
chart(s). See *Medical record(s)*.

chart back(s), 72
charting, in medical records, 37
childbirth, insurance coding for, 202–203
chronological order, 74
chronological resume, 245, 248, *250*
circulatory system, 41–44, *45*
    common disorders of, 43–44
    diseases of, insurance coding for, 203
cirrhosis, 41
civil law, definition of, 17
Civilian Health and Medical Program for Uniformed Services (CHAMPUS), 87, 88
Civilian Health and Medical Program for Veterans Administration (CHAMPVA), 87, 88
claim(s), audit of, 194
claims processing. See *Insurance claim(s), processing of*.
clerk(s), discharge analysis, 78
    for medical records, description of, 146, *147*
    ward. See *Health Unit Coordinator(s)*.
clinical model of health care delivery, 10
clinical psychologist(s), 195
closing(s), in letter writing, spacing of, 118
CMA (certified medical assistant), 3
CMT (certified medical transcriptionist), requirements for, 108
CNA (certified nurse assistant), 195
COB (coordination of benefits), 194, 208, 218
COBRA (Consolidated Omnibus Budget Reconciliation Act of 1985), 194
code(s), Current Procedural Terminology. See *Current Procedural Terminology (CPT) code(s)*.
    International Classification of Diseases. See *International Classification of Diseases (ICD) code(s)*.
coding. See also *Insurance coding*.
    careers in, 88
    in filing systems, 74, 79, 145, 149–150
coding book(s), for insurance claims, *196*, 196–198
coding specialist(s), certified, 3, 88
    medical, 187, *188*
coding system(s). See *Filing system(s)*; *Insurance coding*.
coinsurance, 194
cold(s), 44
collection procedure(s), laws concerning, 209
colon, anatomy of, 41, *43*
color-coding filing system(s), 74, 76, 151
combining form(s), definition of, 33
combining vowel(s), definition of, 33
communication(s), written, processing of, by health unit coordinator, 226–228
communication system(s), 57–62
    at nursing stations, health care coordinator and, 225–226
    intercom use in, 60
    paging system in, 60
    telephone system in, 58–60
    written, 60–62
complement, in grammar, 117
complementary closing(s), in letter writing, spacing of, 118
complete blood count(s) (CBC), insurance coding for, 200
compliance, 9
compound fracture(s), 40–41
compromise and release, for Workers' Compensation, 218
computer system(s), access to, security for, 100, 103
    applications in, 102–103
    auxiliary equipment for, 102
    components of, 101–102
    mainframe, 100, 101
    problems with, resolutions of, 103
condition of admission, 145
    form for, 156, *158–171*
confidential, definition of, 64
confidentiality, in mail processing, 61
congestive heart failure, 43
consent, informed, 17, 20–21, *21*, 93, 94
consent form(s), 95, 227, 228
    for surgery, 156, *172–179*

Index

Consolidated Omnibus Budget Reconciliation Act of 1985 (COBRA), 194
constipation, 39, 41
consultation form(s), 228
consultation report(s), 156, *158–171*
   transcription of, 132, *137–138*
contusion(s), insurance coding for, 200
coordination of benefit(s) (COB), 194, 208, 218
copay, in insurance, 194
copy notation(s), in letter writing, 120
copying machine(s), 62
coronary artery(ies), 41
coronary occlusion, 43
coronary vein(s), 41
correspondence. See also *Letter(s)*.
   types of, 60–61
counselor(s), marriage, family, and child, 195
cover letter(s), 245, 247, *248*
covered expense(s), 194
CPT (Current Procedural Terminology) code(s). See *Current Procedural Terminology (CPT) code(s)*.
CPU (central processing unit), 101, *102*
cranial, 35
credentialing, in quality assurance, 96, 98
CRNA (certified registered nurse anesthetist), 195
CRT, 110
Current Procedural Terminology (CPT) code(s), 87, 88–90
   correct use of, 89, *89*
   in Health Care Procedural Coding System, 89
   in medical records, 154
   reference book for, 197
   use of modifiers in, 193
Current Procedural Terminology (CPT) coding, 190–193
   anesthesia codes in, 191
   block and asterisk procedures in, 191
   evaluation and management codes in, 190–191
   medicine codes in, 193
   qualifying circumstances in, 191–192
   radiology, pathology and lab codes in, 192–193
   surgery codes of, 191
Custodian of Record(s), 145, 152
CVA (cerebrovascular accident), insurance coding for, 200
cystitis, 47

data, statistical, 75
data abstracting, from medical records, 83–85, 145
data entry source document(s), 83
data slip(s), for medical records, 181, 182
database, computer, 100, 101
date line(s), in letter writing, spacing of, 118
DC (Doctor of Chiropractic), 195
DDS (Doctor of Dental Surgery), 195
death(s), processing of, 72
   by health unit coordinators, 231
   record of, 156, *172–179*
death certificate(s), processing of, 84, 85
declarative sentence(s), 117
deductible(s), 87, 90, 194
defamation, definition of, 17
deficiency(ies), in medical records, 74, 78
deficiency slip(s), 145, 148, *148*
Deficit Reduction Act of 1984 (DEFRA), 194
delinquent report(s), 181, 182
demography, definition of, 64
   patient information form for, 65, *66–67*
dependent patient role, in illness, 13
descending colon, 41, *43*
diabetes, 50
   diet for, 239
   insurance coding for, 200, 205
diagnosis, definition of, 96
   major categories of, 187
   principal, 154, 180, 189

Diagnosis Related Group(s) (DRGs), 154, 187
   in quality assurance, 96, 97–98
diagnostic procedure(s), medical orders for, transcription of, 239–240
diaphragm, anatomy of, 44, *46*
diarrhea, 39, 41
diastolic blood pressure, 42–43
dictation, 110
diet(s), types of, 239
diet sheet(s), 226
digestive system, 41, *43*
   diseases of, insurance coding for, 202
disability, insurance claims for, for Worker's Compensation, 218
discharge(s), 64, 72
   abstracting patient data from, 83–84
   after death, 72
   against medical advice, 72, 223, 231
   medical records with, 72
   processing of, by health unit coordinator, 231
      medical records with, 155–156
discharge analysis clerk(s), 78
discharge planning, quality assurance in, 96, 97–98
discharge report(s), transcription of, 132, *140–141*
discharge summary(ies), 145
   analysis of, 180
   forms for, 156, *158–171*
   physician's, 227
disease(s), definition of, 9
   insurance coding for, 200–206
   wellness and, 11–12, *11–12*
Disease and Operative Index, use of, 75
disinfectant(s), 26
disk(s), for computers, 100, 102
dislocation(s), 41
display terminal(s), for computers, 100, 101, *102*
diuretic(s), 236
DO (Doctor of Osteopathy), 195
Doctor of Chiropractic (DC), 195
Doctor of Dental Surgery (DDS), 195
Doctor of Optometry (OD), 195
Doctor of Osteopathy (DO), 195
Doctor of Podiatry (DPM), 195
doctor's box method, of completing medical records, 182
doctor's note(s), in chart assembly, 80
doctor's order form(s), 227. See also *Physician's order(s)*.
document(s), data source entry, 83
documentation, requirements for, 95
Dorland's Illustrated Medical Dictionary, in medical transcription, 126, *127*
downtime, computer, 103
DPM (Doctor of Podiatry), 195
DRG (diagnosis related groups). See *Diagnosis related group(s) (DRGs)*.
duodenum, anatomy of, 41, *43*
duty of care, definition of, 17, 19
dysentery, 39
dysmenorrhea, 51

E-code(s), in insurance coding, 189
   supplementary listing of, 198
ecological model of health care delivery, 10, *10*
electronic claims processing, 211, *214*
eligibility period(s), 194
emergency(ies), common, 29–31
emergency admission(s), 223, 229–230
emergency medical technician (EMT), 195
emergency room, records from, 156, *158–171*
emergency room report(s), 228
emphysema, 44
employment agency(ies), 246
employment application(s), 247
employment interview(s), 251–252
   thank-you note after, *253*
EMT (emergency medical technician), 195

enclosure(s), in letter writing, 120
endocrine disease(s), insurance coding for, 205
endocrine system, 47–48, 50, *50*
enema(s), 237
envelope(s), typing of, 120, *124–125*
environmental safety, 29
EOMB (explanation of Medicare benefits), 194
erythrocyte(s), 41
esophagus, anatomy of, 41, *43*
ethical behavior, 18
ethics, definition of, 17, 18
eudamonistic model of health care delivery, 10–11
eversion, 36
exclamatory sentence(s), 117
exclusion(s), in insurance coverage, 194
exophthalmic goiter, 48, 50
expense(s), covered, 194
explanation of Medicare benefit(s) (EOMB), 194
express mail, 61
extended benefit(s), 194

face sheet(s), 156, *158–171*
fainting, emergency care for, 29–30
fall(s), emergency care for, 31
fallopian tube(s), anatomy of, 51
family(ies), illness and, 14
family member(s), reception of, 57–58
facsimile (fax), 57, 60
FDA (Food and Drug Administration), 234
FECA (Federal Employee Compensation Act), 87
Federal Employee Compensation Act (FECA), 87
fertilization, 51
fetal death certificate(s), processing of, 84, 85
file guide(s), 74, 77
filing system(s), 74
  color-coded, 74, 76, 151
  establishment of, for medical records, 149
  Kardex in, 226–227
  management of, 75–78
  preparation of records for, 150
  rules of, 76–77
  storage in, 77–78
  types of, 74
first-class mail, 61
flexion, 36
floppy disk(s), 100, 102
FOIA (Freedom of Information Act), 145, 150, 152
font(s), computer, 100, 101
Food and Drug Administration (FDA), 234
form(s). See *specific types and procedures*.
formatting, of computer disks, 100, 102
fourth-class mail, 61
fracture(s), 39, 40–41
fraud, in insurance claims, 209
Freedom of Information Act (FOIA), 145, 150, 152
frontal plane(s), 35
functional resume, 245, 248, *251*

gastritis, 41
glaucoma, insurance coding for, 200
goiter, 48, 50
Good Samaritan Act, 17, 19–20
grammar, in medical transcription, 112–113
graphic sheet(s), 156, *172–179*, 228
  in chart assembly, 80
greenstick fracture(s), 41
greeting(s), in letter writing, spacing of, 118

hallucination(s), insurance coding for, 200, 206
handwashing, 25–26, *26*
hard copy, from computer, 100, 101
  in medical transcription, 110, 111
hard disk(s), 100, 102

hardware, for computers, 100, 101, *102*
HCFA (Health Care Financing Administration), 87
  and coding for Medicare, 187, 189
HCFA-1500 insurance form(s), *91*
  processing of, 211, *212*
HCPCS (Health Care Procedural Coding System), 89–90
HCPS (Health Care Procedural Coding System), levels of, 89–90
health, and illness or disease, 11–12, *11–12*
  beliefs about, 12–13
  definition of, 9
  holistic, 12, *13*
  personal definition of, 11
  trends in, 15–16
health behaviors, 12–13
health care delivery, models of, *10*, 10–11
Health Care Financing Administration (HCFA), 87
  and coding for Medicare, 187, 189
  insurance form of, *91*
    processing of, 211, *212*
Health Care Procedural Coding System (HCPCS), 89–90
  coding levels of, 89–90
health care worker(s), licensing of, 19
health insurance, programs of, 88
health maintenance organization(s) (HMOs), 3, 194
health unit coordinator(s), 223–232, *234*
  and medical records, 227–228
  and nurse's station communications, 225–228
  nursing station management by, 228–229
  patient admissions by, 229–230, *230*
  patient discharges by, 231
  patient transfers by, 231–232
  processing of deaths by, 231
  role of, 224–225
    in renewing or stopping orders, 240, *240*
    in requisitions, 223, 226–227
  transcription of medical orders by, 235–241
    interpreting physician's orders in, 235–236
    pharmacology in, 236–237
  use of paging systems by, 225–226
heart disease(s), insurance coding for, 200, 203–204
heart failure, congestive, 43
heart murmur(s), 43–44
heartburn, 41
hemorrhage, 43
hepatitis, 41
hinge joint(s), 40
HMO(s) (health maintenance organizations), 3, 194
Hodgkin's disease, 43
holistic health, 12, *13*
hormone(s), 47–48, 50
hospital(s), organization of, 6, 6t
hospital chart(s). See also *Medical record(s)*.
  contents of, 79–81
Hospital Formulary(ies), in medical orders transcription, 236
hospitalization, effects of, 14–15
host(s), 24, *25*
hydrocephalus, insurance coding for, 200, 203
hypertension, 43
  insurance coding for, 200, 203
hypnotic(s), 236
hypoglycemic(s), 236
hypothyroidism, insurance coding for, 200, 205

ICD (International Classification of Diseases) code(s). See *International Classification of Diseases (ICD) code(s)*.
ileum, anatomy of, 41, *43*
illness, and families, 14
  and wellness, 11–12, *11–12*
  definition of, 9
  stages of, 13–14
  trends in, 15–16
imperative sentence(s), 117
imprinter(s), at nurse's stations, 226
incontinence, 39

Index

index, 74, 75
    of patient information, 75
infection(s), definition of, 23
    insurance coding for, 205
    nosocomial, 23
    susceptibility to, 24
infection control, 24–27
infection cycle, 24–25, *25*
influenza, 44
informational correspondence, 60
informed consent, 17, 20–21, *21*, 93, 94
injury(ies), prevention of, body alignment in, 29, *30*
    risk management and, 98–99
inpatient record analysis checklist, *81*
insulin, 50
insurability, evidence of, 194
insurance benefit(s), 194
    coordination of, 194, 208, 218
    exclusions in, 194
    extended, 194
    major medical, 194
insurance claim(s), adjustments in, 194
    assignment of benefits in, 208, 210
    coding books for, *196*, 196–198
    deductibles in, 90
    electronic, processing of, 211, *214*
    patient history forms in, 210
    patient information for, obtaining, 209–210, *210*
    peer review of, 194
    processing of, 211, *212–214*, 214–216, *217*
        coordination of benefits in, 194, 208, 218
        for Medicaid, 215–216, 216, *217*
        for Medicare, 211, 214–215
        for Worker's Compensation, 216, 218
        forms for, 90, *91*, 211, *212–214*, 214–216, *217*
        laws concerning, 208–210
        manual, 211, *212–213*
    processing of changes for, 90
    release of information forms in, 208, 210
insurance coding. See *Coding system(s)*.
    careers in, 88
    changes in, processing of, 90
    Current Procedural Terminology codes in, 88–90, 190–193
    for cardiovascular diseases, 203–204
    for digestive diseases, 202
    for disorders of sensory organs, 204
    for endocrine diseases, 205
    for infectious diseases, 205
    for integumentary diseases, 201
    for mental health disorders, 204
    for musculoskeletal diseases, 201–202
    for neoplasms, 205
    for neurologic disorders, 204
    for pregnancy and childbirth, 202–203
    for reproductive diseases, 202
    for respiratory diseases, 204
    for urinary diseases, 202
    forms for, 90, *91*
    function of, 188
    International Classification of Diseases system in, 188–190, *190*
    of diseases, 200–206
    provider abbreviations for, 194–196
    reference materials for, 196–198
    terminology for, 194–196
    use of Health Care Procedural Coding System in, 89–90
insurance form(s), 90, *91*
    completion of, 90
insurance policy(ies), 194
insurance program(s), 88
insured, 87
integumentary system, 46, *49*
    diseases of, insurance coding for, 201
intercom(s), 57, 60
    use of, by health unit coordinators, 225–226

interface, 100, 101
International Classification of Diseases (ICD) code(s), 87, 89, 154, 187, 188–190, *190*
    components of, 189
    reference book for, 196–197
    volume I categories of, *190*
    working with, 189–190
    World Health Organization and, 89
interoffice mail, 61
interrogative sentence(s), 117
interview(s), for employment, 251–252
    thank-you note after, *253*
intestine(s), anatomy of, 41, *43*
intravenous (IV) fluid(s), administration of, 237
invasion of privacy, 93
    laws concerning, 93–94
inversion, 36
involuntary muscle(s), 40, *42–43*
ischemic heart disease, insurance coding for, 200, 203
islets of Langerhans, 50
IV (intravenous) fluid(s), administration of, 237

JCAHO (Joint Commission on Accreditation of Healthcare Organizations). See *Joint Commission on Accreditation of Healthcare Organizations (JCAHO)*.
jejunum, anatomy of, 41, *43*
job application(s), 247
joint(s), 40
Joint Commission on Accreditation of Healthcare Organizations (JCAHO), 74, 78, 96, 145
    medical records analysis by, 157
    role of in quality assurance, 97, 149

Kardex, 223, 226–227
kidney(s), 46–47, *49*
kyphosis, insurance coding for, 200

laboratory form(s), for reports, 228
laboratory report(s), 156, *158–171*
    in chart assembly, 81
laboratory sheet(s), 226
large intestine, anatomy of, 41, *43*
larynx, anatomy of, 44
last quarter carry-over(s), 194
lateral, 34
law(s), and medical transcriptionists, 108
    civil, 17
    concerning collection procedures, 209
    concerning insurance claims, 208–210
    concerning liability, 94, 95
    concerning medical records, 21–22, 150
    concerning patient privacy, 93–95
    medical, 18
lawsuit(s), risk management and, 98–99
LCSW (licensed clinical social worker), 195
ledger card(s), 208
length of stay (LOS), average of, calculation of, 84
    quality assurance in, 96, 97–98
letter(s), cover, 245, 247, *248*
    styles of, 120, *121–124*
    writing of, components of, 118
        medical transcriptionists and, 118, 119, *121–124*
leukocyte(s), 41
leukorrhea, 51
liability, definition of, 93
    laws concerning, 94, 95
    medical record maintenance and, 94–95
    personal, 17, 20
liability provision(s), third party, 195
licensed clinical social worker(s) (LCSW), 195
licensed practical nurse(s) (LPN), 195
licensed vocational nurse (LVN), 195
licensing, of health care workers, 19
life expectancy, trends in, 15

ligament(s), 39, 40
line of gravity, 23, *28*, 28–29
litigation, definition of, 93
live birth(s), processing of, by medical records clerks, 183–184
    registering of, 84–85
LOS (length of stay). See *length of stay (LOS)*.
LPN (licensed practical nurse), 195
lung(s), anatomy of, 44
LVN (licensed vocational nurse), 195
lymphatic system, 42

mail, processing of, 60–62
    release of information via, 152
        screening of, 57, 61
        types of, 61
mailgram, 61
mailing notation(s), in letter writing, 120
mainframe computer(s), 100, 101
major diagnostic categories (MDC), 187
major medical insurance(s), 194
malpractice, definition of, 17, 19–20, 96
MAR (medication administration record), 223, 228, 234
marriage, family, and child counselor (MFCC), 195
Master Patient Index (MPI), 155, *155*
    use of, 75
MD (medical doctor), 195
MDC (major diagnostic categories), 187
medial, 34
Medicaid, 87
    billing for, 216
    in medical records standards, 78
    processing claims for, 215–216, *217*
    receipt of payment from, 216
Medicaid claim(s), Health Care Procedural Coding System for, 89–90
Medicaid fraud, and credentialing, 98
medical abbreviation(s), 33–34, 193–196, 255–260
medical abstract(s), 74, 75, 78–79
medical asepsis, 23, 25–27, *26*
medical care contact stage of illness, 13
medical claim(s). See *insurance claim(s)*.
medical coding specialist(s), 187, *188*
medical doctor (MD), 195
medical ethics, definition of, 17, 18
medical history, forms for, 65, *69*, 80
    transcription of, 127, *128–131*
medical law, 18
medical order(s), renewing or stopping, by health unit coordinator, 240, *240*
    transcription of, by health unit coordinator, 235–241
        for diagnostic procedures, 239–240
        for nursing care, 237, 239
        Physician's Desk Reference in, 236
medical power of attorney, 64, 65
Medical Practice Acts, 17
medical record(s), abstracting data from, 83–85, 145
    analysis of, 145, 148, 154, 157, 180
        by Joint Commission on Accreditation of Healthcare Organizations, 157
        nurse's notes in, 157, 180
        physician's orders in, 154, 157, 180
        progress notes in, 157, 180
    as legal documents, 95
    assembly and analysis of, 78–81, *79–81*, 148, *148*
    assembly of, 156–157, *158–179*, 180
    charting in, 37
    consultation reports in, transcription of, 132, *137–138*
    Current Procedural Terminology codes in, 154
    custodian of, 145, 152
    deficiencies in, 74, 78
    definition of, 145
    discharge reports in, transcription of, 132, *140–141*
    filing systems for, 74–78, 145, 149–150

health unit coordinator and, 227–228
history and physical reports in, transcription of, 126–127, *128–131*
incomplete, physician notification of, 181, 182–183
laws concerning, 21–22, 93–95, 150
maintenance of, 147–148
    by health unit coordinator, 235
management of, 74–81
pathology and radiology reports in, transcription of, 127, 132, *133–136*
patient access to, 150, 152
processing of, 154–156
quality assurance in, 99, 148
release of, 152
    authorization for, 145, 150, *151*
retrieval of, 154, 156, *156*
standards for, 78
transcription of, 126–127, *128–131*, 132, *133–141*
understanding, 36–37
updating of, 183
utilization review of, 149
Medical Record Technician (MRT), 3
medical records clerk(s), advanced duties of, 181–184
    and physician notification of incomplete records, 182–184
    description of, 146, *147*
    role of, in increasing physician productivity, 183
        in meeting accreditation standards, 181–182
        in processing birth certificates, 183–184
medical records department, filing systems for, 149–150
    responsibilities of, 147–148
medical records team, members of, 146
medical service(s), administrative. See *administrative medical service(s)*.
medical social worker (MSW), 195
medical specialty(ies), 6, 7t
medical symbol(s), 255–260
medical technologist (MT), 195
medical terminology, 33–36
    anatomical position in, 34–35, *35*
    for insurance coding, 194–196
medical transcription, by health unit coordinator, 235–241
    capitalization in, 114
    computer hard copy in, 110, 111
    equipment for, 110–111, *111*, *112*
    grammar in, 112–113
    letter writing in, spacing in, 118, 120
        styles of, 120, *121–124*
    numbers in, 114–115
    of consultation reports, 132, *137–138*
    of discharge reports, 132, *140–141*
    of hospital charts, 126–127, *128–131*, 132, *133–141*
    of radiology and pathology reports, 127, 132, *133–136*
    proofreading in, 118, *119*
    punctuation in, 113
    references materials for, 126, *127*
    sentence structure in, 117–118
    typing envelopes in, 120, *124–125*
medical transcriptionist(s), laws concerning, 108
    role of, 107–108, *108*
    training of, 108
Medicare, 87
    assignment of benefits in, 215
    Health Care Procedural Coding System for, 89–90
    in medical record standards, 78
    insurance claims for, processing of, 211, 214–215
    providers of, 214–215
    receipt of payment from, 215
medication administration record(s) (MAR), 223, 228, 234
medication order(s), transcription of, by health unit cooordinators, 236–237
medication record(s), 156, *172–179*
memory, of computer, 100, 101
menstruation, 51
mental health disorder(s), insurance coding for, 204

Index

menu(s), in medical transcription, 110
   of computer, 100
Merck Manual, 197
mesentery, anatomy of, 41, *43*
metastasis, insurance coding for, 200–201, 205
meter(s), for postage, 61
MFCC (marriage, family, and child counselor), 195
MI (myocardial infarction), 43
   insurance coding for, 201, 203
midwife, 195
modem(s), 100, 101
   in medical transcription, 110
modified block style, 120, *122, 123*
monitoring, of quality of care, 96
morbidity, definition of, 9
mouth, anatomy of, 41, *43*
movement(s), types of, 36
MPI (Master Patient Index), 75, 155, *155*
MRT (Medical record technician), 3
MSW (medical social worker), 195
MT (medical technologist), 195
muscoskeletal system, 39–41, *40, 42–43*
musculoskeletal system, 23
   common disorders of, 40–41
   diseases of, insurance coding for, 201–202
myocardial infarction (MI), 43
   insurance coding for, 201, 203

NA (nurse assistant), 195
narcotic(s), 236
nasal cavity(ies), 44
National Association of Health Unit Clerks-Coordinator(s) (NAHUC), 3, 223
National Formulary (NF), in medical orders transcription, 236
nausea, 41
NCR, 154
negligence, 93, 94
   definition of, 17, 19
   risk management and, 98–99
neoplasm(s), insurance coding for, 205
nephritis, 47
network(s), computer, 100, 101
neurologic disorder(s), insurance coding for, 204
NF (National Formulary), in medical orders transcription, 236
nondisability claim(s), for Worker's Compensation, 218
nosocomial infection, 23
NP (nurse practitioner), 195
number(s), cardinal, 110
   in medical transcription, 114–115
   ordinal, 110
number paging system(s), 226
numerical filing system(s), 74, 76, 150
nurse(s), registered, 196
nurse assistant (NA), 195
nurse practitioner (NP), 195
nurses' accountability record(s), 228
nurses' note(s), 80, 156, *172–179*, 227–228
   in medical records analysis, 157, 180
nursing assignment admission summary(ies), 156, *158–171*
nursing care, medical orders for, transcription of, 237, 239
nursing station(s), communications in, health unit coordinator and, 225–228
   management of, by health unit coordinator, 228–229

occupational therapist (OT), 196
OD (doctor of optometry), 195
office machine(s), 62
offline, 100
online, 100
orchitis, 51
order(s), medical. See *medical order(s)*.
ordinal number(s), 110
OT (occupational therapist), 196
outguide(s), 74, 77–78

out-of-pocket expense(s), 194
ovary(ies), anatomy of, 51
   physiology of, 50

PA (physician's assistant), 196
page heading(s), in letter writing, 120
paging system(s), 57, 60, 226
   use of, by health unit coordinator, 225–226
paramedic(s), 195
parasitic disease(s), insurance coding for, 205
parathyroid gland, 50, *50*
pathogen(s), definition of, 23
pathology, definition of, 126
pathology report(s), 156, *172–179*
   transcription of, 127, 132, *133*
patient(s), death of, discharge process in, 72
   medical records of, access to, 150, 152
   new, information form for, 65, *68*
   privacy of, 29
      laws concerning, 93–95
   reception of, 57–58
patient care, quality of, 97. See also *quality assurance*.
patient care plan(s), 228
patient census, determination of, 84
patient consent form(s), 95, 156, *172–179*, 227, 228
patient demographic form(s), 65, *66–67*
patient history form(s), 65, *69*, 156, *158–171*, 227
   for insurance claims, 210
patient information, indexing of, 75
   transcription of, 127, *128–131*
patient information form(s), 80
   for insurance claims, 209–210
patient physical form(s), 227
patient record(s). See *medical record(s)*.
patient registration data form(s), 156, *158–171*
patient transfer form(s), 156, *158–171*
Patient's Bill of Rights, 17, 18–19, 93–95
patient's status report sheet(s), 226
payment(s), from Medicaid, 216
   from Medicare, 215
   incoming, processing of, 62
PC (personal computer), 100. See also *computer system(s)*.
PDR (Physician's Desk Reference), 126, *127*, 197. See *Physician's Desk Reference (PDR)*.
peer review(s), in quality assurance, 96, 98
   of insurance claims, 194
penis, anatomy of, 51, *51*
peripheral device(s), for computers, 100, 101
permanent disability, in Worker's Compensation, 218
personal belongings list(s), 227
personal computer (PC), 100. See also *computer system(s)*.
personal liability, 17, 20
pharmacology, in transcription of medical orders, 236–237
pharynx, anatomy of, 41, *43*, 44
phenylketonuria (PKU), insurance coding for, 201
phlebitis, 44
photocopier(s), 62
physical examination(s), reports of, transcription of, 127, *128–131*
physical examination form(s), 156, *158–171*, 227
physical therapist(s) (PT), registered, 196
physical therapy record(s), 228
physical therapy report(s), 156, *158–171*
physician(s), attending, in medical record analysis, 154, 157, 180
   increased productivity of, role of medical records clerks in, 183
   notification of, for incomplete medical records, 182–184
   privileges of, 154, 157
   suspension of, for incomplete medical records, 154, 157
physician's assistant (PA), 196
Physician's Current Procedural Terminology (CPT) code(s). See *Current Procedural Terminology (CPT) code(s)*.
   reference book for, 197
Physician's Desk Reference (PDR), 126, *127*, 197
   in medical orders transcription, 236
physician's discharge summary(ies), 227

Physician's Index, use of, 75
physician's order(s), forms for, 156, *172–179*
   in chart assembly, 80
   medical records analysis of, 154, 157, 180
   transcription of, by health unit coordinator, 235–236
physician's order form(s), 227
physician's progress note(s), 156, *172–179*, 227
pineal gland, 50
pituitary gland(s), 47–48, *50*
PKU (phenylketonuria), insurance coding for, 201
platelet(s), 41
pleura, anatomy of, 44
pleurisy, 44
plurality, definition of, 83
pneumatic tube system(s), 226
pneumonia, 44
pocket pagefinder system(s), 226
policy(ies), insurance, 194
position(s), anatomical, 34–35, *35*
postage meter(s), 61
postanesthesia record(s), 228
posterior, 34
posture(s), anatomical, 23, 27, *27*, 29, *30*, 35–36, *36*
power of attorney, medical, 64, 65
predicate, 117
preexisting exclusion(s), 195
preferred provider organization(s) (PPOs), 195
pregnancy, insurance coding for, 202–203
principal diagnosis, 154, 180, 189
privacy, in mail processing, 61
   invasion of, 93
      laws concerning, 93–94
   of patients, 29
   in insurance claims, 208–209
Privacy Act of 1974, 145, 150
privileged information, definition of, 17
PRN, 234
progress note(s), in chart assembly, 80
   medical records analysis of, 157, 180
   physician's, 156, *172–179*, 227
pronation, 36
proofreading, in medical transcription, 118, *119*
prostate gland, anatomy of, 51, *51*
prostatism, 39
psychologist(s), clinical, 195
psychosis, insurance coding for, 200, 201, 204
PT (physical therapist), 196
pulse, 42
punctuation, in medical transcription, 113

quality assurance, 96–99, 145, 181
   credentialing in, 98
   in discharge planning, 97–98
   in medical records, 99, 148
   quality assessment in, 98
   risk management in, 98–99
   role of Joint Commission on Accreditation of Healthcare Organizations in, 97, 149
   utilization review in, 97–98

radiology report(s), 156, *158–171*
   transcription of, 132, *134–136*
reasonable care, definition of, 17, 19
record(s), medical. See *Medical record(s)*.
record analysis checklist(s), 78, *81*
record of death(s), 156, *172–179*
Record Technician, accredited, 3, 146, *146*, 187
records management, 74–81
   filing and, 75–78
rectum, anatomy of, 41, *43*
reference initial(s), in letter typing, spacing of, 120
referral(s), to physicians, 60
register(s), for vital statistics, 83, 84
registered mail, 61

registered nurse (RN), 196
registered physical therapist (RPT), 196
registered record administrator (RRA), 3
registrar(s), 83
registry, of tumors, 145, 148
Regulation Z, in patient collections, 209
rehabilitation stage of illness, 13–14
Relative Value Study (RVS), reference book for, 197
release and care agreement(s), *67*
release of information, and patient privacy laws, 93–95
   authorization for, 145, 150, *151*, 152
release of information form(s), 150, *151*
   in insurance claims, 208, 210
reminder(s), in correspondence, 60–61
report(s), pathology, 156, *172–179*
      transcription of, 127, 132, *133*
   radiology, 156, *158–171*
      transcription of, 132, *134–136*
   surgical, 156, *172–179*
   therapy, 156, *158–171*
reproductive system, 50–51, *51*
   diseases of, insurance coding for, 202
requisition(s), role of health unit coordinator in, 223, 226–227
respiratory system, 44, 46, *46*
   diseases of, insurance coding for, 204
respiratory therapy report(s), 156, *158–171*, 228
responsible party(ies), 64, 65
resume(s), 245, 247–251, *249–251*
retrieval, of records, 154, 156, *156*
risk management, in quality assurance, 96, 98–99
RN (registered nurse), 196
role performance model of health care delivery, 10
root word(s), 261–264
rotation, 36
RPT (registered physical therapist), 196
RRA (registered record administrator), 3
rule of personal liability, 17, 20
RVS (Relative Value Study), reference book for, 197

safety, environmental, 29
sagittal plane(s), 35
salpingitis, 51
salutation(s), in letter writing, spacing of, 118
scrotum, anatomy of, 50, *51*
second-class mail, 61
security, for computer access, 100, 103
sedative(s), 236
seizure(s), emergency care for, 30–31
sensory organ(s), disorders of, insurance coding for, 204
sentence(s), structure of, 117–118
sick role, behavior in, 14–15
signature(s), in letter typing, spacing of, 120
simple fracture(s), 40
simplified style, of letter writing, 120, *124*
sinus(es), 44
skeletal system, 39–41, *40*
skilled nursing facility (SNF), 87
skin, 46, *49*
skip, 208
small intestine, anatomy of, 41, *43*
smooth muscle(s), 40, *42–43*
SNF (skilled nursing facility), 87
SOAP note(s), 37
Social Security Administration, insurance programs of, 88
software, 100–102
   applications of, 102–103
   in medical transcription, 110
special delivery mail, 61
specialty(ies), medical, 6, 7t
spina bifida, insurance coding for, 201, 203
sprain(s), 41
   emergency care for, 31
spreadsheet(s), 100, 101
standard of care, 93, 94

Index

STAT, 234
statement(s), for billing, 57, 61
statistic(s), vital, 83
statistical data, 75
status report sheet(s), 226
sterility, 51
sterilization, 26–27
stomach, anatomy of, 41, *43*
storage, of filing systems, 77–78
subject, in grammar, 117
subject line(s), in letter writing, 120
subpoena(s), of medical records, 145, 152
subrogation, 195
subscriber(s), 87
suctioning, 237
supination, 36
supplies, at nursing station, health unit coordinator and, 229
surgery, consent forms for, 156, *172–179*
   schedule sheets for, 226
surgical asepsis, 25–26, *26*
surgical checklist(s), 228
surgical patient(s), processing of, by health unit coordinator, 232
surgical report(s), 156, *172–179*
susceptibility, to infection, 24
susceptible, definition of, 23
suspension(s), of physicians, for incomplete medical records, 154, 157
suspension list(s), 181, 182
symbol(s), medical, 255–260
symptom experience stage of illness, 13–14
systolic blood pressure, 42–43

Tabor's Cyclopedic Dictionary, in medical transcription, 126, *127*
tagging, 75, 77
Tax Equity and Fiscal Responsibility Act of 1982 (TEFRA), 208
taxpayer identification number (TIN), 195
teeth, anatomy of, 41, *43*
TEFRA (Tax Equity and Fiscal Responsibility Act of 1982), 208
telephone(s), release of information via, 152
   use of, by health unit coordinators, 225
telephone system(s), etiquette in, 59
   holding in, 59
   screening in, 57, 59
   transferring in, 59
temporary disability claim(s), for Worker's Compensation, 218
tendon(s), 40
terminology, medical, 33–36
   anatomical position in, 34–35, *35*
   for insurance coding, 194–196
testes, anatomy of, 50–51, *51*
   physiology of, 50
thank-you note(s), after employment interview, *253*
therapy report(s), 156, *158–171*
third party(ies), 145
   liability provisions for, 195
third-class mail, 61
thymus gland, 50
thyroid gland(s), 48, 50, *50*
TIN (taxpayer identification number), 195
tort, 93, 94
trachea, anatomy of, 44
transcription, definition of, 107
   medical. See *medical transcription*.
transcriptionist(s), medical. See *medical transcriptionist(s)*.
transfer(s), in telephone system, 57, 59
   of patients, 64, 65–66, *71*, 231–232
    forms for, *71*, 156, *158–171*
transport sheet(s), 226
transportation schedule(s), 226

transverse colon, 41, *43*
transverse plane(s), 35
treatment area(s), maintenance of, 31
treatment order(s), 234
Truth in Lending Act, 209
tuberculosis, 46
tumor(s), insurance coding for, 205
tumor registry(ies), 145, 148
typewriter(s), for medical transcription, 110–111, *111*

UB (uniform bill), 208, 211, *213*
UB-92 form(s), for hospital billing, processing of, 211, *213*
UCR (usual, customary and reasonable), in insurance coding, 195
ulcer(s), insurance coding for, 201
   of digestive system, 41
uniform bill (UB), 208, 211, *213*
unit clerk(s). See *health unit coordinator(s)*.
unit record(s), 145
United States Pharmacopeia (USP), in medical orders transcription, 236
universal precautions, 27
UR (utilization review). See *utilization review (UR)*.
uremia, 47
ureter(s), 47, *49*
urethra, anatomy of, 47, *49*
urinalysis, 201
urinary catherization(s), 234
urinary system, 46–47, *49*
   diseases of, insurance coding for, 202
urinary tract infection (UTI), 201
USP (United States Pharmacopeia), in medical orders transcription, 236
usual, customary and reasonable (UCR), in insurance coding, 195
uterus, anatomy of, 51
UTI (urinary tract infection), 201
utilization review (UR), 96, 97–98, 145, 181
   of medical records, 149
   quality assurance in, 97–98

vagina, anatomy of, 51
V-code(s), in insurance coding, 189
   supplementary listings of, 197–198
vein(s), 41, *45*
   varicose, 44
visitor(s), reception of, 57–58
vital signs sheet(s), 223, 228
vital statistic(s), 83
   reporting requirements for, 84–85
voice paging, 226
voluntary muscle(s), 40, *42–43*
vulva, anatomy of, 51

ward clerk(s). See *health unit coordinator(s)*.
wellness, definition of, 9
   disease and, 11–12, *11–12*
WHO (World Health Organization), and International Classification of Diseases, 89
witness(es), for forms, 95
word element(s), definition of, 33
word processing, 100, 102
   in medical transcription, 110
work assignment sheet(s), 226
Worker's Compensation, 87
   processing of claims for, 216, 218
World Health Organization (WHO), and International Classification of Diseases codes, 89
written communication(s), 60–62
   processing of, by health unit coordinator, 226–228